Discrete Mathematics Through Applications

Discrete Mathematics Through Applications

Nancy Crisler,
Patience Fisher,
and Gary Froelich

W. H. Freeman and Company
New York

Library of Congress Cataloging-in-Publication Data

Crisler, Nancy.
Discrete mathematics through applications / Nancy Crisler, Patience Fisher, and Gary Froelich.
p. cm.
Includes index.
ISBN 0-7167-2577-0
1. Mathematics. 2. Computer science—Mathematics. I. Fisher, Patience. II. Froelich, Gary W. III. Title.
QA39.2.C75 1994
510—dc20
94-2597
CIP
AC

Printed in the United States of America

1 2 3 4 5 6 7 8 9 0 RRD 9 9 8 7 6 5 4

CONTENTS

PREFACE

This book is intended to introduce students to discrete mathematics and its importance in today's world.

Whenever discrete mathematics is a topic of conversation, someone eventually asks What is discrete mathematics? Often the query is accompanied by intentional or unintentional witticisms derived from feigned or real confusion over the difference between the words "discrete" and "discreet."

A good short answer contrasts "discrete" topics with those that are "continuous." Real numbers, for example, are a continuous set because between any two of them there are infinitely many more. Therefore, the traditional study of functions that goes on in algebra, trigonometry, and calculus is not discrete mathematics because functions studied in these courses are defined on the real numbers. On the other hand, discrete topics often exist in traditional courses, but they are relatively isolated. Many algebra texts, for example, include enough material on counting and probability to ensure an understanding of the binomial theorem.

Unfortunately, the traditional mathematics curriculum ignores much of the mathematics that is important today. Early twentieth-century America's place in the world was secured by science and engineering, and calculus was by far the most important mathematical tool. Today, however, mathematics is being applied to problems in the social sciences and, of course, to problems that arise in the design of efficient computer systems. Most of these applications involve discrete mathematics. Moreover, an understanding of many important problems in discrete mathematics requires very little background, which makes the subject accessible to nearly all students. Compare this to calculus, which traditionally has required several years of preparation prior to its study.

The case for including discrete mathematics in the experience of all students is a difficult one to argue against. Indeed, the National Council of Teachers of Mathematics (NCTM) included a separate standard on discrete mathematics in its 1989 *Curriculum and Evaluation Standards for School Mathematics,* thereby elevating discrete mathematics to the same position as algebra, geometry, and calculus. Moreover, after the

Standards was circulated, the NCTM convened a task force of mathematics educators to write *Discrete Mathematics and the Secondary Mathematics Curriculum,* a report that provided guidelines for the integration of topics from discrete mathematics into existing courses and for the establishment of a separate discrete mathematics course. The report was followed by a project in which teachers experienced in teaching discrete mathematics taught it to other teachers at several sites around the country.

The authors of this book were among the members of that NCTM task force. We felt that a textbook that adequately addressed the content outlined in our report did not exist and discussed the need for such a book with Dr. Solomon Garfunkel, the executive director of the Consortium for Mathematics and Its Applications (COMAP), a nonprofit educational corporation that has long had an interest in the inclusion of applications of discrete mathematics in mathematics curricula. COMAP offered financial support for the writing of the book, and we began work in earnest.

Throughout the writing we tested the material in our own classes and in the NCTM teacher education project mentioned earlier. Many other mathematics educators read the material and tested parts of it in their classes. We incorporated numerous suggestions offered by these people into the final manuscript.

As teachers, we have found a great deal of satisfaction in teaching discrete mathematics. There are so many applications that we have seldom, if ever, heard students ask What is this good for? Indeed, the mathematics most often derives from the applications rather than the applications appearing only after mathematical topics have been "mastered."

As authors, we have found even greater satisfaction in writing a book that reflects our own interest in these topics and the way in which we teach them. The features of the book that reflect the approach to discrete mathematics that has worked best for us are:

- The initial treatment of a topic usually consists of an investigation or "Explore This" problem or lesson. We believe that exploration of a problem is important to an understanding of the problem and its solutions.

- The lessons are written in an informal style that is designed to be read by the student. There are few formal definitions or other formal treatments of mathematical topics.

◂ The exercises often contain new ideas. Students "learn by doing." In most cases, students are expected to do all the exercises.

In addition, there are several supplements to the text that we have found helpful. They are

◂ An *Instructor's Manual.* It includes our suggestions for teaching individual lessons, comments on exercises, assessment tips, masters for creating transparencies, and a complete answer key.

◂ A disk of software for the first two chapters. It includes three programs, *Election Machine, Preference Schedule,* and *Weighted Voting,* which are designed for Chapter 1, and a fourth program, *Moving Knife,* which is useful in Chapter 2. The disk is available in Apple II, IBM, and Macintosh formats.

◂ Videos. In particular the video *Geometry: New Tools for New Technologies* has several segments useful in Chapters 4 and 5. Other videos that are mentioned in the Instructor's Manual are available from either the Annenburg Corporation for Public Broadcasting, 1-800-LEARNER, or the Consortium for Mathematics and Its Applications, 1-800-77COMAP.

For information about these supplements, contact W. H. Freeman, 41 Madison Ave., New York, NY 10010, 1-800-347-9405.

Finally, we would like to thank everyone who has contributed to the preparation of *Discrete Mathematics Through Applications.* There are, however, simply too many mathematics educators who have provided encouragement and advice, who have read and commented on the manuscript, and who have taught various parts of the book in their own classes to be able to list them all. Indeed, some of the feedback you have given us has come through others and we may not even know your names. Rest assured that we are deeply grateful to all of you.

We also are indebted to the many students who have patiently studied discrete mathematics from a manuscript containing the inevitable typographical errors and less-than-perfect lessons. The lessons still aren't perfect, but they are much better because of you. Our thanks, in particular, to the three of you who wrote programs that appear on the supplementary disk: Kari Bauer, Dan Froelich, and Tom Hehre.

We owe a great deal to the Consortium for Mathematics and Its Applications for its technical and financial support. Dr. Solomon Garfunkel's faith in our ability to finish the project is particularly appreciated. We also want to thank the COMAP production staff: Laurie Holbrook, Phil McGaw, and Roger Slade for producing an excellent working manuscript, and Marc Kaufman for happily conducting the research that produced most of the newsclips.

Our appreciation, too, to W. H. Freeman and Company and Jerry Lyons for their willingness to take a risk, and to the members of the editorial and production staff for their outstanding efforts and many long days that kept the book on schedule. Among them are: Kay Ueno, Christine Hastings, Michele Barry, Sheila Anderson, Alice Fernandes-Brown, John Hatzakis, Bill Page, Larry Marcus, and Travis Amos. In addition, we would like to thank Tom Durfee for his humorous illustrations.

For many years we have shared our interest in discrete mathematics with others at institutes, seminars, and conferences. We know that many of you have returned to your schools to incorporate discrete topics into your own courses and curricula. That knowledge has made the many weeks and months on the road worthwhile, and we hope that, in turn, this book proves useful to you in your efforts.

Nancy Crisler
Patience Fisher
Gary Froelich

February 1994

FOREWORD

Discrete mathematics and its applications is one of the most rapidly expanding areas in the mathematical sciences. The modeling and understanding of finite systems is central to the development of the economy, computer science, the natural and physical sciences, and mathematics itself. This same rapid growth of discrete topics and applications has made the definition and development of course work in discrete mathematics a more difficult task than the development of materials and courses of study in more established areas of the mathematical sciences.

Some versions of discrete mathematics materials have attempted to address business applications, others applications for the computer sciences. The present text provides a sound introduction to social choice, to matrices and their uses, to graph theory and its applications, and to counting and finite probability. As such, it directly responds to the call made for discrete mathematics by the National Council of Teachers of Mathematics' *Curriculum and Evaluation Standards for School Mathematics* (1989). But, more than that, it provides a solid introduction to the processes of optimization, existence, and algorithm construction that characterize discrete studies.

Central to students' development in mathematics is mastery of the process of problem-solving, communication, reasoning, and modeling. All of these processes receive ample treatment in this text. However, teachers must expend special effort to see that students are actually required to develop their abilities in these areas. Discrete mathematics provides an excellent platform for allowing students to develop these skills, as the fundamental structure for analyzing situations in the positive integers. As such, students are free to analyze specific subcases and generalize their hypotheses to larger settings. Work with matrices, graphs, and combinatorics provides rich settings for these processes to develop and blossom.

The initial work with social choice allows students to begin their study of discrete mathematics in a realm rich with applications, modeling real world events. When used as the initial entry to the course, this helps break the mold of studying and learning mathematics only to do more of the same. Students suddenly see mathematics as a useful tool in human decision making. They also come to see that not all problems have immediate, if any, solutions. This realization sets the course not only for the text but also for the students' future careers as users of mathematics in their personal and professional lives.

John A. Dossey
Department of Mathematical Sciences
U.S. Military Academy, West Point, New York
February 1994

Discrete Mathematics Through Applications

E.D A.D
LA JUNTA
DE REGISTRO
REGISTER
TO
VOTE
It's that simple.
LLENE EST
FORMULA
E.D
BOARD
OF
REGISTRY
MEETS
HERE

Election Theory

Throughout our lives we are faced with decisions. As students, we must decide which courses to take and how to divide our time among schoolwork, activities, social events, and, perhaps, a job. As adults, we are faced with many new decisions, including whether to vote for one candidate or another. Many of these decisions are not made in isolation. Our decisions, and those of others, must be combined to achieve a common result. Your school's administrators, for example, compile all the students' selections of courses into a master schedule.

How are the wishes of many individuals combined to yield a single result? Do the methods for doing so always treat each person fairly? If not, is it possible to improve on these methods? This chapter examines ways of combining preferences of individuals into a single result that are used in such areas as elections and athletic power polls. The answers to these and many other questions related to voting can be found in an area of discrete mathematics called election theory.

Lesson 1·1

An Election Activity

How is the winner of an election decided? How do sportswriters select the best college football team in the country? In this lesson, we consider how individual preferences can be combined into a single result.

Explore This

Take a piece of paper and write down the names of the following soft drinks, in the order given:

- Coke
- Dr. Pepper
- Mountain Dew
- Pepsi
- 7-Up

Beside the name of the soft drink you like best, write 1. Beside the name of your next favorite soft drink, write 2. Continue until you have ranked all five soft drinks.

At the direction of your instructor, collect the papers from all the members of your class. Write the rankings of each individual on the classroom board. Because everyone has written the names in the same order, this should require only writing the digits from each ballot. List these rankings on a sheet of paper. Now, as your instructor directs, divide your class into groups of three or four people.

Your group's task is to determine a first-, second-, third-, fourth-, and fifth-rated soft drink for the entire class. Use any method that is acceptable to all the members of your group. You should be able to explain your method to the other groups when you have finished. Allow about 15 minutes of class time for the groups to reach their decisions.

After all the groups have finished, a spokesperson for each group should present the group's decision to the class and explain the method used to reach the decision. The spokesperson should write the group's ranking of the five soft drinks on the board before explaining the method. Record each group's results in your notebook.

Exercises

1. Which soft drink was ranked first by the most people? Did all the groups rank the same soft drink first?
2. Repeat Exercise 1 for the soft drink ranked second.
3. Repeat Exercise 1 for the soft drink ranked third.
4. Repeat Exercise 1 for the soft drink ranked fourth.
5. Repeat Exercise 1 for the soft drink ranked fifth.
6. Write a description of the method used by your group. Make it clear enough so that someone else could use the method. You may want to break down the method into steps and number each one.

7. Did any of the other groups use the same method that your group used?

8. Did your method result in any ties between soft drinks? How could you modify your method to break a tie?

9. A **preference schedule** that is shown below displays four choices, called A, B, C, and D. The preference schedule indicates that the individual whose preference it represents liked B best, C second, D third, and A fourth.

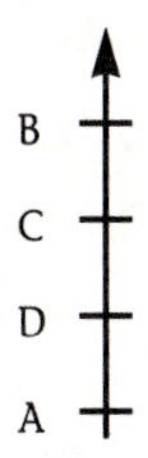

The number written under a preference schedule indicates the number of individuals who expressed that preference.

A set of preferences for a group of 26 people is shown below.

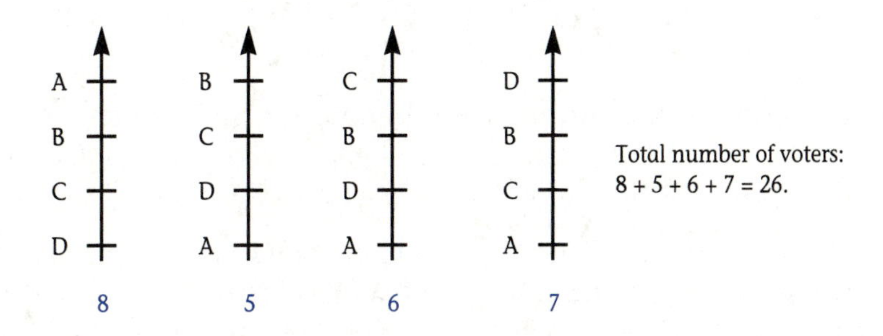

a. Apply your group's method for determining the class's soft drink ranking to this set of preferences. List the first, second, third, and fourth choices that your method produces for the group of 26. If your method is too inexact to produce a group ranking, discuss this with your group members and revise the method so that it can be used here.

b. Do the members of your group approve of the results? Were any of the choices questionable? In other words, do the first, second, third, and fourth choices produced by your method seem reasonable, or are there reasons that one or more of the rankings seem unfair?

10. You have three items to rank. Call them A, B, and C. Below are the six possible preferences you could express.

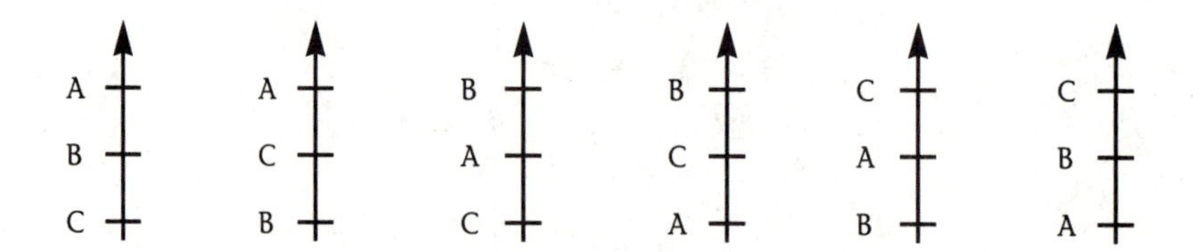

A fourth choice, D, enters the picture. If D is attached to the bottom of each of the previous schedules, there would be six schedules with D at the bottom. Similarly, there would be six schedules with each of A, B, and C at the bottom, or a total of $4(6) = 24$ schedules.

Thus, the possible number of schedules with four choices is four times the possible number of schedules with three choices. There are 24 possible schedules with four choices. How many are possible with five choices? With six?

11. Let S_n mean the number of schedules possible when there are n choices. You have seen that $S_n = nS_{n-1}$. Write an English translation of the mathematical sentence $S_n = nS_{n-1}$.

12. The mathematical sentence in Exercise 11 is an example of a **recurrence relation,** a verbal or symbolic statement that describes how one number in a list can be derived from the previous numbers. If, for example, the first number in a list is 7 and the recurrence relation states that to obtain any number in the list you must add 4 to the previous number, the second number in the list will be $7 + 4$, or 11. This recurrence relation is stated symbolically as $T_n = 4 + T_{n-1}$. Another example of a recurrence relation is

$$T_n = n + T_{n-1}.$$

Complete the following table for the recurrence relation $T_n = n + T_{n-1}$:

n	T_n
1	3
2	$2 + 3 = 5$
3	
4	
5	

13. Complete the following table for the recurrence relation $A_n = 3 + 2A_{n-1}$:

n	A_n
1	4
2	$3 + 2(4) = 11$
3	
4	
5	

Group-ranking Methods and Algorithms

If the soft drink data collected from your class are typical, you will know that the problem of establishing a group ranking is not without controversy. The reason is that there is seldom complete agreement on the correct way to do this. We will examine several common methods of determining group preferences from a set of individual preferences. As we do so, consider whether any of the methods discussed are similar to the one your group used.

Consider the preferences of Exercise 9 of the previous lesson, which are shown again in Figure 1.1.

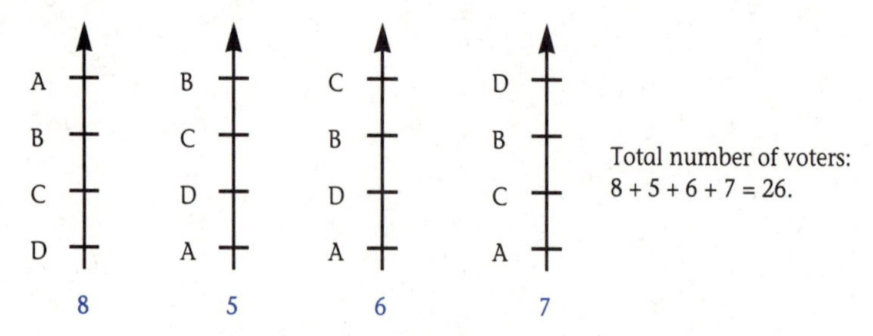

Figure 1.1 Preferences of 26 voters.

Many group-ranking situations, such as elections in which only one office is to be filled, require the selection of a first choice only. In the set of preferences shown in Figure 1.1, choice A is ranked first on 8 schedules, more often than any other choice. If A wins on this basis, A will be known as the **plurality winner.** A, however, is first on only 8 out of 26 schedules, or on 30.8% of the schedules. Had A been first on over half the schedules, A would also be a **majority winner.** The plurality winner is based on first-place rankings only. The winner is the choice that received the most votes:

A: 8 B: 5 C: 6 D: 7.

The Borda Method

Did any of the groups in your class determine their group ranking by assigning points to the first, second, third, and fourth choice and obtaining a point total for each soft drink? If so, these groups used a variation of a **Borda count.** This method is named after Jean-Charles de Borda (1733–1799), a French cavalry officer, naval captain, mathematician, and scientist, who devised it because of his dissatisfaction with the plurality method.

The most common way of applying this method to a ranking of n choices is to assign n points to first place, $n - 1$ to second, $n - 2$ to third, . . . , and 1 point to last. The group ranking is established by totaling each choice's points.

In our example, A was ranked first by 8 people and fourth by the remaining 18. A's point total is $8(4) + 18(1) = 50$, and B's total is $8(3) + 5(4) + 6(3) + 7(3) = 83$. Similar calculations give totals of 69 for C and 58 for D. Thus, Borda ranks B first, C second, D third, and A fourth. In this case, the plurality method winner does not fare well under the Borda system.

The Borda winner is based on point totals:

$$\text{A: } 8(4) + 5(1) + 6(1) + 7(1) = 50.$$
$$\text{B: } 8(3) + 5(4) + 6(3) + 7(3) = 83.$$
$$\text{C: } 8(2) + 5(3) + 6(4) + 7(2) = 69.$$
$$\text{D: } 8(1) + 5(2) + 6(2) + 7(4) = 58.$$

The Runoff Method

Runoff elections are common but often expensive because of the cost of holding another election and time-consuming because they require a

NEWS CLIP

Top 25 Coaches' Poll

Monday, September 27, 1993, USA TODAY CNN

1. Florida State (4-0) ▸ Last week's ranking: 1. ▸ Poll points: 1,546 points (58 first-place votes). ▸ Result: Did not play. ▸ Next: Saturday vs. Georgia Tech.

2. Alabama (4-0) ▸ Last week: 2. ▸ Poll points: 1,480 (2). ▸ Result: Beat Louisiana Tech 56–3. Sherman Williams ran for three touchdowns, and David Palmer caught two TD passes. ▸ Next: Saturday at South Carolina.

3. Miami (Fla.) (3-0) ▸ Last week: 3. ▸ Points: 1,403 (2). ▸ Result: Beat then-No. 12 Colorado 35–29. Frank Costa had two TD passes, and Donnell Bennett rushed for two scores. ▸ Next: Saturday vs. Georgia Southern.

4. Notre Dame (4-0) ▸ Last week: 4. ▸ Points: 1,305. ▸ Result: Beat Purdue 17–0. Defensive end Brian Hamilton returned fumble 28 yards for touchdown in third quarter, and Irish scored 10 points in final four minutes. ▸ Next: Saturday at Stanford.

5. Nebraska (4-0) ▸ Last week: 5. ▸ Points: 1,252. ▸ Result: Beat Colorado State 48–13. Tommie Frazier passed for two touchdowns and ran for one. ▸ Next: Oct. 7 at Oklahoma State.

6. Florida (3-0) ▸ Last week: 6. ▸ Points: 1,200. ▸ Result: Did not play. ▸ Next: Saturday vs. Mississippi State.

7. Ohio State (3-0) ▸ Last week: 7. ▸ Points: 1,155. ▸ Result: Did not play. ▸ Next: Saturday vs. Northwestern.

8. Penn State (4-0) ▸ Last week: 8. ▸ Points: 1,109. ▸ Result: Beat Rutgers 31–7. Kerry Collins, who replaced John Sacca as starting quarterback last week, was 18-for-25 for 222 yards and four touchdowns. ▸ Next: Saturday at Maryland.

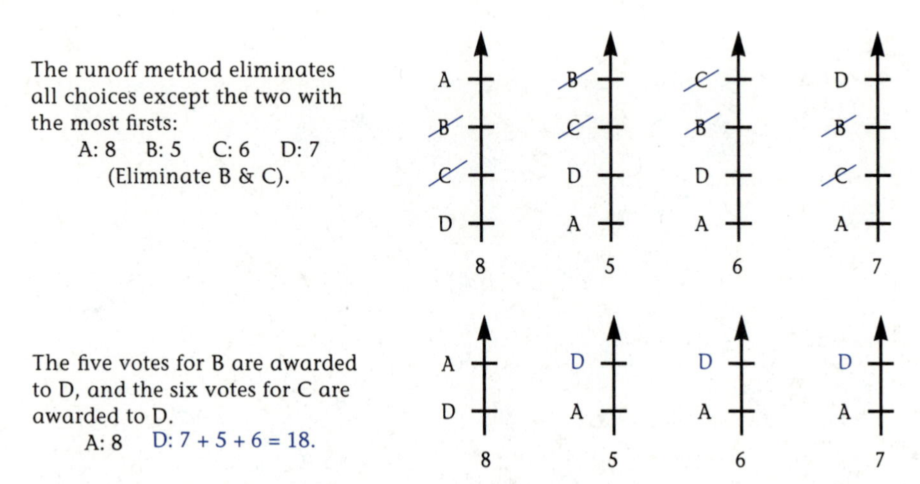

Figure 1.2 The runoff method.

second trip to the polls. With the use of preference schedules, however, much of this inconvenience can be avoided.

To conduct a runoff, determine the number of firsts for each choice. In our example, A is first eight times, B is first five times, C is first six times, and D is first seven times.

Eliminate all but the two highest totals: Choices B and C are eliminated and A and D are retained. Now consider each of the preference schedules on which the eliminated choices were chosen first. B was first on the second schedule. Of the two remaining choices, A and D, D is ranked higher than A and is awarded these 5 votes. Similarly, D is awarded the 6 votes from the third schedule. The totals are now 8 for A and $7 + 5 + 6 = 18$ for D, and so D is the runoff winner (see Figure 1.2).

The Sequential Runoff Method

The sequential runoff method differs from the runoff method in that it eliminates choices only one at a time. In our example, B is eliminated first because it is ranked first the fewest times, and so these 5 votes are awarded to C. The point totals now are 8 for A, $5 + 6 = 11$ for C, and 7 for D.

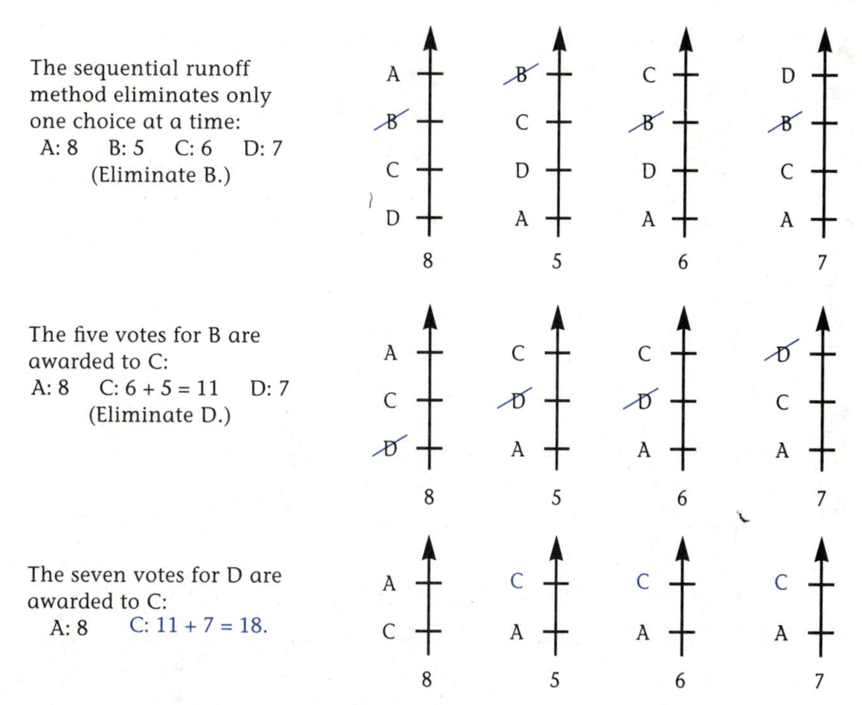

Figure 1.3 The sequential runoff method.

There are now three candidates remaining. D's total is the smallest, so D is eliminated next. The 7 votes are awarded to the remaining choice that is ranked highest by these 7. Thus, C is given an additional 7 votes and so defeats A by 18 to 8 (see Figure 1.3).

Exercises

1. Which of the soft drinks is the plurality winner in your class? Is it also a majority winner?
2. Which soft drink is the Borda winner in your class?
3. Which soft drink is the runoff winner in your class?
4. Which soft drink is the sequential runoff winner in your class?

5. For our example, determine the percentage of schedules on which each choice is ranked first and last.
   a. Enter the results in the following table:

Choice	First	Last
A		
B		
C		
D		

   b. Based on only these percentages, which choice do you think would be the most objectionable? The least objectionable?
   c. Which choice do you think should be ranked first for the group? Explain your reasoning.
   d. Give at least one argument against your choice.

6. Determine the plurality, Borda, runoff, and sequential runoff winners for the set of preferences shown below.

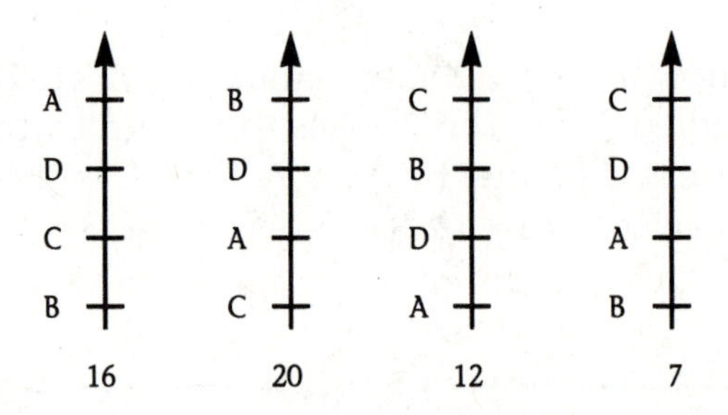

7. The Borda method automatically determines a complete ranking, but the other methods examined here produce only a first choice. Each of these may be extended, however, to produce a complete group ranking. Describe how the runoff method could be extended to determine a second, third, and so forth. Apply this to the example on page 12 and list the second, third, and fourth choices produced by the method.

NEWS CLIP

Sydney Wins! 2000 Summer Olympics go to Australia

Kansas City Star, Friday, September 24, 1993

Sydney, Australia, edged out Beijing Thursday for the right to hold the 2000 Summer Olympic Games. Beijing, which was considered the slight favorite, led in each of the first three rounds of voting but could not gain an overall majority. Here's how the International Olympic Committee voted. A simple majority was required to win.

	First round	Second round	Third round**	Fourth round**
Beijing	32	37	40	43*
Sydney	30	30	37	45
Manchester, England	11	13	11*	
Berlin	9	9*		
Istanbul, Turkey	7*			

* Eliminated
**One member did not vote

8. In the sequential runoff method, the number of choices after a given round is one less than the number after the previous round. Let C_n represent the number of choices after n rounds, and write this as a recurrence relation.

9. A procedure for solving a problem is known as an **algorithm.** This section has presented various algorithms for determining a group preference from individual preferences. Algorithms are often presented in numbered steps in order to make them easy to apply.

   The following is an algorithmic description of the runoff method:

1. For each choice, determine the number of preference schedules on which the choice was ranked first.

2. Eliminate all the choices except the two that were ranked first most often.

3. For each preference schedule, assign all firsts to the remaining choice that ranks highest on that schedule.

4. Count the number of preference schedules on which each of the remaining choices is ranked first.

5. The winner is the choice ranked first on the most schedules.

a. Write an algorithmic description of the sequential runoff method.
b. Write an algorithmic description of the Borda method.

10. The number of first-, second-, third-, and fourth-place votes for each choice in an election can be described in a table, or *matrix,* as shown below.

The preferences:

A	B	C	C
D	D	B	D
C	A	D	A
B	C	A	B
20	10	12	15

The matrix:

	A	B	C	D
1st	20	10	27	0
2nd	0	12	0	45
3rd	25	0	20	12
4th	12	35	10	0

The number of points that a choice receives for first, second, third, and fourth place can be written in a matrix, as shown below.

	1st	2nd	3rd	4th
Points	4	3	2	1

A new matrix that gives the Borda point totals for each choice can be computed by writing this matrix alongside the first, as shown on the next page.

$$[4 \quad 3 \quad 2 \quad 1] \begin{bmatrix} 20 & 10 & 27 & 0 \\ 0 & 12 & 0 & 45 \\ 25 & 0 & 20 & 12 \\ 12 & 35 & 10 & 0 \end{bmatrix}$$

The new matrix is computed by multiplying each entry of the first matrix by the corresponding entry in the first column of the second matrix and finding the sum of these products:

$$4(20) + 3(0) + 2(25) + 1(12) = 142.$$

This number is the first entry in a new matrix that gives the Borda point totals for choices A, B, C, and D:

$$\begin{array}{rcccc} & A & B & C & D \\ \text{Point totals:} & [142 & \_\_ & \_\_ & \_\_] \end{array}$$

Calculate the remaining entries of the new matrix.

Computer Explorations

11. Enter the soft drink preferences from your class into the election machine program on the disk that accompanies this book. Compare the results given by the computer to your answers to the first four exercises of this section. Resolve any discrepancies.

Projects

12. Write a short report on the history of any of the methods discussed in this section. Look into the lives of people who were influential in developing the method. Discuss the factors that led them to their proposals.

13. Write a short report on a method of determining a group ranking that is not included in this section. What, for example, is the Hare method? Apply the method you investigate to our example and to your class's soft drink preferences.

14. Find three different examples of group rankings that are used in our society. Examples are a local school board election and a football power poll. Describe how the group ranking is determined. Compare each method with the five methods described in this section. Is it the same? What are some advantages and disadvantages of the method?

15. Select one or more European countries and report on the methods they use to conduct elections.

More Group-ranking Methods and Paradoxes

Different methods of determining a group ranking often give different results. Similar controversies led the marquis de Condorcet (1743–1794), a French mathematician, philosopher, economist, and good friend of Jean-Charles de Borda, to propose that any choice that could obtain a majority over every other choice should win.

The Condorcet method requires that a choice be able to defeat each of the other choices in a one-on-one contest.

Again consider the set of preference schedules used in the last section (see Figure 1.4).

To examine the data for a Condorcet winner, compare each choice with every other choice. Begin by comparing A with each of B, C, and D. A is ranked higher than B on 8 schedules and lower on 18. Because A

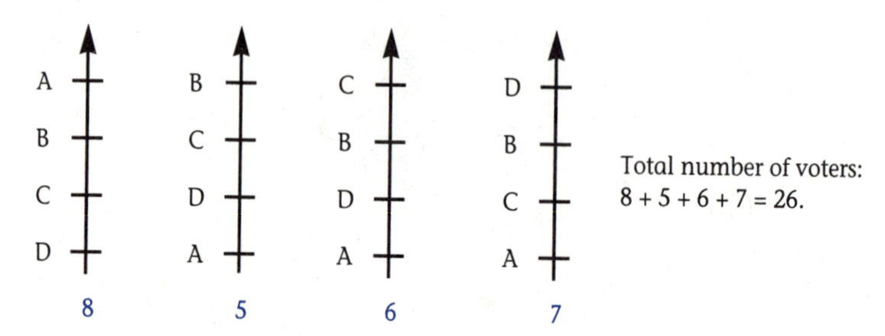

Figure 1.4 Preferences of 26 voters.

cannot obtain a majority against B, it is impossible for A to be the Condorcet winner.

Now consider B. You've already seen that B has a majority over A, so begin by comparing B with C. B is ranked higher on 8 + 5 + 7 = 20 schedules and lower on 6. Now compare B with D. B is ranked higher on 8 + 5 + 6 = 19 schedules and lower on 7. B therefore can obtain a majority over each of the other choices and so is the Condorcet winner. The following table shows the results of all possible one-on-one contests in this example:

	A	B	C	D
A		L	L	L
B	W		W	W
C	W	L		W
D	W	L	L	

To see how a choice does in one-on-one contests, read across the row associated with that choice. A, for example, loses in one-on-one contests with B, C, and D.

Although the Condorcet method may sound ideal, it sometimes fails to produce a winner. Consider the set of schedules shown in Figure 1.5.

Notice that A is preferred to B on 40 of the 60 schedules but that A is preferred to C on only 20. Although C is preferred to A on 40 of the 60, C is preferred to B on only 20. Therefore there is no Condorcet winner.

You might expect that if A is preferred to B by a majority of voters and if B is preferred to C by a majority of voters, then a majority of

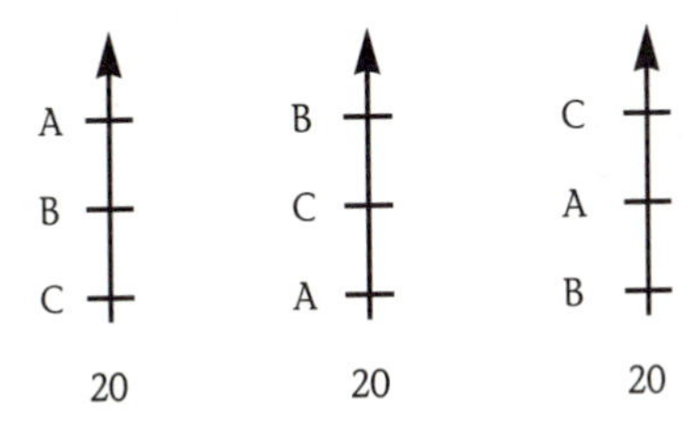

Figure 1.5 Preferences of 60 voters.

voters would prefer A to C. But this example shows that this need not be the case.

In your mathematics classes, you learned that many relationships are *transitive*. The relation "greater than (>)," for example, is transitive because if $a > b$ and $b > c$, then $a > c$.

You have just seen that group rankings may violate the transitive property. Because this intransitivity seems contrary to our intuition, it is known as a **paradox.** This particular paradox is sometimes referred to as the **Condorcet paradox.** There are other paradoxes that can occur when one is trying to determine a group ranking, which you will encounter in this lesson's exercises.

Exercises

1. Determine the Condorcet winner in the soft drink ballot conducted in Lesson 1.1.

2. Suppose the Condorcet method were used to decide the result of an election. Propose a method for resolving situations in which there is no Condorcet winner.

3. The choices in the set of preferences shown at the top of the next page represent three bills to be considered by a legislative body. The members will debate two of the bills and choose one of them. The chosen bill and the third will then be debated and another vote taken. Suppose you are responsible for deciding which two will appear on the agenda first.

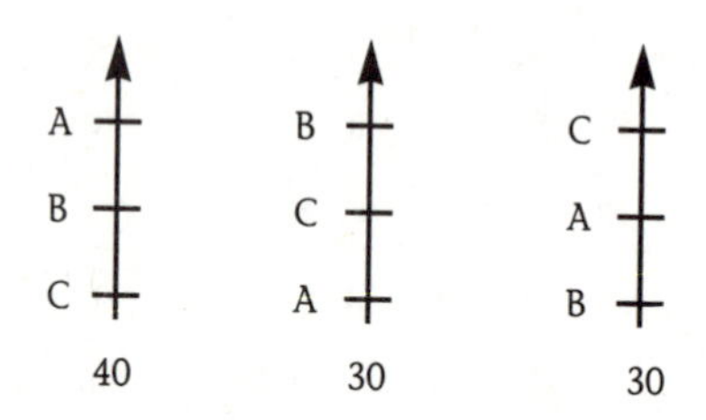

If you strongly prefer Bill C, which two bills would you place first on the agenda? Why?

4. A panel of sportswriters is selecting the best football team in a league, and the preferences are distributed as shown below.

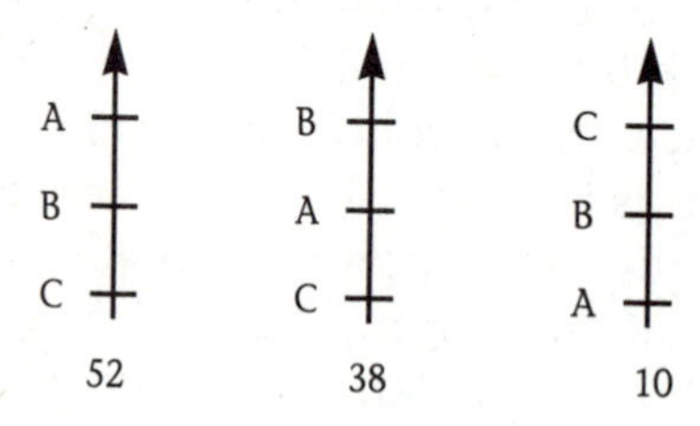

   a. Determine the winning team by using a 3–2–1 Borda count.
   b. The 38 who ranked B first and A second decide to lie in order to improve the chances of their favorite and so rank C second. Determine the winner by using a 3–2–1 Borda count.

5. When people decide to vote differently from the way they actually feel, they are said to be *voting insincerely*. People are often encouraged to vote insincerely because they have some idea of the result beforehand. Explain why advance knowledge of a result is possible in our society.

6. Most political elections in our society are decided by the plurality method. Construct a set of preferences with three choices in which the plurality method would encourage insincere voting. Identify the group of voters that would be encouraged to vote insincerely and explain the effect of their doing so on the election.

7. a. Use a runoff to determine the winner in the set of preferences shown on the next page.

A	C	B	B
B	A	C	A
C	B	A	C
38	30	25	7

b. These schedules represent a situation in which the votes are made public. Because they expect to receive some favors from the winner and because they expect A to win, the seven voters associated with the last schedule decide to change their preference from

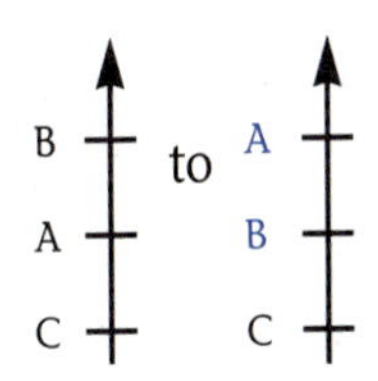

and to "go with the winner." Conduct a new runoff and determine the winner.

c. Explain why the results constitute a paradox.

8. a. Use a 4–3–2–1 Borda count to determine a group ranking for the set of preferences shown below.

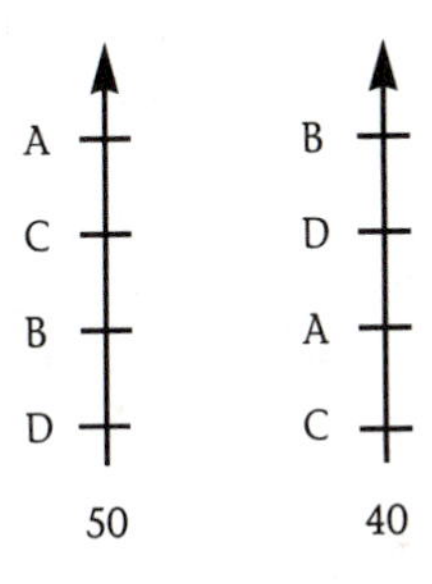

b. These preferences represent the ratings of four college athletic teams, and team C has been disqualified because of a recruiting violation. Write the schedules with team C removed. Now use a 3–2–1 Borda count to determine a group ranking from the original preferences with team C removed.

c. Explain why these results constitute a paradox.

9. Write a brief summary of the five methods of achieving a group ranking that have been discussed in this and the previous section. Include at least one example of why each method can lead to unfair results.

10. The Condorcet method requires that in theory each choice be compared with every other one, although in practice many of the comparisons do not have to be made. Consider what the number of comparisons would be if every comparison were made. If there are only two choices, a single comparison is all that is necessary. In the diagram below, the choices are represented by points, or vertices, and the comparison by a segment, or edge.

a. Add a third choice, C, to the diagram. Connect it to A and to B to represent the number of additional comparisons. How many new comparisons are there? What is the total number of comparisons that must be made?

b. Add a fourth choice, D, to the diagram. Connect it to each of A, B, and C. How many new comparisons are there, and what is the total number of comparisons?

c. Add a fifth choice to the diagram and repeat. Then add a sixth choice and repeat. Complete the following table:

Number of Choices	Number of New Comparisons	Total Number of Comparisons
1	0	0
2	1	1
3		
4		
5		
6		

11. Let C_n represent the total number of comparisons necessary when there are n choices. Write a recurrence relation that expresses the relationship between C_n and C_{n-1}.

Computer Explorations

12. Use the preference schedule program that accompanies this book to find a set of preferences with at least four choices that demonstrate the same paradox found in Exercise 7 when the sequential runoff method is used.

13. Use the preference schedule program to enter several schedules with five choices. Use the program's features to alter your data in order to produce a set of preferences with several different winners. Can you find a set of preferences with five choices and five winners? If so, what is the minimum number of schedules with which this can be done?

Projects

14. Research and report on paradoxes in mathematics. Try to determine whether the paradoxes have been satisfactorily resolved.

15. Research and report on paradoxes outside mathematics. In what way have these paradoxes been resolved?

16. Select an issue of current interest in your community that involves more than two choices. Have each member of your class vote by writing a preference schedule. Compile the preferences and determine the winner by five different methods.

17. Investigate the contributions of Charles Dodgson (Lewis Carroll) to group-ranking procedures. Was he responsible for any of the group-ranking procedures you have studied? What did he suggest doing when the Condorcet method failed to produce a winner?

Arrow's Conditions and Approval Voting

Ten representatives of the five foreign language clubs of Central High School are meeting to select a location for the clubs' annual joint dinner. The committee must choose among a Chinese, French, Italian, or Mexican restaurant. The preferences of the representatives fall into three categories (see Figure 1.6).

Racquel suggests that because the last two dinners have been held at Mexican and Chinese restaurants, this year's dinner should be at either an Italian or a French restaurant. The group votes 7 to 3 in favor of the Italian restaurant.

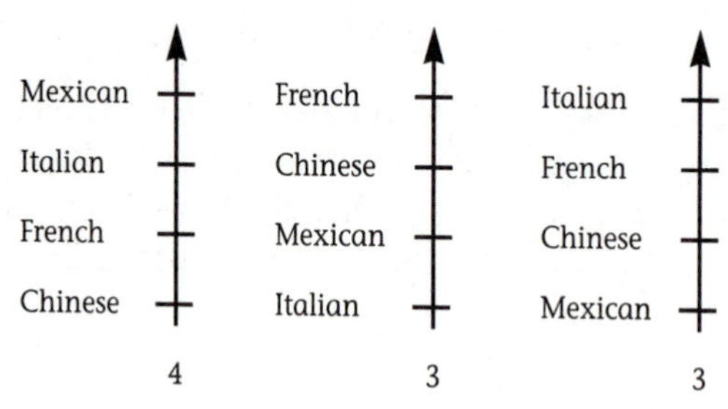

Figure 1.6 Preferences of 10 students.

Martin, who doesn't like Italian food, says that the community's newest Mexican restaurant has an outstanding reputation. He proposes that the group choose between Italian and Mexican. The other members agree and vote 7 to 3 to hold the dinner at the Mexican restaurant.

Sarah, whose parents own a Chinese restaurant, says that she can obtain a substantial price discount for the event. The group votes between the Mexican and Chinese restaurants and selects the Chinese by a 6 to 4 margin.

Look carefully at the group members' preferences. Note that French food is preferred to Chinese by all, yet the voting has resulted in the selection of the Chinese restaurant!

In 1951, paradoxes such as this led Kenneth Arrow, an American economist, to formulate a list of conditions that he considered necessary for a fair method of determining a group ranking. These conditions are today known as **Arrow's conditions.**

Arrow settled on five criteria. One of them states that if every member of a group prefers one choice to another, the group ranking should do the same. According to this criterion, the choice of the Chinese restaurant when all members rated French food more favorably is unfair. Thus, a group-ranking scheme that allows pairs of choices to be compared in any order is undesirable.

Arrow's Conditions

1. Nondictatorship: The preferences of a single individual should not become the group ranking without considering the other individuals' preferences.

2. Individual sovereignty: Each individual is allowed to order the choices in any way and can indicate ties.

3. Unanimity: If every individual prefers one choice to another, the group ranking should do the same.

4. Freedom from irrelevant alternatives: The group ranking between any pair of choices does not depend on the individuals' preferences for the remaining choices.

5. Uniqueness of the group ranking: The method of producing the group ranking should give the same result whenever it is applied to a given set of preferences. The group ranking must also be transitive.

Arrow inspected many of the common methods of determining a group ranking for their adherence to his five criteria. He also looked for new methods that would meet all five. After doing so, he arrived at a surprising conclusion.

In the following exercises, you will examine a number of group-ranking methods for their adherence to Arrow's criteria. You will also learn Arrow's surprising result.

Exercises

1. Your teacher decides to order soft drinks based on the soft drink vote conducted at the beginning of this chapter but in doing so selects the preference schedule of a single student (the teacher's pet). Which of Arrow's conditions are violated by this method of determining a group ranking?

2. Instead of selecting the preference schedule of a favorite student, your teacher places all of the individual preferences in a hat, draws one, and orders the soft drinks based on only that one. If this method were repeated, would the same group ranking result? Which of Arrow's conditions does this method violate?

3. Do any of Arrow's conditions require that the voting mechanism include a secret ballot? Is a secret ballot desirable in all group-ranking situations?

4. Examine the paradox demonstrated in Exercise 8 of Lesson 1.3 on page 23. Which of Arrow's conditions are violated?

5. Construct a set of preference schedules with three choices, A, B, and C, showing that the plurality method violates Arrow's fourth condition. In other words, construct a set of preferences in which the outcome between A and B depends on whether or not C is on the ballot.
6. There often are situations in which insincere voting results. Do any of Arrow's conditions state that insincere voting should not be part of a fair group-ranking procedure?
7. Suppose that there are only two choices in a list of preferences and that the plurality method is used to decide the group ranking. Which of Arrow's conditions could be violated?
8. After failing to find a group-ranking method for three or more choices that always obeyed all of his fairness conditions, Arrow began to suspect that such a method might be impossible. He applied logical reasoning to the problem and proved that no method could do so. In other words, any group-ranking method will violate at least one of Arrow's conditions in some situations.

   Arrow's proof demonstrates how mathematical reasoning can be applied to areas outside mathematics. This and other achievements resulted in Arrow's receiving the Nobel Prize for economics.

   Although Arrow's work means that we cannot achieve perfection in group rankings, it doesn't mean that we can't improve our methods. Recent studies have led experts to believe that a system called **approval voting** violates Arrow's criteria less often than does any other known method.

   In approval voting, you may vote for as many choices as you like, but you do not rank them. You mark all those of which you approve. For example, if there are five choices, you may vote for as few as none or as many as five.

   a. Write a soft drink ballot like the one you used in Lesson 1.1 of this chapter. Place an "X" beside each of the soft drinks that you feel is acceptable. At the direction of your instructor, collect the ballots from each member of the class. Count the number of votes for each soft drink and determine the winner.
   b. Determine a complete group ranking.
   c. Was the approval winner the same as the earlier plurality winner in your class?

d. How does the group ranking in part b compare with the earlier Borda ranking?

9. Examine Exercise 4 of Lesson 1.3 on page 22. Would the sports-writers who supported team B be encouraged to vote insincerely if approval voting were used?

10. What is the effect on a group ranking of casting approval votes for all the choices? Of casting approval votes for none of them?

11. Approval voting offers a voter many more choices than does the system used in most elections. If there are three candidates for a single office, for example, the present system allows four choices: Vote for any one of the three candidates or for none of them. Approval voting allows one to vote for none, any one, any two, or all three. In how many ways can you vote when approval voting is used? If there are two choices, A and B, there are four possibilities. We indicate voting for none by writing { }, voting for A by writing {A}, voting for B by writing {B}, and voting for both by writing {A,B}.
    a. List all ways of voting under the approval system when there are three choices.
    b. List all ways of voting under the approval system when there are four choices.
    c. Generalize the pattern by letting V_n represent the number of ways of voting under the approval system when there are n choices and writing a recurrence relation that describes the relationship between V_n and V_{n-1}.

12. Listing all the ways of voting under the approval system can be difficult if not approached systematically. The following algorithm describes how to find all the ways of voting for two items. The results shown are applied to a ballot with five choices.

	List 1	List 2
1. List all choices in order.	A B C D E	
2. Draw a line through the first choice, and write it once for each remaining choice that doesn't have a line through it.	A̸ B C D E	

3. In the second list, write the choice that you just drew a line through beside each choice that doesn't have a line through it.

AB AC AD AE

4. Repeat Steps 2 and 3 until each choice has a line through it. The items in the second list show all the ways of voting for two items.

Write an algorithm that describes how to find all the ways of voting for three choices. You may use the results of the previous algorithm to begin the new one.

13. Many patterns can be found in the various ways of voting when the approval system is used. The table below shows the number of ways of voting for exactly one item when there are several choices on the ballot. For example, in Exercise 11, you listed all ways of voting when there are three choices on the ballot. Three of these, {A}, {B}, and {C}, are selections of exactly one item.

Number of Choices on the Ballot	Number of Ways of Selecting Exactly One Item
1	1
2	2
3	3
4	___
5	___

Complete the table.

14. Let $V1_n$ represent the number of ways of selecting exactly one item when there are n choices on the ballot, and write a recurrence relation that expresses the relationship between $V1_n$ and $V1_{n-1}$.

15. The table below shows the number of ways of voting for exactly two items when there are from one to five choices on the ballot. For example, your list in Exercise 11 shows that when there are three

choices on the ballot, there are three ways of selecting exactly two items: {A,B}, {A,C}, and {B,C}.

Number of Choices on the Ballot	Number of Ways of Selecting Exactly Two Items
2	1
3	3
4	____
5	____

Complete the table.

16. Let $V2_n$ represent the number of ways of selecting exactly two items when there are n choices on the ballot, and write a recurrence relation that expresses the relationship between $V2_n$ and $V2_{n-1}$. Can you find more than one way to do this?

Projects

17. Investigate the number of ways of voting under the approval system for other recurrence relations (see Exercises 13 through 16).
18. Arrow's result is an example of an *impossibility theorem.* Investigate and report on other impossibility theorems.
19. Research and report on Arrow's theorem. The theorem is usually proved by an indirect method. What does this mean? How is it applied in Arrow's case?
20. Design a computer program that lists all possible ways of voting when approval voting is used. Use the letters A, B, C, . . . , to represent the choices. The program should ask for the number of choices and then display all the possible ways of voting for one choice, two choices, and so forth.

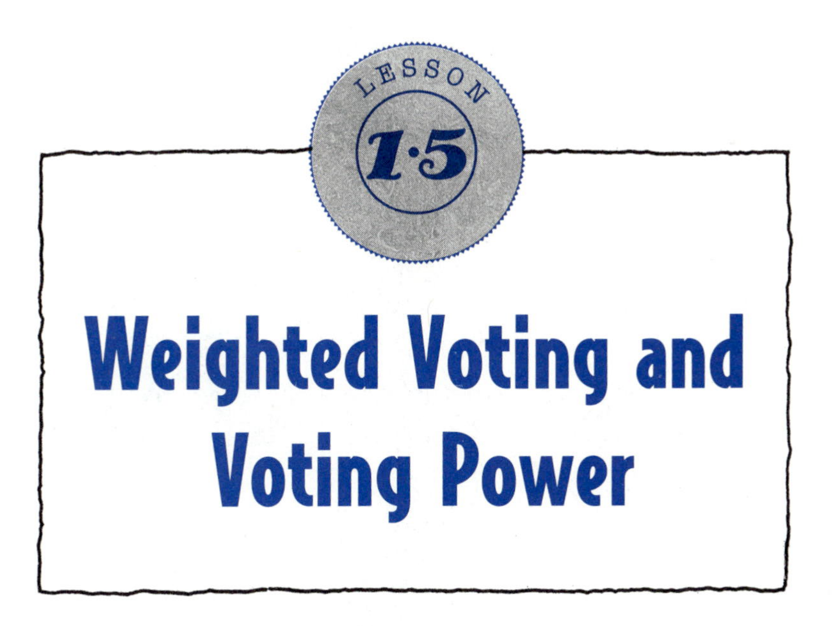

Weighted Voting and Voting Power

A small high school has 110 students. Because of recent growth in the size of the community, the sophomore class is quite large. It has 50 members, and the junior and senior classes each have 30 members.

The school's student council is composed of a single representative from each class. Each of the three members is given a number of votes proportionate to the size of the class represented. Accordingly, the sophomore representative has five votes, and the junior and senior representatives each have three. The passage of any issue that is before the council requires a simple majority of six votes. This procedure is known as **weighted voting.**

> Weighted voting occurs whenever some members of the voting body have more votes than the others do.

Consider the number of ways that voting could occur in this example. It is possible that an issue could be favored by none of the members, one of them, two of them, or all three. In which cases would the issue be passed? Listed below are all the possible ways of voting for an issue and the associated number of votes.

{;0} {So;5} {Jr;3} {Sr;3} {So,Jr;8}
{So,Sr;8} {Jr,Sr;6} {So,Jr,Sr;11}

For example, {Jr,Sr; 6} indicates that the junior and senior representatives could vote for an issue and that they have a total of 6 votes between them.

Each of these collections is known as a *coalition.* Those that are successful are known as **winning coalitions.** The winning coalitions in this example are those with 6 or more votes and are listed below along with their respective vote totals.

{So,Jr;8} {So,Sr;8} {Jr,Sr;6} {So,Jr,Sr;11}

A coalition is any group of voters, and a winning coalition is any group of voters that has enough votes to pass an issue.

Of the winning coalitions, the last is different from the other three in one important respect: If any one of the members decides to vote differently, the coalition will still win. None of the members is essential to the coalition. When all of the members are essential to a winning coalition, it is called a *minimal winning coalition.* The minimal winning coalitions in this example are listed below.

{So,Jr;8} {So,Sr;8} {Jr,Sr;6}

A winning coalition that will become a losing coalition if any of the members is removed is called a minimal winning coalition.

Notice that the sophomore representative is essential to two of the winning coalitions. This is also true of the junior and senior representatives. In other words, in about the same number of times, each of the representatives can expect to cast a key vote in passing an issue.

A paradox: Although the votes have been distributed to give greater power to the sophomores, the actual outcome is that all members have the same amount of power! Since distributing the number of votes in a way that reflects the distribution of the population does not result in a fair distribution of power, mathematical procedures can be used to try to find a way to measure actual power when weighted voting is used.

A measure of the power of a member of a voting body is called a **power index,** or the number of winning coalitions to which each member of that body is essential. In our example, each representative is essential to two winning coalitions and thereby has a power index of 2. In other words, if the sophomore representative is removed from either of the marked coalitions, the coalition will no longer win. Therefore, the sophomore power index is 2.

*{So,Jr;8} {Jr,Sr;6}
*{So,Sr;8} {So,Jr,Sr;11}

Many people think that votes should be distributed so that the members' power indices reflect the differences in the population. John Banzhaff III, a law professor at George Washington University, has initiated several legal actions that have upheld this belief.

A Power Index Algorithm

1. List all coalitions of voters that are winning coalitions.

2. Select any voter, and record a 0 for that voter's power index.

3. From the list in Step 1 select any coalition of which the voter is a member. Subtract from the coalition's vote total the number of votes that this voter has. If the result is less than the number of votes necessary to pass an issue, add 1 to the voter's power index.

4. Repeat Step 3 until you have checked all the coalitions for which the voter is a member.

5. Repeat Steps 2 through 4 until you have checked all the voters.

Consider a group of three members in which A has seven votes, B has three, and C has three. The coalitions and winning coalitions are listed below.

All coalitions: { ;0} {A;7} {B;3} {C;3} {A,B;10}
{A,C;10} {B,C;6} {A,B,C;13}
Winning coalitions: {A;7} {A,B;10} {A,C;10} {A,B,C;13}

In this situation, A is the only one with any power. All four of the winning coalitions will become losing coalitions if A is removed. B and C have no power because every winning coalition will remain winning if either is removed.

If one member has over half of the votes, that member alone is essential to the winning coalitions and so is known as a *dictator*. When a member is essential to no winning coalitions, that member is known as a *dummy*. In this case, A is a dictator and B and C are dummies.

Exercises

1. Consider a situation in which A, B, and C have 3, 2, and 1 votes, respectively, and in which 4 votes are needed to pass an issue.

a. List all possible coalitions, all winning coalitions, and all minimal winning coalitions.
b. Determine the power index for each voter.
c. Suppose that the number of votes needed to pass an issue is increased from 4 to 5. Determine the power index of each voter.

2. The examples in this lesson depicted a situation with three voters that resulted in equal power for all three and another that resulted in total power for one member. In Exercise 1, the power was distributed differently. What other distributions of power are possible with only three members? Find a distribution of votes that results in a power distribution different from the three that you have already examined.

3. In the student council example in this section, can the votes be distributed so that the members' power indices will follow the ratios of the class sizes?

4. In the student council example in this section, suppose that the representatives of the junior and senior classes always differ on issues and never vote alike. Does this make any practical difference in the power of the three representatives?

5. (See Exercise 11 of Lesson 1.4, p. 30.) Let C_n represent the number of coalitions that can be formed in a group of n voters. Write a recurrence relation that describes the relationship between C_n and C_{n-1}.

6. One way to determine all winning coalitions and all minimal winning coalitions in a weighted-voting situation is to work from a list of all possible coalitions. Use A, B, C, and D to represent the individuals in a group of four voters, and list all possible coalitions.

7. Weighted voting is commonly used to decide issues at meetings of corporate stockholders. Each member is given a vote for each share of stock held. Suppose that a company has four stockholders: A, B, C, and D. They own 26%, 25%, 25%, and 24% of the stock, respectively, and more than 50% of the vote is needed to pass an issue.
   a. Determine the power index of each stockholder. Use your results from Exercise 6 as an aid.

b. The company has four stockholders, and they own 47%, 41%, 7%, and 5% of the stock. Find the power index of each stockholder.

c. Compare the percentage of stock held by the smallest stockholder in parts a and b. Do the same for the power index of the smallest stockholder.

8. In 1964, the Nassau County, New York, Board of Supervisors had 6 members. The number of votes given to each was 31, 31, 21, 28, 2, and 2.

   a. Determine the power index for each member.

   b. The Nassau County Board was composed of representatives of five municipalities: Hempstead, North Hempstead, Oyster Bay, Glen Cove, and Long Beach. The respective populations were 728,625; 213,225; 285,545; 22,752; and 25,654. The members with 31 votes both represented Hempstead. The others each represented the municipality listed in the same order as above.

NEWS CLIP

Nassau Districting Ruled Against Law

Special to THE NEW YORK TIMES, *January 15, 1970*

ALBANY—The Court of Appeals ruled today that the present "weighted voting" plan of the Nassau County Supervisors was unconstitutional but that a new plan was not necessary until after the 1970 Federal census.

In a unanimous opinion, the state's highest court said the county's present Charter provision is a clear violation of the one-man-one-vote principal in that it specifically denied the Town of Hempstead representation that reflected its population.

The town, the court pointed out, constituted 57.12 per cent of the county's population, but because of the weighted voting plan its representatives on the board could cast only 49.6 per cent of the board's vote.

Compare the power indices of the municipalities with their populations.

9. Would defining a voter's power index as the number of minimal winning coalitions to which the member belongs be equivalent to the definition given in Lesson 1.5?

Computer Explorations

10. Use the weighted-voting program on the disk that accompanies this book to experiment with different weighted-voting systems when there are three voters. Change the number of votes given to each voter and the number of votes required to pass an issue. How many different power distributions are possible? Do the same for weighted-voting systems when there are four voters.

Projects

11. The Security Council of the United Nations is composed of 5 permanent members and 10 others that are elected to 2-year terms.

For a measure to pass, it must obtain at least 9 votes that include all 5 of the permanent members. Determine the power index for a permanent member and for a temporary member.

12. Research and report on power indices. What, for example, is the Banzhaf power index? What is the Shapley–Shubik power index? Which is similar to the index used in this section?

13. The president of the United States is legally chosen by the electoral college. What does this system do to the power of voters in different states in selecting the president? Research the matter and report on the relative power of voters in different states.

14. Research and report on court decisions involving voting power in situations in which weighted voting is used.

1. Consider the set of preferences shown below.

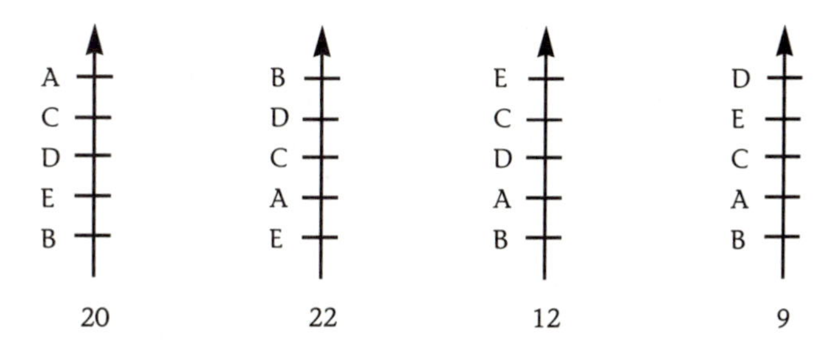

   a. Determine the winner using a 5–4–3–2–1 Borda count.
   b. Determine the plurality winner.
   c. Determine the runoff winner.
   d. Determine the sequential runoff winner.
   e. Determine the Condorcet winner.
   f. Suppose that this election is conducted by the approval method and that all the voters decide to approve of the first two choices on their preference schedules. Determine the approval winner.

2. Complete the following table for the recurrence relation $B_n = 2B_{n-1} + n$.

n	B_n
1	3
2	$2(3) + 2 = 8$
3	
4	
5	

3. In this chapter, you have encountered many paradoxes involving group-ranking methods.
   a. One of the most amazing paradoxes occurs when a winning choice becomes a loser when its standing actually improves. In which group-ranking method(s) did this occur?
   b. Discuss at least one other paradox that occurs with group-ranking methods.

4. In the 1912 presidential election, polls showed that the preferences of voters were as shown below.

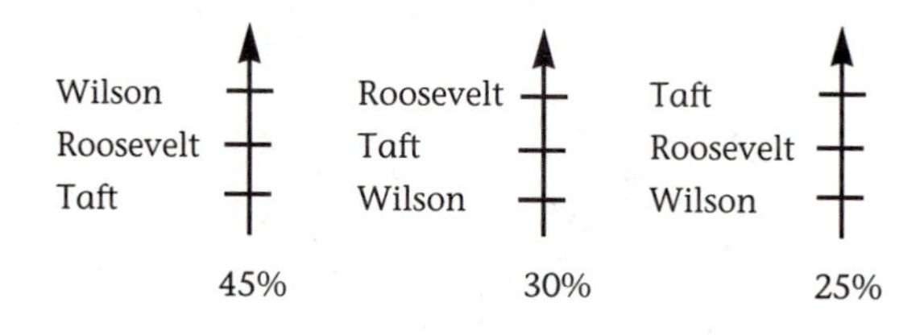

   a. Who won the election? Was he a majority winner?
   b. How did the majority of voters feel about the winner?
   c. How might one of the groups of voters have changed the results of the election by voting insincerely?
   d. Discuss who might have won the election if a different method had been used.

5. Your class is ranking soft drinks, and someone suggests that the names of the soft drinks be placed in a hat and the group ranking be determined by drawing them from the hat. Which of Arrow's conditions does this method violate?

6. State Arrow's theorem. In other words, what did Arrow prove?

7. Consider a situation in which voters A, B, C, and D have 4, 3, 3, and 2 votes, respectively, and 7 votes are needed to pass an issue.
   a. List all winning coalitions and their vote totals.
   b. Find the power index for each voter.
   c. Do the power indices reflect the number of votes assigned to each voter?
   d. Suppose the number of votes necessary to pass an issue is increased from 7 to 8. How does this change the voters' power indices?

Bibliography

Brams, Steven J. 1976. *Paradoxes in Politics.* New York: Free Press.

Brams, Steven J., and Peter Fishburn. 1983. *Approval Voting.* Boston: Birkhauser.

Bunch, Brian. 1982. *Mathematical Fallacies and Paradoxes.* New York: Van Nostrand Reinhold.

COMAP. 1991. *For All Practical Purposes: Introduction to Contemporary Mathematics.* 2nd ed. New York: Freeman.

Davis, Morton. 1980. *Mathematically Speaking.* New York: Harcourt Brace Jovanovich.

Falletta, Nicholas. 1983. *The Paradoxican.* Gardon City, NY: Doubleday.

Hoffman, Paul. 1988. *Archimedes' Revenge.* New York: Norton.

Lucas, William F. 1992. *Fair Voting: Weighted Votes for Unequal Constituencies.* Lexington, MA: COMAP.

Lum, Lewis, and David C. Kurtz. *Voting Made Easy: A Mathematical Theory of Election Procedures.* Greensboro, NC: Guilford Press.

Meyer, Rochelle Wilson, and Walter Meyer. 1990. *Play It Again Sam: Recurrence Equations in Mathematics and Computer Science.* Lexington, MA: COMAP.

Pakhomov, Valery. 1993. "Democracy and Mathematics." *Quantum,* January–February, pp. 4–9.

Fair Division

Fascination with cakes of all kinds and shapes extends from artists like Thiebaud who express their interest with paint and canvas, to mathematicians who explore cake-cutting algorithms. As children, we have all experienced the feeling of unfairness that occurs when someone else gets the piece of cake that we think is better than all the others.

How do you divide a cake among a group of people so that all of them will think the result is fair? Is the meaning of fairness when a cake is divided among children different from the meaning of fairness when an estate is divided among heirs or when seats in congress are divided among states? Are the methods that are commonly used to divide cakes, estates, and legislatures necessarily the fairest methods? Discrete mathematics plays an important role in answering these questions.

LESSON 2·1

A Fair Division Activity

There are many different circumstances in which the division of an object in a fair way is important to the people involved. Three of the most common are the division of food among children, a house in an estate among heirs, and the seats in a governmental body among districts. Each has characteristics that make it different from the others. In this lesson you and the other people in your class will propose your own solutions to these three types of problems.

Explore This

At the direction of your instructor, divide your class into groups of three people. Write the numbers 1, 2, or 3 on each of several slips of paper. Have each group draw one of the slips from a bag or box. Each group should consider the fair division problem listed below that corresponds to the number drawn by the group. Allow about 15 minutes for each group to discuss the problem.

After all groups have finished their discussions, a spokesperson for each group should present the group's decision to the class. Each group that discussed Problem 1 should report first, and so on. Each member of the class should record the method used by each group.

1. Martha and Ray want to divide the last piece of the cake that their mother baked yesterday. Propose a method of dividing the piece of cake that will satisfy both Martha and Ray.

2. Juan and Mary are the only heirs to their mother's estate. The only object of significant value is the house in which they were raised. Propose a method that fairly disposes of the house.

3. The sophomore, junior, and senior classes of Central High School have 333, 288, and 279 members, respectively. The school's student council is composed of 20 members divided among the three classes. Determine a fair number of seats on the council for each class.

Exercises

1. a. Did any groups resolve Problem 1 by relying on the mother's authority? In what way?
   b. Does the decision of such a problem by the mother or other authority figure usually result in a solution that seems fair to both children?
   c. Cite at least two examples of situations in which fair division problems are resolved by an authority.

2. a. Did any of the groups use a random event such as a coin flip to resolve Problem 1? In what way?
   b. Does the use of randomness in such a problem usually result in a solution that seems fair to both children?
   c. Cite at least two examples of situations in which randomness is used to resolve an issue.

3. a. Did any of the groups use a means of measuring the piece of cake to resolve Problem 1? In what way?

b. Does the use of measurement in such problems produce a solution that seems fair to both children?
c. Give an example of a situation in which measurement is likely to result in an agreeable solution to a fair division problem.

4. A common way of resolving this problem is to have one of the children cut the cake into two pieces and to have the other choose first.
   a. Suppose that Martha cuts the cake. How will she feel about the two pieces?
   b. If Ray gets to choose one of the two pieces that Martha cut, how will he feel about the two pieces?
   c. If you were one of the participants in this scheme, would you rather be the cutter or the chooser? Why?
5. Write a description of what you consider to be desirable results of a process that fairly divides a cake among any number of people.
6. Did any of the groups resolve Problem 2 by selling the house and dividing the cash? Why might this be unsatisfactory to one or more of the heirs?
7. Did any of the groups use a method that considers the possibility that the heirs might not agree on the value of the house? In what way?
8. Juan thinks the house is worth $60,000, and Mary feels it is worth $70,000.
   a. Who do you think should receive the house?
   b. How might the person who doesn't get the house be compensated?
9. Write a description of what you consider to be desirable results of a process that fairly divides a house among several heirs.
10. How might the possibility of lying about the value of the house affect the result of a division process?
11. Did all of the groups that discussed Problem 3 divide the seats among the classes in the same way? If not, describe the differences.
12. If some of the groups that discussed Problem 3 obtained different results, which of the methods do you think is the fairest? If all

groups produced the same result, do you agree that the result is fair? Why or why not?

13. Write a description of what you consider to be desirable results of a process that fairly divides the seats in a student council among a school's classes.

Estate Division

A fair division problem can be either *discrete* or *continuous*. The problem of dividing a house among heirs and that of dividing a student council among classes are examples of the discrete case. Discrete division is involved whenever objects of division cannot be meaningfully separated into pieces. Dividing a cake is an example of the continuous case because the cake can be divided into any number of parts.

In this lesson, we look at the fair division of an estate among heirs. The issue of fairly dividing a student council is examined in a later lesson. We consider first an algorithm for dividing an estate among heirs that results in an appealing paradox: Each of the heirs receives a share that is larger than he or she thinks is fair.

1. Each heir submits a bid for each item in the estate. Bids are not made on cash in the estate because it is a continuous medium that can be divided equally without controversy.

2. A fair share is determined for each heir by finding the sum of his or her bids and dividing this sum by the number of heirs.

3. Each item in the estate is given to the heir who bid the highest on that item.

4. Each heir is given an amount of cash from the estate that is equal to his or her fair share (from Step 2) less the amount that the heir bid on the objects he or she received. If this amount is negative, the heir pays that amount into the estate.

5. The remaining cash in the estate is divided equally among the heirs.

Example

Amanda, Brian, and Charlene are heirs to an estate that includes a house, a boat, a car, and $75,000 in cash. Each submits a bid for the house, boat, and car. The bids are summarized in the following table, or matrix.

	House	Boat	Car
Amanda	$40,000	$3,000	$5,000
Brian	$35,000	$5,000	$8,000
Charlene	$38,000	$4,000	$9,000

The entries in Amanda's row, for example, indicate the value to Amanda of each item in the estate.

A fair share is determined for each heir.

Amanda: ($40,000 + $3,000 + $5,000 + $75,000)/3 = $41,000.
Brian: ($35,000 + $5,000 + $8,000 + $75,000)/3 = $41,000.
Charlene: ($38,000 + $4,000 + $9,000 + $75,000)/3 = $42,000.

The house is given to Amanda, the boat to Brian, and the car to Charlene. Cash equal to the difference between the fair share and the value of the awarded items is given to each heir.

Amanda: $41,000 − $40,000 = $1,000.
Brian: $41,000 − $5,000 = $36,000.
Charlene: $42,000 − $9,000 = $33,000.

NEWS CLIP

Puccini's Inheritance Part of a Murky Tale That's, Well, Operatic

By Paul Hofmann, *Special to* THE NEW YORK TIMES, Sept. 20, 1990

ROME, Sept. 19—When Giacomo Puccini died in a Brussels clinic in 1924, he left a fortune estimated at more than $100 million at today's value. His estate would grow further over the years through substantial royalties from "La Bohème," "Tosca," "Madama Butterfly" and other operas. . . .

Puccini's wife, Elvira, was in Viapreggio when the composer died of throat cancer at the age of 66. His will named their only son, Antonio, as sole heir. The younger Puccini died in 1946, with his widow, Rita, considered the only heir. After her death in 1979, what had remained of the Puccini fortune went to her brother, Baron Livio dell'Anna, a lawyer. The Baron died in 1986, leaving no offspring or other known relatives.

Meanwhile, a would-be heir to the Puccini wealth had entered the scene. She was Simonetta Giurumello, who asserted she was a daughter of the composer's son and fought for her claim through the courts. When Italy's highest tribunal in civil matters, the Court of Cassaion, upheld her claim, she changed her name to Simonetta Puccini, as was her right. . . .

The Italian judiciary will, among other things, have to determine the legal status of the Puccini Foundation, which formally owns the Torre del Lago villa, and whether the property may be sold to satisfy the claims of Simonetta Puccini and of tax officials. . . .

The cash given to the heirs totals $70,000, and the remaining $5,000 is divided equally. Thus, each heir receives $1666.67, more than a fair share. The results of this division may be summarized in a matrix:

	Amanda	Brian	Charlene
Total of bids and cash	$123,000	$123,000	$126,000
Fair share (total ÷ 3)	$41,000	$41,000	$42,000
Items received	House	Boat	Car
Value of items rec'd.	$40,000	$5,000	$9,000
Cash received (fair share: value of items received)	$1,000	$36,000	$33,000
Share of remaining cash	$1,666.67	$1,666.67	$1,666.67

Exercises

1. The application of any fair division algorithm requires certain assumptions, or axioms. For example, the success of the estate division algorithm requires that each heir place a value on each object in the estate. If any heir considers an object priceless, the algorithm will fail. List at least one other axiom that you think is necessary for the success of this algorithm.

2. Suppose that Garfield and Marmaduke are heirs to an estate that contains only a house. Garfield bids $70,000, and Marmaduke bids $60,000.
   a. What does Garfield feel is a fair share? Marmaduke?
   b. What is the difference between Garfield's fair share and Garfield's bid for the house?
   c. Because the value of the house is more than Garfield's fair share, Garfield must pay cash into the estate. How much cash must Garfield pay?
   d. Marmaduke is given an amount of cash from Garfield's payment equal to Marmaduke's fair share. How much does Marmaduke receive? If the remaining cash is divided equally, what will be the final value of Marmaduke's settlement? Of Garfield's?
   e. Garfield must borrow money in order to pay into the estate, and

the interest on this loan is $2,000. Do you think this should be considered when arriving at a settlement? If so, suggest how the settlement should be revised.

f. If the division between Garfield and Marmaduke were settled in the usual way, Marmaduke would be given half of Garfield's bid. Compare the final settlements for Garfield and Marmaduke if this method is used with the settlements of part d. Which result do you think is best?

3. Amy, Bart, and Carol are heirs to an estate that consists of a valuable painting, a car, a New York Yankees season ticket, and $5,000 in cash. They submit the bids shown in the matrix below. Use the method of this lesson to divide the estate among the heirs. For each heir, state the fair share, the items received, the amount of cash, and the final settlement. Summarize your results in a matrix.

	Painting	Car	Ticket
Amy	$2,000	$4,000	$500
Bart	$5,000	$2,000	$100
Carol	$3,000	$3,000	$300

Hint: When a division requires that some heirs pay into the estate, be sure to keep an accurate record of the estate's cash:

Cash	$5,000
Received from Amy	____
Received from Bart	____
Paid to Carol	____
Remaining cash	____

4. Suppose that in the division of Exercise 3, Amy had received previous financial support from the estate. A will states that she is to receive only 20% of the estate, whereas Bart and Carol are to receive 40% each. Adapt the algorithm of this lesson to this situation, and describe a fair division of the estate.

5. Suppose two heirs submit an identical highest bid for an item. How would you resolve this?

6. Alan, Betty, and Carl are heirs to an estate and have submitted the bids shown below.

	House	Boat	Car
Alan	\$55,000	\$3,000	\$8,000
Betty	\$60,000	\$4,000	\$6,000
Carl	\$56,000	\$3,000	\$7,000

The awarding of the items in the estate can be indicated in a matrix, as shown below.

	Alan	Betty	Carl
House	0	1	0
Boat	0	1	0
Car	1	0	0

The numbers in Alan's column indicate the items that he received. For example, the 1 in Alan's column and the car's row indicates that Alan received the car. Each of the other entries in Alan's column is a 0, which indicates that Alan received neither the house nor the boat.

A new matrix can be computed by writing the second matrix alongside the first, as shown below.

$$\begin{bmatrix} \$55{,}000 & \$3{,}000 & \$8{,}000 \\ \$60{,}000 & \$4{,}000 & \$6{,}000 \\ \$56{,}000 & \$3{,}000 & \$7{,}000 \end{bmatrix} \begin{bmatrix} 0 & 1 & 0 \\ 0 & 1 & 0 \\ 1 & 0 & 0 \end{bmatrix}$$

The new matrix is computed by multiplying each entry in the first row of the first matrix by the corresponding entry in the first column of the second matrix and finding the sum of these products:

$$\$55{,}000(0) + \$3{,}000(0) + \$8{,}000(1) = \$8{,}000.$$

Because this number was obtained from the first row of the first matrix and the first column of the second matrix, it is written in the first row and the first column of the new matrix.

The entry for the first row and the second column of the new matrix is found by performing a similar calculation with the first row of the first matrix and the second column of the second matrix:

$$\$55{,}000(1) + \$3{,}000(1) + \$8{,}000(0) = \$58{,}000.$$

a. Calculate the remaining entries of the new matrix.
b. The \$8,000 in the first row and the first column of the new matrix can be interpreted as the value to Alan of the items he received. Write an interpretation of the number in the first row and the second column of the new matrix.
c. Write an interpretation of the number in the second row and the second column of the new matrix.

Computer Explorations

7. Use a computer spreadsheet to perform the type of estate division done in this lesson. A sample output and the formulas that generated it are shown in the figures below. In this case, the results are those of Exercise 3. Once your spreadsheet is complete, use it to answer the questions that follow.

	A	B	C	D	E	F
1		Estate Division Spreadsheet				
2						
3		Amy	Bart	Carol		
4	Painting	2000.00	5000.00	3000.00		
5	Car	4000.00	2000.00	3000.00		
6	Ticket	500.00	100.00	300.00		
7						
8		Amy	Bart	Carol		Cash:
9	Bid total	11500.00	12100.00	11300.00		5000.00
10	Share	0.333333	0.333333	0.333333		666.67
11	Fair share	3833.33	4033.33	3766.67		966.67
12	Object value	4500.00	5000.00	0.00		-3766.67
13	Cash received	-666.67	-966.67	3766.67		
14	Extra cash	955.56	955.56	955.56		2866.67
15						
16	Final total	4788.89	4988.89	4722.23		
17						

	A	B	C	D	E	F
1		Estate Division Spreadsheet				
2						
3		Amy	Bart	Carol		
4	Painting	2000	5000	3000		
5	Car	4000	2000	3000		
6	Ticket	500	100	300		
7						
8		Amy	Bart	Carol		Cash:
9	Bid total	=SUM(B4:B6)+$F$9	=SUM(C4:C6)+$F$9	=SUM(D4:D6)+$F$9		5000
10	Share	=1/3	=1/3	=1/3		=−B13
11	Fair share	=B9*B10	=C9*C10	=D9*D10		=−C13
12	Object value	=B5+B6	=C4	=0		=−D13
13	Cash received	=B11−B12	=C11−C12	=D11−D12		
14	Extra cash	=$F$14*B10	=$F$14*C10	=$F$14*D10		=SUM(F9:F12)
15						
16	Final total	=SUM(B12:B14)	=SUM(C12:C14)	=SUM(D12:D14)		

a. What would happen if the amount of cash in the estate were 0? Change the amount in F9 to 0 and see.

b. What would happen if Bart lied about the value he placed on the car and said he thought it was worth $5,000? Change the amount in C5 to 5,000 and see. (Change the cash back to 5,000 before doing this one.)

c. What would happen if Bart really did feel that the car was worth $5,000 but that he had accepted a $2,000 bribe from Amy to bid $2,000? How do you think this would change the value of the final settlements for Bart and Amy?

d. How would you change the formulas in the spreadsheet to account for the situation in Exercise 4?

Projects

8. Matrix calculations of the type shown above can be useful in programming computers to do tedious calculations. Research and report on the use of matrix applications in computer science.

9. Research division procedures that are used at auctions. What are Dutch and English auctions? Why do some auctions award the contract to the second-highest bidder? When are closed and open bids used? Why?

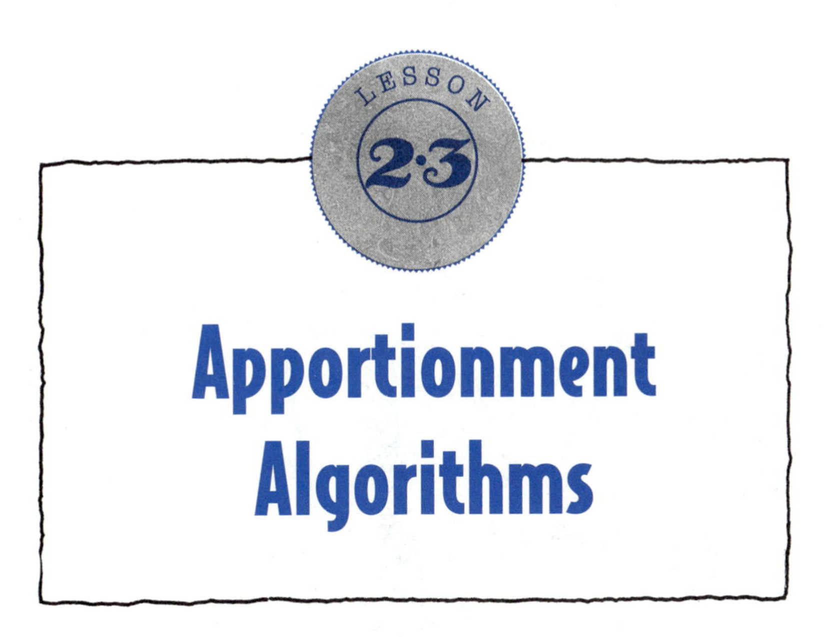

Apportionment Algorithms

Fair division controversies are common. Parties that do not agree often settle their differences in court. In this country, political and legal disputes over the division of legislative representation are as old as the nation itself. The first veto by an American president occurred when George Washington vetoed an apportionment bill advocated by Alexander Hamilton in favor of a method championed by Thomas Jefferson. In this lesson, we will consider the two methods of dividing representation advocated by these well-known statesmen.

Central High School has sophomore, junior, and senior classes of 464, 240, and 196 students, respectively. The 20 seats on the school's student council are divided among the three classes. Divide the total student population of 900 by 20 seats to obtain a quotient of 45 students per seat. Ideally, each council member would represent 45 students. This number is called the **ideal ratio** of students per seat. In cases of political representation, it is often called the *ideal district size.*

$$\text{Ideal ratio} = \frac{\text{Total population}}{\text{Number of seats}}$$

Because the sophomore class has 464 members, it deserves 464 ÷ 45 = 10.31 seats. Accordingly, 10.31 is called the sophomore class **quota.** Similarly, the junior and senior quotas are 5.33 and 4.36 seats, respectively.

The Quotas

Sophomores:	10.31
Juniors:	5.33
Seniors:	4.36

$$\text{Quota} = \frac{\text{Class Size}}{\text{Ideal Ratio}}$$

The seats on the council are discrete and cannot be subdivided. That is, a single seat cannot be split to give 0.36 of it to the seniors, 0.33

NEWS CLIP

Supreme Court Upholds Method Used in Apportionment of House

By Linda Greenhouse, *Special to* THE NEW YORK TIMES, April 1, 1992

WASHINGTON, March 31—In a decision that dashed Montana's hope of retaining two seats in the House of Representatives, the Supreme Court today upheld the constitutionality of the method Congress has used for 50 years to apportion seats among the states.

The unanimous ruling overturned a decision issued last fall by a special three-judge Federal District Court in Montana. That court, ruling in a lawsuit brought by the state, had ordered the use of a different method under which Montana would have kept the two House seats it has had since 1910 and the State of Washington would have lost one of its nine.

While the decision affected only those two states, it was of much broader interest because it was the Court's first look at Congressional apportionment in light of the strict one-person, one-vote requirement the Court now applies to legislative districting.

The Court insists on virtual mathematical equality for Congressional districts in a state. But such equality is not possible for districts in different states, because districts may not cross state lines and the Constitution requires at least one Representative for each state.

of it to the juniors, and 0.31 of it to the sophomores. How, then, should the 20 seats be distributed?

The methods favored by Hamilton and Jefferson have one thing in common: Each ignores the decimal part of the quota and assigns a number of seats equal to the whole number part of the quota. Regardless of whether the quota is 10.31 or 10.91, each method awards 10 seats. Ignoring the decimal part of a number in this way is called *truncating*.

As its initial step, the **Hamilton method** gives 10, 5, and 4 seats to the sophomore, junior, and senior classes, respectively. This accounts for only 19 of the 20 seats. This method gives the remaining seat to the class whose quota has the largest decimal part. The decimal part of the senior quota, 0.36, is larger than either of the other two decimal parts, and so the senior class is given the extra seat. The results are summarized in the following table:

Class Size	Quota	Hamilton Apportionment
464	10.31	10
240	5.33	5
196	4.36	5

Thomas Jefferson favored a more complicated method. It would also give 10, 5, and 4 seats to the sophomore, junior, and senior classes, respectively. The difference between the two methods lies in how the remaining seat is treated. The **Jefferson method** requires the quota for each class to increase until the sophomore quota reaches 11, the junior quota reaches 6, or the senior quota reaches 5. How does the Jefferson method increase these quotas? Because the quota is found by dividing the class size by the ideal ratio, the quotas will increase if the ideal ratio decreases. The Jefferson method changes the ideal ratio so that the desired apportionment is reached.

Consider, for example, what will happen if the ideal ratio is decreased to 40 students per seat. The new quotas are $464 \div 40 = 11.60$, $240 \div 40 = 6$, and $196 \div 40 = 4.9$. The Jefferson method would apportion 11, 6, and 4 seats. This is a total of 21 seats, so the ideal ratio is now too small. The results are summarized in the table below.

Class Size	Quota with Ideal Ratio of 45	Quota with Ideal Ratio of 40
464	$464 \div 45 = 10.31$	$464 \div 40 = 11.6$
240	$240 \div 45 = 5.33$	$240 \div 40 = 6.0$
196	$196 \div 45 = 4.36$	$196 \div 40 = 4.9$
Seats	$10 + 5 + 4 = 19$	$11 + 6 + 4 = 21$

The sophomore class, for example, receives 10 seats when the ideal ratio is 45 and 11 seats when the ideal ratio is 40. There must be some ideal ratio between 45 and 40 that causes the sophomore apportionment to increase from 10 seats to 11. This can be found by dividing 464 by 11 to obtain 42.18. The number 42.18 is called the **Jefferson adjusted ratio** for the sophomore class.

We can compute the ideal ratio that causes the sophomore quota to pass 11, by dividing 464 by 11 to obtain an adjusted ratio of 42.18:

$$\text{Jefferson adjusted ratio} = \frac{\text{Class size}}{\text{Truncated quota} + 1}$$

Similarly, the junior class quota passes 6 when the ratio drops below $240 \div 6 = 40$, and the senior class quota passes 5 when the ratio drops below $196 \div 5 = 39.2$.

If we gradually decrease the ratio from the ideal 45, it will meet 42.18 before it meets 40 or 39.2. Therefore, the sophomore class is given the extra seat.

The printout from a computer spreadsheet (Table 1) shows how the quotas for each class change as the ideal ratio is gradually decreased. The sophomore quota passes the next integer before either the junior or the senior quota does. Therefore, the extra seat is assigned to the sophomores.

The Jefferson method can be summarized in algorithmic form:

1. Divide the total population by the number of seats to obtain the ideal ratio.
2. Divide the population of each class (state, district, etc.) by the ideal ratio to obtain the class quota.
3. Assign a number of seats to each class equal to its truncated quota.
4. If the number of seats assigned matches the total number of seats available, then stop.
5. If the number of seats assigned is smaller than the total number seats available, then divide the size of each class by one more than the number of seats assigned to it in Step 3, to obtain an adjusted ratio.
6. Give an extra seat to the class with the largest adjusted ratio.

TABLE 1

Adjusted Ratio	Sophomore	Junior	Senior
45.00	10.31	5.33	4.36
44.80	10.36	5.36	4.38
44.60	10.40	5.38	4.39
44.40	10.45	5.41	4.41
44.20	10.50	5.43	4.43
44.00	10.55	5.45	4.45
43.80	10.59	5.48	4.47
43.60	10.64	5.50	4.50
43.40	10.69	5.53	4.52
43.20	10.74	5.56	4.54
43.00	10.79	5.58	4.56
42.80	10.84	5.61	4.58
42.60	10.89	5.63	4.60
42.40	10.94	5.66	4.62
42.20	11.00	5.69	4.64
42.00	11.05	5.71	4.67
41.80	11.10	5.74	4.69
41.60	11.15	5.77	4.71
41.40	11.21	5.80	4.73
41.20	11.26	5.83	4.76
41.00	11.32	5.85	4.78
40.80	11.37	5.88	4.80
40.60	11.43	5.91	4.83
40.40	11.49	5.94	4.85
40.20	11.54	5.97	4.88
40.00	11.60	6.00	4.90
39.80	11.66	6.03	4.92
39.60	11.72	6.06	4.95
39.40	11.78	6.09	4.97
39.20	11.84	6.12	5.00
39.00	11.90	6.15	5.03

This form of the Jefferson algorithm applies only to situations in which one extra seat is to be awarded. In some cases, the Jefferson method can fail to award more than a single seat. This lesson's exercises examine such a case.

Although the method that Hamilton favored may seem more appealing because of its simplicity, it has been abandoned. The exercises show why this has happened.

Exercises

1. The student council at Central High has had difficulty deciding a number of issues because of conflicts between the sophomore representatives and the representatives of the other two classes. The vote has been a 10–10 tie. The council has decided to add a seat in order to prevent ties on future issues.
   a. Based on the data in this lesson, which class do you think should have the extra seat?
   b. Find the new ideal ratio of students per seat for the 21-seat council.
   c. Use the new ratio to determine the quota for each of the three classes.
   d. Use the Hamilton method to allocate the 21 seats on the new council to the three classes.
   e. Compare the Hamilton apportionment for the 21-seat council with that of the 20-seat council, and explain why the results constitute a paradox.

2. A senior council member who recently studied apportionment in the school's discrete mathematics course is unhappy over the loss of one of the senior seats and proposes the apportionment be made by a different method.
   a. Find an adjusted ratio for each class as described in the Jefferson algorithm of this lesson.
   b. Decrease the 21-seat ideal ratio until all 21 seats are allocated. State the number of seats given each class by the Jefferson method.

3. Compare the 21-seat Jefferson apportionment with the 20-seat Jefferson apportionment. Does the Jefferson method produce a paradox similar to the one described in Exercise 1?

4. Revise the Jefferson apportionment algorithm given in this lesson to account for situations in which more than one additional seat must be distributed.

5. The paradox observed in Exercise 1 occurs because increases in a divisor do not produce equal changes in quotients. When the size of a representative assembly increases and the total population remains the same, the ideal ratio decreases. For example, two classes (states, districts, etc.) have populations of 100 and 230. An increase in the size of the council has caused the ideal ratio to decrease from 22 to 21.

   a. Complete the following table, and explain why it could result in the shift of a council member from one class to the other:

Class Size	Quota with Ideal Ratio of 22	Quota with Ideal Ratio of 21
100		
230		

   b. Will the paradox observed in Exercise 1 result in the loss of a seat for a small class or a large one? Why?

6. The student council members at Central High, aware of the strange results that slight differences can make, decide to monitor the council's apportionment. At the end of the first quarter of the school year, the class numbers have changed somewhat.

Sophomores	459
Juniors	244
Seniors	197

   a. Use the Hamilton method to divide the council's 21 seats among the classes.

At the end of the first semester the classes have changed again.

Sophomores	460
Juniors	274
Seniors	196

At the council's first meeting of the new semester, the members are amazed when one of the representatives of the senior class, the only class that has decreased in size, demands that the council be reapportioned.

b. Use the Hamilton method to reapportion the council.
c. Explain why the results constitute a paradox.
d. Use an analysis similar to that of Exercise 5 to explain why this type of paradox occurs. Will it have an adverse effect on small classes or large classes?

Projects

7. Research and report on the occurrence of paradoxes in the apportionment of the United States House of Representatives. What methods of apportionment have been used?

More Apportionment Algorithms and Paradoxes

Widespread dissatisfaction over the paradoxes that occur when the Hamilton method is used has led to the method's falling from favor as a means of apportionment. The Jefferson method, however, favors large states. Next, we consider other methods of apportionment and some significant recent results in the debate over which method is the fairest.

The Jefferson method is one of several *divisor methods* of apportionment. The term *divisor* is used because the methods determine quotas by dividing the population by an ideal ratio or an adjusted ratio. This ratio is the divisor. The two apportionment methods given the most attention today are a divisor method named after Daniel Webster and another named after Joseph Hill, an American statistician.

The Webster and Hill methods differ from the Jefferson method in the way that they round quotas. Recall that the Jefferson method truncates a quota and apportions a number of seats equal to the integer part of the quota. Quotas of 11.06 and 11.92 both receive 11 seats under the Jefferson method.

The United States House of Representatives is apportioned by the Hill method, although this has not always been the case. At various times it has been apportioned by the Jefferson method, the Hamilton method, and the Webster method.

The **Webster method** uses the rounding method that is most familiar. A quota above or equal to 11.5 receives 12 seats, and a quota below 11.5 receives 11 seats; 11.5 is sometimes called the **arithmetic mean** of 11 and 12.

The **Hill method** is slightly more complicated. It computes the **geometric mean** of the integers directly above and below the quota and rounds up if the quota exceeds the geometric mean, down if it doesn't. The geometric mean of two numbers is the square root of their product. For example, a quota between 11 and 12 would have to exceed $\sqrt{11 \times 12} = 11.4891$ to receive 12 seats (see Figure 2.1).

The following table summarizes the apportionment of the Central High student council by means of each of these methods:

Class Size	Quota	Initial Apportionment: Hamilton	Jefferson	Webster	Hill
464	10.31	10	10	10	10
240	5.33	5	5	5	5
196	4.36	5	4	4	4

The Jefferson, Webster, and Hill methods all fail to assign one of the seats and so require an adjusted ratio. The following table lists the adjusted ratio necessary for each class to gain a seat by means of each of the methods. Recall that the adjusted Jefferson ratio for the sophomore class is determined by dividing the class size by 11. The adjusted Webster ratio for the sophomore class is found by dividing the class size by 10.5. The adjusted Hill ratio for the sophomore class is found by dividing the class size by the geometric mean of 10 and 11, or 10.4881.

Class Size	Adjusted Ratio for: Jefferson	Webster	Hill
464	$464 \div 11 = 42.1818$	$464 \div 10.5 = 44.1905$	$464 \div \sqrt{10 \times 11} = 44.2407$
240	$240 \div 6 = 40.0000$	$240 \div 5.5 = 43.6364$	$240 \div \sqrt{5 \times 6} = 43.8178$
196	$195 \div 5 = 39.2000$	$196 \div 4.5 = 43.5556$	$196 \div \sqrt{4 \times 5} = 43.8269$

Each of these three methods requires that the ideal ratio of 45 be decreased until it is smaller than exactly one of the adjusted ratios. If

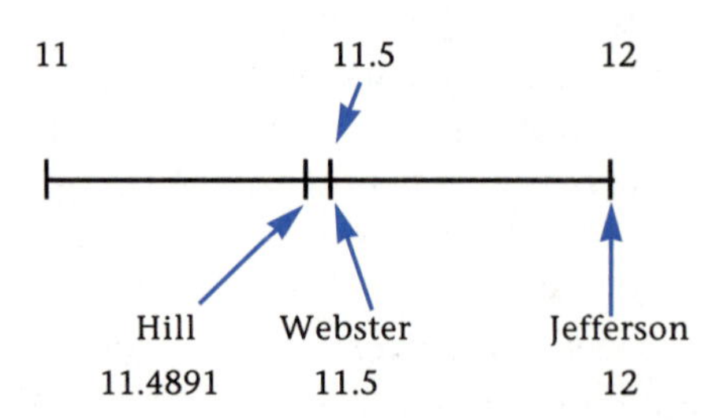

Figure 2.1 The Hill, Webster, and Jefferson roundup points for quotas between 11 and 12.

the adjusted ratios are listed in decreasing order, the Jefferson method will list the sophomore ratio first, the junior ratio second, and the senior ratio third. The extra seat is awarded to the sophomore class. Similarly, the Webster and Hill methods list the sophomore class first and award the extra seat to that class. Note, however, that the Hill method lists the senior class second rather than third.

The following exercises consider the Webster and Hill methods and some surprising results.

Exercises

1. a. Complete the following apportionment table for the 21-seat Central High student council described in the exercises of the previous lesson (see page 65):

Class Size	Quota	Initial Apportionment: Hamilton	Jefferson	Webster	Hill
464	____	____	____	____	____
240	____	____	____	____	____
196	____	____	____	____	____

b. The Jefferson method distributes only 19 seats. Check your results from the exercises of the previous lesson to determine which classes are given an additional seat. Give the final distribution of the 21 seats according to the Jefferson method.

Both the Webster and Hill methods apportion 22 seats. In such cases the ideal ratio must be increased until one of the classes loses a seat. The sophomore class, for example, will lose a seat under the Webster method if its quota drops below 10.5. This requires an adjusted ratio of $464 \div 10.5 = 44.1905$. For the sophomore class to lose a seat under the Hill method, its quota must drop below $464 \div 10.4881 = 44.2407$.

c. Complete the following table of adjusted ratios for the Webster and Hill methods:

Class Size	Adjusted Ratio For Webster	Hill
464	$464 \div 10.5 = 44.1905$	$464 \div \sqrt{10 \times 11} = 44.2407$
240	____	____
196	____	____

d. List the adjusted ratios for the Webster method in increasing order. The ideal ratio must be increased until it passes only the first ratio in your list. This class will lose one seat. Give the final Webster apportionment.

e. List the adjusted ratios for the Hill method in increasing order. The ideal ratio must be increased until it passes only the first ratio in your list. This class will lose one seat. Give the final Hill apportionment.

f. If the council has 21 seats, which method(s) would be favored by each class?

2. Is it possible to measure the fairness of an apportionment method? One way to do this is to measure the discrepancy between a quota and the actual apportionment. If a quota is 11.25 and 11 seats are apportioned, the unfairness is 0.25 seats. If 12 seats are apportioned, the unfairness is 0.75 seats.

   a. Use the apportionments for the 20-seat council to measure the discrepancy for each class by means of each method. Record the results in the following table:

Class Size	Quota	Amount of Discrepancy: Hamilton	Jefferson	Webster	Hill
464	10.31	____	____	____	____
240	5.33	____	____	____	____
196	4.36	____	____	____	____
Total discrepancy		____	____	____	____

   b. According to this measure, which method is fairest?

3. None of the divisor methods is plagued by the paradoxes that caused the demise of the Hamilton method. Divisor methods, however, can cause problems. Mountain High School has four classes: freshman, sophomore, junior, and senior. The freshman class has 1,105 members, the sophomore class has 185, the junior class has 130, and the senior class has 80. The school student council has 30 members distributed among the four classes by means of the Webster method of apportionment.
   a. What is the ideal ratio?
   b. Complete the following Webster apportionment table:

Class Size	Quota	Initial Webster Apportionment
1,105	22.1	22
185	____	____
130	____	____
80	____	____

   c. Because the Webster method apportions too many seats, the ideal ratio must be decreased. Calculate the adjusted ratio necessary for each class to lose a seat, and complete the following table:

Class Size	Adjusted Ratio
1,105	$1105 \div 21.5 = 51.3953$
185	____
130	____
80	____

   d. Increase the ideal ratio until the extra seat is removed. State the final apportionment.
   e. Explain why the results would be considered unfair to the freshman class.

This situation is known as a **violation of quota.** It occurs whenever a class (district, state) is given a number of seats that is not equal to the integer part of its quota or one more than that. It can

TABLE 2

Adjusted Ratio	Freshman	Sophomore	Junior	Senior
50.00	22.10	3.70	2.60	1.60
50.20	22.01	3.69	2.59	1.59
50.40	21.92	3.67	2.58	1.59
50.60	21.84	3.66	2.57	1.58
50.80	21.75	3.64	2.56	1.57
51.00	21.67	3.63	2.55	1.57
51.20	21.58	3.61	2.54	1.56
51.40	21.50	3.60	2.53	1.56
51.60	21.41	3.59	2.52	1.55
51.80	21.33	3.57	2.51	1.54
52.00	21.25	3.56	2.50	1.54

occur with any divisor method and is considered a flaw of divisor methods.

The printout from a computer spreadsheet (Table 2) shows how the quotas for each class change as the ideal ratio is gradually increased. The freshman quota drops much more rapidly than do the others.

Around 1980, two American mathematicians, Michel L. Balinski (left below) and H. Peyton Young (right below), proved an important impossibility theorem: Any method of apportionment sometimes produces at least one of three undesirable results. These

are the violation of quota and the two paradoxes that occur with the Hamilton method.

4. Compare the Hill roundoff point with the Webster roundoff point for quotas of various sizes. How, for example, do the two compare for quotas between 2 and 3, between 10 and 11, between 100 and 101? Can you prove any relationship between the two?

Projects

5. Research and report on the work of Balinski and Young. What method of apportionment do they recommend? Is it currently used to apportion the United States House of Representatives? Why or why not?

6. Research and report on the methods that have been used to apportion the United States House of Representatives. Have any of the divisor methods ever resulted in a violation of quota?

7. Investigate the divisor methods for other ways in which the violation of quota can occur. Can you find, for example, four classes for which the number of apportioned seats falls two short of the size of the council and results in the awarding of both seats to the same class?

8. Investigate measures of fairness such as the one given in Exercise 2. Does this measure of fairness favor a particular apportionment method? Are there other ways of measuring fairness?

LESSON 2.5

Fair Division Algorithms: The Continuous Case

In this lesson, we return to the problem of fairly dividing a cake, which was initially discussed in the first lesson of this chapter. You may want to review your answers to Exercises 1 through 5 of that lesson before continuing (see pages 48–49).

A cake is continuously divisible; it may be divided into any number of parts. This is not true, however, of the discrete objects in an estate or seats in a legislature.

Our investigation of the fair division of a cake in Lesson 2.1 revealed a number of important considerations. For example, the resolution of the problem by an outside authority often does not result in a solution that is fair in the eyes of both individuals. In addition, the appeal to a random event such as a coin toss does not result in a division that is fair in the eyes of both parties. On the basis of these observations, the division of a cake between two people will be called fair only if each person feels that he or she has received at least one-half of the cake. An acceptable solution to the problem of dividing an object such as a cake among several individuals is not possible without a *definition of fairness.*

A division among n people is called fair if each person feels that he or she has received at least $1/n$th of the object.

Lesson 2.1 showed that an appeal to a measurement scheme such as weighing may not be adequate, because an individual's evaluation may be based on more than just size. Cake icing, for example, could be important. A successful solution to the cake division problem also depends on several *assumptions.*

Our fair division assumptions are as follows:

1. Each individual is capable of dividing a portion of the cake into several portions that he or she feels are equal.

2. Each individual is capable of placing a value on any portion

of the cake. The total of the values placed on all parts of a cake by an individual is 1, or 100%.

3. The value that each individual places on a portion of the cake may be based on more than just the size of the portion.

Let's now use these assumptions and the definition of fairness as an aid in examining the solution to the problem of dividing a piece of cake fairly among two people. Recall that this required one person to cut the cake and the other to choose the first of the two portions.

The requirement that one person cut the cake assumes that the individual will cut it into two portions that he or she feels are equal. This is the first assumption in the above list. The second assumption ensures that the person who doesn't cut will place values on the portions that total 1. This individual, however, may not feel that the pieces are equal and thereby will choose a piece that he or she feels is more than half the cake. Even if the chooser feels that his or her piece is more than half the cake, the division is fair because the definition requires only that each person feel that his or her piece is at least half of the cake.

How is a cake fairly divided among three, four, or more people? No solution to this problem is adequate unless it adheres to the definition of a fair division. A fair division among three people, for example, requires that each individual place a value of at least one-third on the received portion.

In mathematics, there is often more than one way to solve a problem. When this is the case, the solution is not *unique*. That is, the problem of dividing a cake fairly among three or more people has more than one solution.

A solution to a complicated problem can often be based on the solution to a simpler one. The following solution to the three-person division problem uses the solution to the two-person problem:

Call the three individuals Ann, Bart, and Carl. The solution is described in algorithmic form:

1. Ann cuts the cake into two pieces that she feels are equal.

2. Bart chooses one of the pieces; Ann gets the other.

3. Ann cuts her piece into three pieces that she considers equal; Bart does the same with his.

4. Carl chooses one of Ann's three pieces and one of Bart's.

To see that the division is fair, it is necessary to show that each person places a value of at least one-third on the portion that he or she received.

Consider Ann. In Step 1, Ann feels that each piece is one-half of the cake. She therefore feels that the piece she received in Step 2 is half the cake. She feels that each piece she cut in Step 3 is one-third of half the cake, or one-sixth of the cake. She therefore feels that she receives two-sixths or one-third of the cake in Step 4.

Bart's case is similar except that he may feel that the portion he chose in Step 2 is more than half the cake. Thus, he may feel that each piece he cut in Step 3 is more than one-sixth of the cake and that his final portion is more than two-sixths or one-third.

Now consider Carl. He may feel that the two pieces that Ann cut in Step 1 are not equal. His value could be, for example, 0.6 for one piece and 0.4 for the other. Likewise, he may not feel that the cuts made in Step 3 are equal. He could, for example, decide that the piece he valued at 0.6 was divided into pieces he values at 0.3, 0.2, and 0.1. Similarly, he could decide that the piece he valued at 0.4 was divided into pieces he values at 0.2, 0.1, and 0.1. Because he chooses first, however, he will pick the largest piece from each: the 0.3 from the division of the piece he valued at 0.6 and the 0.2 from the piece he valued at 0.4. Thus, he feels that the value of the portion he receives is $0.3 + 0.2 = 0.5$.

But will this always result in a fair division for Carl? The previous example does not demonstrate that Carl will be satisfied in all possible circumstances. Suppose Carl's value for one of the two pieces that Ann cuts is x. His value for the other piece must be $1 - x$. Although Carl may not feel that the piece he values at x has been divided into equal thirds, he will choose the largest of the three pieces and value it as at least one-third of x, or $\frac{1}{3}x$. Similarly, he will value the piece he chooses from the part he values at $1 - x$ as at least $\frac{1}{3}(1 - x)$. The total value of his two pieces is therefore at least $\frac{1}{3}x + \frac{1}{3}(1 - x) = \frac{1}{3}x + \frac{1}{3} - \frac{1}{3}x = \frac{1}{3}$. This method is referred to as the **cut-and-choose method.**

Exercises

1. In the division among Ann, Bart, and Carl, who will evaluate his or her share at exactly one-third? Who might feel that he or she received more than one-third?

2. Does the division among Ann, Bart, and Carl result in three pieces or three portions?

In Exercises 3 and 4, suppose Carl feels that Ann's initial division is fair, that Ann's subdivision is also even, but that Bart's subdivision is not. (Give your answers as fractions or decimals rounded to the nearest 0.01.)

3. What value will Carl place on the piece he takes from Ann?

4. Although Carl feels that the piece Bart divided is half the cake, he does not feel that Bart subdivided it equally. He could, for example, place values of 0.3, 0.1, and 0.1, or values of 0.4, 0.06, and 0.04, on the three pieces. The largest value he could place on any of these three pieces is 0.5.
   a. What is the smallest value he could place on the piece he takes from Bart?
   b. What is the largest total value he could place on his two pieces?
   c. What is the smallest total value he could place on his two pieces?

5. In mathematics, a fundamental principle of counting is that if there are m ways of performing one task and n ways of performing another, then there are $m \times n$ ways of performing both. For example, a tossed coin may land in two ways, and a rolled die may land in six ways. Together they may land in a total of $2 \times 6 = 12$ ways.
   a. If two people each have a piece of cake and each cuts his or her piece into three pieces, how many pieces will result?
   b. If k people each have a piece of cake and each cuts his or her piece into $k + 1$ pieces, what are two equivalent expressions for the total number of pieces that result?

c. If $k + 1$ boxes each contain $k + 5$ toothpicks, what are two equivalent expressions for the total number of toothpicks?

d. Two offices are to be filled in an election: mayor and governor. If there are three candidates for governor and four for mayor and conventional voting procedures are used, in how many ways may one vote?

6. Consider the following division of a cake among three people: Arnold, Betty, and Charlie. Arnold cuts the cake into three pieces he considers equal. Betty chooses one of the pieces, and Charlie chooses either of the remaining two. Arnold gets the third piece.
   a. Will Arnold feel he has received at least one-third of the cake? Might he feel he has received more?
   b. Will Betty feel she has received at least one-third of the cake? Might she feel she has received more?
   c. Will Charlie feel he has received at least one-third of the cake? Might he feel he has received more?

7. Arnold, Betty, and Charlie decide to divide a cake in the following way: Arnold slices a piece he considers one-third of the cake. Betty inspects the piece. If she feels it is more than one-third of the cake, she will cut enough from the cake so that she feels it is one-third of the cake. The removed portion is returned to the cake. Charlie now inspects Betty's piece and has the option of doing the same. The piece of cake is given to the last person who cut from it.

   One of the remaining two people slices a piece that he or she feels is half of the remaining part of the cake. The other person inspects the piece with the option of removing some of the cake if he or she feels it is more than half the remainder.
   a. Will the person who receives the first piece feel that it is at least one-third of the cake? Could he or she feel it is more than one-third?
   b. Will the person who receives the second piece feel that it is at least one-third of the cake? Could he or she feel it is more than one-third?
   c. Will the person who receives the third piece feel that it is at least one-third of the cake? Could he or she feel it is more than one-third?

   This method is referred to as the **inspection method.**

Computer Explorations

8. Use the moving knife program on the disk that accompanies this book to divide a cake of any shape among yourself and two other people in your class by means of the **moving knife method.**
9. Explain why each of the three people in your group feels that he or she received at least one-third of the cake.
10. Suppose we establish the following criterion for an "optimal" division of a cake among three people: that the division result in exactly three pieces rather than more. Does the cut-and-choose, the inspection, or the moving knife method result in an optimal solution?
11. Could the cut-and-choose, the inspection, or the moving knife method be extended to divide a cake among four people? Explain how this could be done.

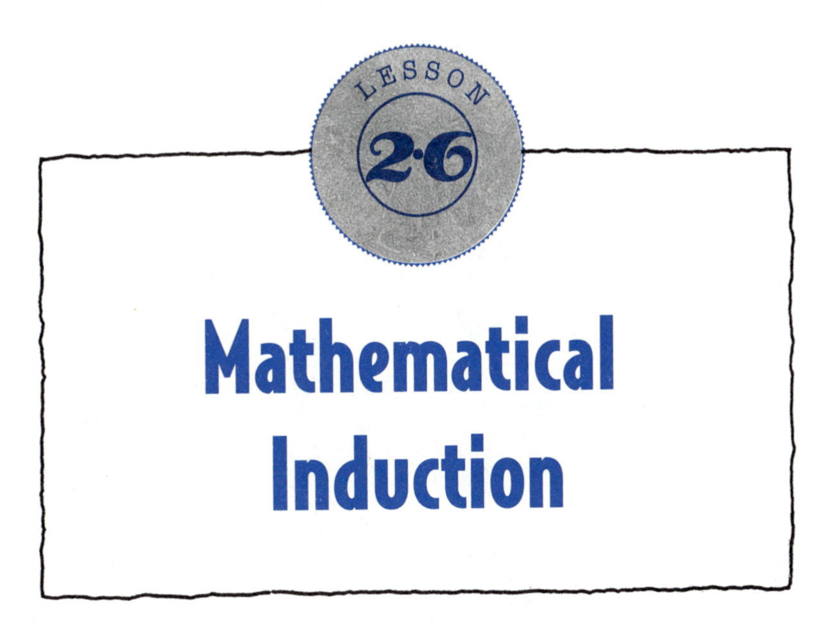

Mathematical Induction

Mathematics is often concerned with discovering patterns in the world around us. The identification and verification of patterns in figures and numbers enable us to analyze and predict trends in science, business, economics, world affairs, and many other areas. Patterns, however, can be elusive, and erroneous predictions often result in the expenditure of unnecessary effort and money. In this lesson, we consider a way of verifying certain kinds of patterns that occur in discrete mathematics, by applying a method of reasoning known as **mathematical induction.**

Mathematical induction is a method of proving that certain types of discrete patterns continue.

For example, notice that the fair division of a cake by the cut-and-choose method can apparently be continued indefinitely. The cut-and-choose method was first applied to the problem of dividing a cake among two people. This was accomplished by having one person cut the cake into two parts that he or she considered equal and letting the other person choose first.

Extending the method to three people required that two of them first apply the two-person method. Then each of these two cut his or her piece into three pieces that he or she considered equal, and the third person selected one of the three pieces from each.

To extend this to four people, three of them apply the three-person solution. Then each of these three people divides his or her portion into four portions that he or she considers equal. The fourth person chooses one portion from each of the other three people. In order for the four-person solution to be fair, each of the three cutters must feel that he or she is left with three portions that are each one-fourth of their original share, which was at least one-third of the cake. This is at least $\frac{3}{4} \times \frac{1}{3} = \frac{1}{4}$ of the cake.

Although the fourth person may not feel that each of the three portions is at least one-third of the cake, he or she must feel that the total value of the three portions is 1. Suppose the fourth person assigns values p_1, p_2, and p_3 to the three portions. Then $p_1 + p_2 + p_3 = 1$. Because the fourth person is given first choice of a portion from each of the original three people, he or she will place a value of at least $\frac{1}{4}p_1 + \frac{1}{4}p_2 + \frac{1}{4}p_3$ on the resulting portion. Accordingly, $\frac{1}{4}p_1 + \frac{1}{4}p_2 + \frac{1}{4}p_3 = \frac{1}{4}(p_1 + p_2 + p_3) = \frac{1}{4}(1)$, or one-fourth of the entire cake.

Can this method be extended to yield a 5-person cake division? To a 6-person division? It appears that it can. If, for example, a cake is to be divided fairly among 10 people by this method, the division can be based on a 9-person solution, which is based on an 8-person solution, which is based on. . . .

Mathematical induction generalizes this pattern of solutions by proving that it always is possible to extend the solution to a group that is one larger than the previous. The generalization is achieved by using a variable rather than a specific number.

Suppose that you know how to divide a cake fairly among k people. You need to show that it also is possible to divide a cake fairly among $k + 1$ people. Doing so will demonstrate that not only can the two-person solution be extended to three people and the three-person extended to four, which has already been done, but that the solution can always be extended.

The proof begins by applying the assumption that k people can fairly divide the cake. Then each of the k people divides his or her portion into $k + 1$ portions that he or she feels are equal. The $k + 1$st person then selects one portion from each. It is now necessary to show that this results in a share of at least $1/(k + 1)$ for each of the $k + 1$ people.

In the case of the k people who cut, each will feel that each portion he or she cut is $1/(k + 1)$ of at least $\frac{1}{k}$ of the cake or at least $1/((k + 1)k)$ of the cake. He or she keeps k of the $k + 1$ portions. This gives a total value of at least $k(1/((k + 1)k) = 1/(k + 1)$.

Although the chooser may not feel that all of the original k portions are at least $\frac{1}{k}$ of the cake, he or she must feel that the total value is 1. Suppose that this person assigns values of $p_1, p_2, \ldots, p_k$ to the k pieces. Then $p_1 + p_2 + \cdots + p_k = 1$. Because the chooser is given first choice of a piece from the $k + 1$ pieces cut by each of the other people, he or she will place a value of at least $\frac{1}{k+1}p_1 + \frac{1}{k+1}p_2 + \cdots + \frac{1}{k+1}p_k$ on the resulting portion. Then $\frac{1}{k+1}p_1 + \frac{1}{k+1}p_2 + \cdots + \frac{1}{k+1}p_k = \frac{1}{k+1}(p_1 + p_2 + \cdots + p_k) = \frac{1}{k+1}(1)$, or $\frac{1}{k+1}$ of the entire cake.

The proof is complete because it demonstrates that whenever a cake can be divided fairly among k people, it can also be divided fairly among $k + 1$ people. Notice that the mathematical induction argument is merely a generalization of the one given for the four-person solution.

Mathematical induction is frequently used to verify that an observed formula always works.

Luis and Britt are investigating the number of handshakes that will be made by a group of people if each person shakes hands with every other person. Luis notes that if there is only one person, no handshakes are possible and that if there are two people, only one handshake is possible. Britt draws a graph in which the vertices represent people, and the segments, or edges, represent handshakes (see Figure 2.2). She concludes that a group of three people requires a total of three handshakes.

Luis suggests organizing the data into a table.

Number of People in the Group	Number of Handshakes
1	0
2	1
3	3

In the exercises, mathematical induction is used to prove that a hypothesis for the number of handshakes in a group of people is correct. Other examples of the use of mathematical induction are given as well.

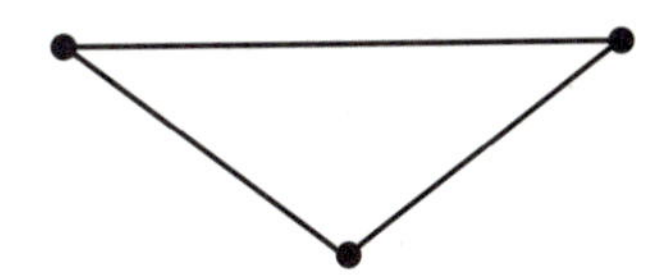

Figure 2.2 Graph representing handshakes among three people.

Exercises

1. To use mathematical induction, you must be able to use symbols to express numeric patterns. Some of the expressions you write in this exercise will be used in the mathematical induction proof.
   a. If there are three people in a group and another person joins the group, there will be four people in the group. If a person leaves the original group of three, there will be two. Write expressions for the number of people if there are k people in a group and another person joins. Do the same if a person leaves the group of k people.
   b. Repeat this exercise for a group of $k + 1$ people. For a group of $2k$ people.

2. Draw a graph like Britt's and a table like Luis's.
   a. Add another vertex to the graph to represent a fourth person, and draw segments to represent the additional handshakes that will result if the group grows to four people. Determine the number of handshakes in a group of four by adding the number of new handshakes to the number for a group of three given in the table. Write in your table the total number of handshakes for a group of four people.
   b. Add a fifth vertex to represent a fifth person, and draw segments to represent the additional handshakes. Add the number of new handshakes to the number for a group of four given in the table. Write in your table the total number of handshakes for a group of five people.

3. a. Suppose that there are seven people in a group and each of them has shaken hands with every other person. If an eighth person enters the group, how many additional handshakes must be made?

b. Suppose that there are k people in a group and each of them has shaken hands with every other person. If a new person enters the group, how many additional handshakes must be made?
c. If H_n represents the number of handshakes in a group of n people, what is the recurrence relation that expresses the relationship between H_n and H_{n-1}? Write a recurrence relation that expresses the relationship between H_{n+1} and H_n.

4. After studying the data for a while, Britt wonders whether the number of handshakes in a group can be found by multiplying the number of people in the group by the number that is 1 less than that and by dividing this product by 2.
   a. If her guess is correct, how many handshakes would there be in a group of 10 people?
   b. Write an expression for the number of handshakes based on Britt's guess if there are k people in a group. Do the same for a group of $2k$ people. Do the same for a group of $k + 1$ people.

Britt's formula, if correct, is sometimes known as a **solution of the recurrence relation** given in part c. One of its advantages is that it allows you to determine the number of handshakes in a group without using the number of handshakes in a smaller group.

To prove that Britt's guess is correct, show that whenever the solution is known to work, it is possible to extend it to a group that is 1 larger. In other words, whenever the conjecture works for a group of k people, it will also work for a group of $k + 1$ people.

   c. Assume that Britt's formula works for a group of k people, and write the formula for such a group.
   d. You need to show that Britt's formula works for a group of $k + 1$ people. Write her formula for $k + 1$ people.
   e. If an additional person enters a group of k people, how many new handshakes are necessary?

An expression for the total number of handshakes in a group of $k + 1$ people can be found by adding the expression for the number of handshakes in a group of k people (part c) to the number of new handshakes (part e):

$$\frac{k(k-1)}{2} + k$$

f. You can conclude that Britt's formula will always work if this expression matches the one in part d. Use algebra to transform the expression until it matches the one you wrote in part d.

5. Although Britt's formula is for the number of handshakes in a group of people, it could also represent the number of potential two-party conflicts in a group.
   a. Use the formula to compare the number of potential conflicts when the size of a group doubles. Does the number of potential disputes also double?
   b. Why do the results of Exercise 4 suggest that some of the costs associated with government, such as that of maintaining a police force, may outpace the growth of population?

In Exercises 1–4 you supplied several of the steps of the mathematical induction proof that began in the lesson. In Exercise 6, you will again supply many of the steps of the induction process, which requires a number of preliminary steps leading to the guessing of a formula, which must then be proved. The preliminary steps are summarized here:

Preliminary Steps

Do the following before using mathematical induction to prove a relationship:

1. Organize a table of data for several small values. For example, how many ways of voting are there with 1, 2, 3, or 4 choices on the ballot?

2. Investigate the problem and the data to describe the pattern of the data with a recurrence relation. For example, how many ways of voting are added when another choice is placed on the ballot?

3. Make up a formula that predicts the outcome for a collection of k items. For example, what is a formula that predicts the number of ways of voting when there are k choices on the ballot?

4. Verify that your formula works for the small values you have tabulated.

6. In Exercise 11 of Lesson 4 in Chapter 1 (page 30), we considered the number of ways of voting under the approval system and wrote a recurrence relation to describe the pattern. We will now use mathematical induction to verify that a suspected formula for the number of ways of voting under the approval system when there are n choices on the ballot is indeed correct. The data you gathered in Chapter 1 are reproduced below.

Number of Choices on the Ballot	Number of Possible Ways of Voting
1	2
2	4
3	8
4	16

a. Collecting these data completes the first of the preliminary steps. The second step requires you to determine the way in which the data change when an additional choice is added to the ballot. That is, what recurrence relation describes the relationship between the number of ways of voting when there are $k + 1$ choices on the ballot (V_{k+1}) and the number of ways of voting when there are k choices on the ballot (V_k)? Be careful—if you do not establish the recurrence relation properly, the proof that comes later will fail. Here's a hint:

With Three Choices There are Eight Ways		The New Ways When a Choice Is Added
{ }		{D}
{A}		{A,D}
{B}		{B,D}
{C}	Append D to each	{C,D}
{A,B}	⟶	{A,B,D}
{A,C}		{A,C,D}
{B,C}		{B,C,D}
{A,B,C}		{A,B,C,D}

The column on the left shows all the ways of voting when there are three choices. The column on the right shows the number of new ways of voting when a fourth choice is added. Both columns together give all the ways of voting with four choices.

b. You are ready for the third and fourth preliminary steps. Look carefully at the data. Do you notice that the values in the second column are the same power of 2 as the numbers in the first? What formula does this suggest for the number of ways of voting when there are n choices? Check the formula with each pair of values in the table.

This completes the preliminary process. It is now time to attempt the proof. The steps you must do to complete the proof are summarized below.

The Proof

Once you have decided on a formula, you can prove that it is correct by using mathematical induction:

1. State the meaning of your formula for a collection of size k. This is the assumption and is similar to the "given" in a geometric proof.

2. State the meaning of your formula for a collection of size $k + 1$, by substituting $k + 1$ for k in the previous formula. This is the goal and is similar to the "prove" in a geometric proof.

3. Use your recurrence relation to describe the effect of an additional object on the formula you stated in the first step.

4. Use algebra to transform the previous expression until it matches the one you stated in the second step.

c. Assume that the formula works for a ballot with k choices: $V_k = 2^k$. You must show that the formula works for a ballot with $k + 1$ choices. Write the formula for a ballot with $k + 1$ choices. This completes the first two steps of the proof process.

d. Write an expression for the total number of ways of voting on a ballot with $k + 1$ choices by applying the recurrence relation you gave in part a to the formula you stated in part b. Use algebra to show that this is equivalent to your answer to part c.

This completes steps 3 and 4 of the proof process and your induction proof is finished.

For each of the following situations, collect and organize data into a table, investigate the data, decide on a formula, and use mathematical induction to prove that the formula is correct. If you need help with any of these steps, use the summaries of the preliminary steps and the proof in Exercise 6.

7. Dominoes come in sets of different sizes. A double-six set, for example, contains a domino that pairs every number of spots from 0 to 6 with itself and with every other number of spots. Find a formula for the number of dominoes in a double-k set.

8. Bowling pins are normally set in a triangular configuration. Find a formula for the number of pins in a triangular configuration of k rows.

9. An ancient legend has it that the inventor of the game of chess was offered a reward of his own choosing for the delight the game gave the king. The inventor asked for enough grains of wheat to be able to place one grain on the first square of the chessboard, two on the second, four on the third, and so forth, doubling the number of grains each time. Find a formula for the total number of grains on a chessboard after the kth square has been filled.

10. It takes 4 toothpicks to make a 1×1 square, and it takes 12 toothpicks to make a 2×2 square that is subdivided into 1×1 squares. Find a formula for the number of toothpicks needed to make a $k \times k$ square that is subdivided into 1×1 squares.

11. In Exercises 15 and 16 of Lesson 1.4 (pages 31–32), you found a recurrence relation for the number of ways of selecting exactly two items when there are several choices on a ballot. Find a formula for the number of ways of selecting exactly two items when there are k choices on the ballot.

12. In the popular song "The Twelve Days of Christmas," one gift is given on the first day, one plus two on the second, and so on. Find a formula for the total number of gifts given on the kth day.

1. Joan, Henry, and Sam are heirs to an estate that includes a vacant lot, a boat, a computer, a stereo, and $10,000 in cash. Each heir has submitted bids for the items in the estate as summarized in the matrix below.

	Joan	Henry	Sam
Vacant lot	$8,000	$7,500	$6,200
Boat	$6,500	$5,700	$6,700
Computer	$1,340	$1,500	$1,400
Stereo	$ 800	$1,100	$1,000

For each heir, find the fair share, the items received, the amount of cash, and the final settlement. Summarize your results in a matrix.

2. Anne, Beth, and Jay are heirs to an estate that includes a computer, a car, and a stereo. Each heir has submitted bids for the items in the estate as summarized in the matrix below.

	Anne	Beth	Jay
Computer	$1,800	$1,500	$1,650
Car	$2,600	$2,400	$2,000
Stereo	$1,000	$ 800	$1,200

For each heir, find the fair share, the items received, the amount of cash, and the final settlement. Summarize your results in a matrix.

3. States A, B, and C have populations of 647, 247, and 106, respectively. There are 100 seats to be apportioned among them.
   a. What is the ideal ratio?
   b. Find the quota for each state.
   c. Apportion the 100 seats among the three states by the Hamilton method.
   d. What is the initial Jefferson apportionment?
   e. Find the Jefferson adjusted ratios for each state.
   f. Apportion the 100 seats by the Jefferson method.
   g. What is the initial Webster apportionment?
   h. Find the Webster adjusted ratios for each state.
   i. Apportion the 100 seats by the Webster method.
   j. What is the initial Hill apportionment?
   k. Find the Hill adjusted ratios for each state.
   l. Apportion the 100 seats by the Hill method.
   m. Suppose that the populations of the states change to 650, 255, and 105, respectively. Reapportion the 100 seats by the means of the Hamilton method.
   n. Explain why the results of the previous question constitute a paradox.
4. Discuss the theorem proved by Michel Balinski and H. Peyton Young. That is, what did they prove?
5. Arnold, Betty, and Charlie are dividing a cake in the following way:

   ◂ Arnold divides the cake into what he considers six equal pieces.

   ◂ The pieces are then chosen in this order: Betty, Charlie, Betty, Charlie, Arnold, Arnold.

   ◂ Who is guaranteed a fair share by his or her own assessment?
6. Four people have divided a cake into four pieces that each considers fair, and then a fifth person arrives. Describe a method of dividing the four existing pieces so that each of the five receives a fair share.

7. For the following situation, collect and organize data into a table, investigate the data, decide on a formula, and use mathematical induction to prove that the formula is correct:

   In a set of concentric circles, a ring is any region that lies between any two of the circles. Make up a formula for the number of rings in a set of k concentric circles.

Bibliography

Balinski, Michel, and H. Peyton Young. 1982. *Fair Representation: Meeting the Ideal of One Man, One Vote.* New Haven, CT: Yale University Press.

Bradberry, Brent A. 1992. "A Geometric View of Some Apportionment Paradoxes." *Mathematics Magazine* 65(1):3–17.

Bunch, Brian. 1982. *Mathematical Fallacies and Paradoxes.* New York: Van Nostrand Reinhold.

Burrows, Herbert, et al. 1989. *Mathematical Induction.* Lexington, MA: COMAP.

COMAP. 1991. *For All Practical Purposes: Introduction to Contemporary Mathematics.* 2nd ed. New York: Freeman.

Eisner, Milton P. 1982. *Methods of Congressional Apportionment.* Lexington, MA: COMAP.

Falletta, Nicholas. 1983. *The Paradoxican.* Garden City, NY: Doubleday.

Lambert, J. P. 1988. *Voting Games, Power Indices, and Presidential Elections.* Lexington, MA: COMAP.

Meyer, Rochelle Wilson, and Walter Meyer. 1990. *Play It Again Sam: Recurrence Equations in Mathematics and Computer Science.* Lexington, MA: COMAP.

Edison's Triple Bargain
TOASTER $6.95 CASH LAMP $6.95 CASH RADIO $15.95 CASH } or 6 MONTHS TO PAY
HERE
BALLS
OUTS
0
AT BAT
STRIKES
TEAMS 1 2 3 4 5 6 7 8 9 10 R. P. TEAMS 1 2 3 4 5 6 7 8 9 10 R. P.
BKLYN 1 1 1 0 0 0 2 1 0
BOSTON 0 0 0 0 0 0 0 0
ST. LOU 0 0 0 0 0 0 1 0 0 1
PITTS. 0 0 1 0 0 2 0 0 3
BATTING ORDER
UMPIRES
PLATE
BASES
STEGMAIER'S
BEER
THE DODGERS USE...
LIFEBUOY
ICE COLD
Coca-Cola
SOLD HERE
WRINKLE-PROOF TIES
HIT SIGN WIN SUIT
ABE STARK
1514 PITKIN AVE.
BROOKLYN'S LEADING CLOTHIER
NEW

CHAPTER 3

Matrix Operations and Applications

It has been said that sports fans are the nation's foremost consumers of statistics and that baseball fans are the most prominent among them. Whether it"s in the information conveyed by a scoreboard, as on this one that commemorates the 1941 National League pennant race, or in the current information on league leaders, baseball records are full of numbers.

How can large collections of data be organized and managed in an efficient way? What calculations provide meaningful information to people who use the data? How can computers and calculators assist them? Baseball statisticians, business executives, and wildlife biologists are among the diverse groups of people who turn to the mathematics of matrices for answers to these questions.

Addition and Subtraction of Matrices

As a first example, suppose that we are planning a pizza and video party for a few of our friends and we have to make some decisions about ordering food. We start by calling the pizza houses that deliver in our neighborhood to ask about prices for large single-topping pizzas, liter containers of cold drinks, and family-sized salads with house dressing. We could record this information in a table such as the following:

	Gina's	Vin's	Toni's	Sal's
Pizza	\$12.16	\$10.10	\$10.86	\$10.65
Drinks	\$1.15	\$1.09	\$0.89	\$1.05
Salad	\$4.05	\$3.69	\$3.89	\$3.85

Or we could write the data in **matrix form,** which simply means writing the numbers in a rectangular array and enclosing them in brackets or parentheses.

$$\begin{array}{r} \\ \text{Pizza} \\ \text{Drinks} \\ \text{Salad} \end{array} \begin{array}{c} \begin{array}{cccc} \text{Gina's} & \text{Vin's} & \text{Toni's} & \text{Sal's} \end{array} \\ \begin{bmatrix} \$12.16 & \$10.10 & \$10.86 & \$10.65 \\ \$1.15 & \$1.09 & \$0.89 & \$1.05 \\ \$4.05 & \$3.69 & \$3.89 & \$3.85 \end{bmatrix} \end{array}$$

For the sake of simplicity, we could also omit the row and column labels and dollar signs in our matrix and write only the values.

$$\begin{bmatrix} 12.16 & 10.10 & 10.86 & 10.65 \\ 1.15 & 1.09 & 0.89 & 1.05 \\ 4.05 & 3.69 & 3.89 & 3.85 \end{bmatrix}$$

If we delete the labels, however, we have to keep in mind that the rows represent prices for pizzas, drinks, and salads, and that the columns represent the various pizza houses.

Each of the individual entries in the matrix is called an **element** or a **component** of the matrix. A matrix such as this one, with three rows and four columns, is said to be a matrix with **order** or **dimensions** 3 by 4 (written as 3×4). It is also customary to refer to this matrix as simply a 3×4 matrix.

In general, if a matrix has m rows and n columns, in which m and n represent counting numbers, it is called an *m by n matrix.*

After looking over our data, we might decide to drop Gina's options from the possible choices, since they are more expensive item by item than are any of the others. If we do this, we will be left with a 3×3 *square matrix.* Notice that in a square matrix the number of rows equals the number of columns, or $m = n$.

$$\begin{array}{r c} & \begin{array}{ccc} \text{Vin's} & \text{Toni's} & \text{Sal's} \end{array} \\ \begin{array}{r} \text{Pizza} \\ \text{Drinks} \\ \text{Salad} \end{array} & \begin{bmatrix} \$10.10 & \$10.86 & \$10.65 \\ \$1.09 & \$0.89 & \$1.05 \\ \$3.69 & \$3.89 & \$3.85 \end{bmatrix} \end{array}$$

Matrices do not always have multiple rows and columns. For example, as we think about which pizza most people prefer, regardless of the price, we might decide that it would be Sal's. If we list only the prices for Sal's offerings, we will have a **column matrix** of order 3×1. Matrices that have only one column are sometimes referred to as **column vectors.**

$$\begin{array}{r c} & \text{Sal's} \\ \begin{array}{r} \text{Pizza} \\ \text{Drinks} \\ \text{Salad} \end{array} & \begin{bmatrix} 10.65 \\ 1.05 \\ 3.85 \end{bmatrix} \end{array}$$

If we choose to look at the pizza prices alone, they can be represented with a 1×3 **row matrix** or **row vector.**

$$\begin{array}{r c} & \begin{array}{ccc} \text{Vin's} & \text{Toni's} & \text{Sal's} \end{array} \\ \text{Pizza} & [\$10.10 \quad \$10.86 \quad \$10.65] \end{array}$$

Notice that when we give the order of a matrix, we write the number of rows followed by the number of columns. The simplest order is that of a 1×1 matrix such as [10.65] containing a single element.

Exercises

1. Define a matrix in your own words, and justify your definition through discussion with other students in your class.
2. Find and bring to class copies of at least two examples of information presented in matrix form in newspapers or magazines.
   a. What are the dimensions of each of your matrices?
   b. What is represented by the rows and columns of your matrices?
   c. Share your matrices with other class members.

Matrices are powerful tools that have numerous business applications such as keeping track of the flow of goods within industries.

3. A trendy garment company receives orders from three boutiques. The first boutique orders 25 jackets, 75 shirts, and 75 pairs of pants. The second boutique orders 30 jackets, 50 shirts, and 50 pairs of pants. The third boutique orders 20 jackets, 40 shirts, and 35 pairs of pants. Display this information in a matrix whose rows represent the boutiques and whose columns represent the type of garment ordered. Label the rows and columns of your matrix accordingly.

4. Although matrices contain many data values, they can also be thought of as single entities. This features allows us to refer to a matrix with a single capital letter, as follows:

$$\begin{array}{cc} & \begin{array}{ccc}\text{Vin's} & \text{Toni's} & \text{Sal's}\end{array} \\ A = \begin{array}{r}\text{Pizza} \\ \text{Drinks} \\ \text{Salad}\end{array} & \begin{bmatrix} \$10.10 & \$10.86 & \$10.65 \\ \$1.09 & \$0.89 & \$1.05 \\ \$3.69 & \$3.89 & \$3.85 \end{bmatrix}\end{array}$$

or

$$\begin{array}{cc} & \begin{array}{ccc}\text{Vin's} & \text{Toni's} & \text{Sal's}\end{array} \\ S = \text{Pizza} & [\$10.10 \quad \$10.86 \quad \$10.65]\end{array}$$

Individual entries in a matrix are identified by row number and column number, in that order. For example, the value 10.65 is the entry in row 1 and column 3 of matrix A and is referenced as A_{13}. Entry A_{13} represents or is interpreted as the cost of a pizza at Sal's. Notice that A_{31} is not the same as A_{13}. Entry A_{31} has the value 3.69 and represents the cost of a salad at Vin's. In the row matrix S, the entries are referenced as S_1, S_2, and S_3.

a. What is the value of A_{21}? Of A_{12}? Of A_{32}?
b. Write an interpretation of each of the entries in part a.
c. Write an interpretation of S_3.

5. For breakfast Patty had cereal, a medium-sized banana, a cup of 2% fat milk, and a slice of buttered toast. She recorded the following information in her food journal. Cereal: 165 calories, 3 g fat, 33 g carbohydrate, and no cholesterol. Banana: 120 calories, no fat, 26 g carbohydrate, and no cholesterol. Milk: 120 calories, 5 g fat, 11 g carbohydrate, and 15 mg cholesterol. Buttered toast: 125 calories, 6 g fat, 14 g carbohydrate, and 18 mg cholesterol.

a. Write this information in a matrix N whose rows represent the foods. Label the rows and columns of your matrix.
b. State the values of N_{23}, N_{32}, and N_{42}.
c. Write an interpretation of N_{23}, N_{32}, and N_{42}.

6. As we continue to plan our pizza party, we discover that the local supermarket has a sale on 2-liter bottles of soft drinks and so decide not to order drinks from a pizza house after all. Write and label a 2×3 matrix that represents the prices for just pizza and salad at Vin's, Toni's, and Sal's.

7. Suppose that when we were calling the pizza houses about prices, we also collected the following information about the cost of additional toppings and salad dressings:

	Vin's	Toni's	Sal's
Additional toppings	$1.15	$1.10	$1.25
Additional dressings	$0.00	$0.45	$0.50

Represent the information from this table in another 2×3 matrix whose rows represent the additional toppings and dressings and whose columns represent the three pizza houses. Label the rows and columns of your matrix.

8. The next step is to compute what it would cost to order pizzas with two toppings and to allow a choice of two salad dressings. This can be done by simply adding corresponding components of our two price matrices. If we let A represent the basic price matrix and B represent the matrix of additional costs, we can add A and B to get a third matrix, C, which will represent the total prices for pizza and salads at each pizza house. So, if

$$A = \begin{bmatrix} 10.10 & 10.86 & 10.65 \\ 3.69 & 3.89 & 3.85 \end{bmatrix}$$

and

$$B = \begin{bmatrix} 1.15 & 1.10 & 1.25 \\ 0.00 & 0.45 & 0.50 \end{bmatrix}$$

then

$$A + B = \begin{bmatrix} 10.10 + 1.15 & 10.86 + 1.10 & \underline{\qquad} \\ \underline{\qquad} & \underline{\qquad} & \underline{\qquad} \end{bmatrix}$$

$$= \begin{bmatrix} \underline{\quad} & \underline{\quad} & \underline{\quad} \\ \underline{\quad} & \underline{\quad} & \underline{\quad} \end{bmatrix} = C.$$

Complete the addition and label the rows and columns of matrix C.

9. In Exercise 8, the entries of matrix C represent the sum of the corresponding entries in matrices A and B. For example, C_{13}, which represents the cost of a pizza with an extra topping at Sal's, equals the sum of A_{13} and B_{13}.
   a. What is the value of A_{21}? Of B_{21}? Of C_{21}?
   b. Write an interpretation of A_{21}, B_{21}, and C_{21}.

In general, if A and B are m by n matrices, then $C = A + B$ is a matrix whose entries represent the sum of the corresponding entries in matrices A and B. In symbols, we have

$$C_{ij} = A_{ij} + B_{ij} \text{ where } 1 \leq i \leq m \text{ and } 1 \leq j \leq n.$$

It is clear from our discussion that we cannot add matrices of unlike dimensions. It also does not make sense to add matrices whose row and column labels represent unlike or incompatible quantities.

10. A physician associates with each patient a row matrix whose components represent that person's age, weight, and height. Would it be appropriate to add together the matrices associated with two different patients? Explain your answer.

American League Team Averages

Boston Globe, *July 4, 1993*

Chicago White Sox

BATTING

	Avg.	AB	R	H	HR	RBI
LVlr	.385	26	3	10	0	3
Johnson	.301	279	42	84	0	23
Thomas	.301	279	47	84	16	60
Raines	.299	117	27	35	7	18
Burks	.288	240	35	69	10	39
Cora	.274	277	45	76	1	23
Guillen	.272	217	26	59	2	27
Krkvc	.266	188	31	50	11	32
Ventura	.245	269	43	66	14	44
Jackson	.241	112	11	27	6	12
Bell	.228	298	27	68	6	42
Grebeck	.226	84	10	19	1	5
Sax	.224	49	11	11	1	3
Pasqua	.191	89	10	17	3	13
Wrona	.167	6	0	1	0	1
Totals	.265	2605	374	691	80	355

PITCHING

	W-L	ERA	Sv.	IP	H
Pall	2-2	2.56	1	38.2	31
Hernandez	1-3	2.68	14	37.0	30
Schwarz	1-1	2.70	0	30.0	17
Fernandez	9-4	2.79	0	126.0	101
Leach	0-0	2.81	1	16.0	15
Alvarez	7-4	3.19	0	104.1	90
McDwl	12-5	3.65	0	128.1	127
Thigpen	0-0	4.26	1	19.0	24
Bare	3-2	4.68	0	42.1	35
Radinsky	2-0	4.88	1	27.2	35
McCaskill	2-7	5.93	0	60.2	84
Bolton	0-4	8.85	0	20.1	33
Cary	0-0	12.27	0	3.2	7
Totals	40-37	3.95	18	688.1	674

California Angels

BATTING

	Avg.	AB	R	H	HR	RBI
Correia	.500	4	0	2	0	2
Easley	.309	175	26	54	2	18
Curtis	.300	260	50	78	2	29
Salmon	.284	257	48	73	15	50
Lovullo	.283	166	20	47	3	18
Gruber	.277	65	10	18	3	9
DSrcn	.256	254	26	65	3	33
Polonia	.249	277	30	69	1	13
R. Gnzls	.249	173	19	43	1	16
Myers	.246	130	15	32	4	16
Snow	.239	234	42	56	11	41
Davis	.230	269	36	62	10	57
Javier	.219	96	14	21	0	9
Orion	.200	90	5	18	1	4
Tingley	.167	30	3	5	0	4
Totals	.258	2534	349	653	56	324

PITCHING

	W-L	ERA	Sv.	IP	H
Davis	0-0	0.00	0	2.0	0
R. Gnzls	0-0	0.00	0	1.0	0
Weberre	0-1	1.24	3	29.0	22
Frey	1-0	1.88	9	24.0	22
Langston	9-2	2.62	0	123.2	101
Finley	9-6	3.00	0	117.0	115
Grahe	2-1	3.57	6	22.2	22
Sanderson	7-8	3.94	0	112.0	118
Patterson	0-1	6.53	0	30.1	34
Linton	0-1	6.61	0	16.1	17
Valera	3-6	6.62	4	53.0	77
Butcher	0-0	6.75	0	5.1	4
Hathaway	1-0	7.36	0	11.0	12
Springer	0-3	8.87	0	23.1	34
Nielson	0-0	9.64	0	9.1	16
Totals	37-40	4.36	22	681.2	713

11. The manager of a convenience store associates with each customer a column vector whose components represent the customer's purchases. Would it make sense to add together the vectors representing the purchases of two or more customers? Explain your answer.

12. Through June 7, 1991, the three baseball players with the highest batting averages in the American League had the following batting statistics:

	At bats	Runs	Hits	HRs	RBIs	Pct.
C. Ripken (Baltimore)	204	32	72	12	38	.353
E. Martinez (Seattle)	174	33	60	4	20	.345
P. Molitor (Milwaukee)	213	39	73	3	15	.343

A week later the following statistics for the same three players were published.

	At bats	Runs	Hits	HRs	RBIs	Pct.
C. Ripken (Baltimore)	228	34	81	12	39	.355
E. Martinez (Seattle)	190	35	65	4	23	.342
P. Molitor (Milwaukee)	234	45	81	5	20	.346

a. Subtract these two matrices to find and label another matrix that displays the changes in the statistics over the week's time.
b. Compute the batting averages (Pct.) for each player for the week, and add this information as a seventh column in your matrix. (Divide the number of hits by the times at bat to find the batting average for the week.)
c. Which player appears to have had the best week? Why?

13. Define *matrix subtraction,* and justify your definition through discussion with your classmates.

14. The matrices below (U.S. Department of Education, 1991) give the average times in minutes for physical fitness test endurance runs for American youths in 1980 and 1989. The runs were three-quarter mile for ages 10 and 11 and one mile for ages 12 through 17.

	1980		1989	
	Boys	Girls	Boys	Girls
10- and 11-year-olds	6.5	7.4	7.3	8.0
12- and 13-year-olds	8.4	9.8	9.1	10.5
14- to 17-year-olds	7.5	9.6	8.6	10.7

a. Find and label the matrix that represents the change in times in minutes for each age group from 1980 to 1989.
b. In which age group and sex was there the greatest increase in average time for the endurance runs? The smallest increase?
c. How would you explain this increase in times across age groups?

15. In statistics, a correlation matrix is a matrix whose entries represent the degree of relationship among variables. The values in a correlation matrix range from -1 to 1, where 0 indicates that there is no relationship, a negative value indicates that as one variable increases the other decreases, and a positive value indicates that as one variable increases the other one also increases. In a study of the relationship among ACT scores, high school class rank, and college grade point average, the following correlation matrix was generated. Notice that the row labels and the column labels are the same in a correlation matrix.

	ACT Comp.	ACT Eng.	ACT Math	ACT Soc. St.	ACT Sci.	H.S. Rank	Coll. GPA
ACT Composite	1.00	0.80	0.79	0.81	0.82	0.59	0.51
ACT English	0.80	1.00	0.54	0.58	0.55	0.53	0.48
ACT Math	0.79	0.54	1.00	0.42	0.52	0.57	0.42
ACT Social studies	0.81	0.58	0.42	1.00	0.61	0.39	0.39
ACT Science	0.82	0.55	0.52	0.61	1.00	0.44	0.36
High school rank	0.59	0.53	0.57	0.39	0.44	1.00	0.45
College GPA	0.51	0.48	0.42	0.39	0.36	0.45	1.00

Source: Aksamit, Mitchell, and Pozehl, 1986.

The ones in this matrix are located along what is called the **main diagonal,** in which the row and column numbers are the same. If we call this matrix R, then these diagonal elements are referenced as R_{ii} ($i = 1, 2, \cdots, 7$). We say that this matrix is **symmetric,** since $R_{ij} = R_{ji}$ ($i, j = 1, 2, \cdots, 7$).

In a symmetric matrix we need only know the values along the main diagonal and either the triangle above the main diagonal (the *upper* triangle) or below it (the *lower* triangle). Because of this feature, correlation matrices are often written with blanks in either the upper or the lower triangle.

a. Why do you think all the values along the main diagonal of a correlation matrix are ones?
b. Could a matrix that is not square be symmetric? Why?
c. Why are the values in a correlation matrix symmetrical about the main diagonal?
d. Which variable had the highest correlation with college GPA?
e. Which subject area test had the highest correlation with high school rank?

16. In algebra, we learn that the commutative and associative properties hold for addition over the set of real numbers. That is, for all real numbers a, b, and c, $a + b = b + a$, and $a + (b + c) = (a + b) + c$.
    a. Do you think that the *commutative* and *associative properties* hold for the addition of matrices? Why?
    b. Use the matrices below to test your answer in part a.

$$A = \begin{bmatrix} 4 & -2 \\ 3 & 1 \end{bmatrix} \quad B = \begin{bmatrix} 1 & 3 \\ -2 & 5 \end{bmatrix} \quad C = \begin{bmatrix} 2 & 4 \\ 1 & -1 \end{bmatrix}$$

17. Do you think that the commutative and associative properties hold for the subtraction of matrices? Why? Test your reasoning using matrices A, B, and C.

18. When all entries of a matrix are the number zero it is called a *zero matrix* and is denoted using a capital letter O alone or with subscripts ($O_{m \times n}$).

Let $O = \begin{bmatrix} 0 & 0 \\ 0 & 0 \end{bmatrix}$ and let A be any other 2×2 matrix.

a. Show that $A + O = O + A = A$ and that $A - A = O$.
b. Show that $A + (-A) = (-A) + A = O$, where the matrix $-A$, called the negative of A, is obtained by negating each entry in A.

LESSON 3.2

Multiplication of Matrices, Part 1

In the previous lesson, matrix addition and subtraction were defined by looking at some matrix models of real-world situations. In this lesson, we approach matrix multiplication in much the same manner.

For our first example, we return to the data in our pizza problem. One decision that we must make is how many of each type of pizza and salad we can afford to order. Let's start this decision-making process by computing the cost of ordering four each of the pizzas and salads represented by matrix C in Exercise 8 on page 105. To do this, we multiply each element in matrix C by 4 to get a new matrix, T, equal to $4C$. An operation of this type is called **multiplication of a matrix by a scalar.** Multiplication of a matrix by a scalar is analogous to multiplication of integers, in that $4C$ could also be interpreted as repeated addition, or $4C = C + C + C + C$.

$$4C = 4 \times \begin{bmatrix} 11.25 & 11.96 & 11.90 \\ 3.69 & 4.34 & 4.35 \end{bmatrix} = \begin{bmatrix} 4(11.25) & 4(11.96) & 4(11.90) \\ 4(3.69) & 4(4.34) & 4(4.35) \end{bmatrix}$$

$$= \begin{bmatrix} 45.00 & 47.84 & 47.60 \\ 14.76 & 17.36 & 17.40 \end{bmatrix} = T.$$

Labeling the rows and columns of the matrix, we have

$$T = \begin{array}{l} \\ \text{Pizza} \\ \text{Salad} \end{array} \begin{array}{c} \begin{array}{ccc} \text{Vin's} & \text{Toni's} & \text{Sal's} \end{array} \\ \begin{bmatrix} \$45.00 & \$47.84 & \$47.60 \\ \$14.76 & \$17.36 & \$17.40 \end{bmatrix} \end{array}$$

In our next example, we look at a situation that illustrates the **multiplication of a row matrix times a column matrix.** In this example, Jon, a student at Galileo High, runs out to the nearby Super X to buy some junk food to stock up his locker for between-class snacks. He chooses four small bags of chips, five candy bars, a box of cheese crax, three packs of sour drops, and two bags of cookies. Jon's purchases can be represented by a row matrix Q.

$$\begin{array}{cccccc} & \text{chips} & \text{candy} & \text{crax} & \text{drops} & \text{cookies} \\ Q = [& 4 & 5 & 1 & 3 & 2 \;] \end{array}$$

Suppose that chips are 30 cents a bag, candy bars are 35 cents each, crax are 50 cents a box, sour drops cost 20 cents a pack, and cookies sell for 75 cents a bag. These prices are represented in column matrix P below.

$$P = \begin{array}{l} \\ \text{chips} \\ \text{candy} \\ \text{crax} \\ \text{drops} \\ \text{cookies} \end{array} \begin{array}{c} \text{cents} \\ \begin{bmatrix} 30 \\ 35 \\ 50 \\ 20 \\ 75 \end{bmatrix} \end{array}$$

The obvious question to ask now is, How much did Jon pay for all these goodies? To find the answer, we "multiply" the price vector P by the quantity vector Q, as follows:

$$Q \times P = [4 \quad 5 \quad 1 \quad 3 \quad 2] \begin{bmatrix} 30 \\ 35 \\ 50 \\ 20 \\ 75 \end{bmatrix}$$

$$\begin{aligned} &= [4(30) + 5(35) + 1(50) + 3(20) + 2(75)] \\ &= [120 + 175 + 50 + 60 + 150] \\ &= [555] \text{ cents} = [\$5.55]. \end{aligned}$$

This matrix computation is, of course, exactly what the clerk at the Super X would do to figure Jon's bill. The price of each item was multiplied by the number purchased, and the products were summed. In order to do this computation, it is obvious that the number of items and the number of prices must be the same.

> In general: If Q is a row matrix and P is a column matrix, each having the same number of components, then the product QP is defined. Q times P is found by multiplying the corresponding components and summing the results.

To illustrate this, suppose that another student, Trilby, goes along with Jon to the Super X. Her purchases are a bag of chips, two candy bars, two packs of gum that cost 25 cents each, and a medium-sized drink for 75 cents.

1. Write and label a row matrix Q to represent the quantity of each item that Trilby purchased and a column matrix P to represent the prices for each item.

2. Perform the multiplication, Q times P, to find the total cost of Trilby's purchases.

3. Compare your work with that of other students in your class. Was your final matrix = [$2.25]?

As you can see from these examples, when we multiply a row matrix times a column matrix, the answer is a single value. In other words, when we multiply a $(1 \times k)$ row matrix Q by a $(k \times 1)$ column matrix P, the result is a (1×1) single-value matrix C. Schematically this product looks like

$$\underset{(1 \times k)}{Q} \times \underset{(k \times 1)}{P} = \underset{(1 \times 1)}{C}$$

Same

Dimensions of the product.

The previous examples showed how the multiplication of a matrix by a scalar and the multiplication of a row matrix by a column matrix are defined. Our next step is to define the **multiplication of a row**

matrix with more than one column by a row matrix. We model this type of matrix multiplication by examining some additional options in our pizza problem.

Suppose we decide to order five pizzas and three salads, and we want to know the total cost at each of the pizza houses for this combination. If we do the calculations involved here without using matrices, we proceed by multiplying the pizza price by 5 and adding the result to 3 times the salad price for each pizza house, as follows:

Cost at Vin's: 5($11.25) + 3($3.69) = $56.25 + $11.07 = $67.32

Cost at Toni's: 5($11.96) + 3($4.34) = $59.80 + $13.02 = $72.82

Cost at Sal's: 5($11.90) + 3($4.35) = $59.50 + $13.05 = $72.55

To use matrix multiplication to solve this problem, we represent the number of pizzas and salads we plan to order with a 1×2 row matrix A.

$$A = \begin{array}{c} \text{Pizzas} \quad \text{Salads} \\ [\;5 \qquad\quad 3\;]. \end{array}$$

The prices for pizzas and salads at each of the three pizza houses are modeled by the 2×3 matrix C.

$$C = \begin{array}{r} \\ \text{Pizzas} \\ \text{Salads} \end{array}\!\!\begin{array}{c} \begin{array}{ccc} \text{Vin's} & \text{Toni's} & \text{Sal's} \end{array} \\ \begin{bmatrix} 11.25 & 11.96 & 11.90 \\ 3.69 & 4.34 & 4.35 \end{bmatrix} \end{array}.$$

Now when we multiply matrix A times matrix C, we expect the result to be another matrix whose components give the total cost for five pizzas and three salads at each pizza house. To do this, it makes sense to multiply the row matrix A by each of the columns in matrix C, as follows:

$$A \times C = [5 \quad 3] \times \begin{bmatrix} 11.25 & 11.96 & 11.90 \\ 3.69 & 4.34 & 4.35 \end{bmatrix}$$

$$\begin{aligned} &= [5(11.25) + 3(3.69) \qquad 5(11.96) + 3(4.34) \\ &\quad\; 5(11.90) + 3(4.35)] \\ &= [56.25 + 11.07 \qquad 59.80 + 13.02 \qquad 59.50 + 13.05] \\ &= [67.32 \qquad 72.82 \qquad 72.55]. \end{aligned}$$

Now, labeling our final product, we have

Vin's	Toni's	Sal's
[$67.32	$72.82	$72.55].

Notice that in doing this matrix computation, we proceeded in the same manner as we did when we computed the costs without using matrices. In the matrix multiplication, each of the components of the row matrix [5 3] was multiplied by the corresponding component in the columns of matrix C. These products were then summed to give the components of the product matrix.

We can see from this model that matrix multiplication of this sort is defined only when the number of entries in the row matrix equals the number of rows in the multidimensional matrix. In addition, the product is a row matrix with the same number of entries as there are columns in the second matrix.

> In general, if a $(1 \times k)$ row matrix, A, is multiplied times a $(k \times n)$ matrix, C, the result will be a $(1 \times n)$ row matrix, P.

We can represent this result schematically, as follows:

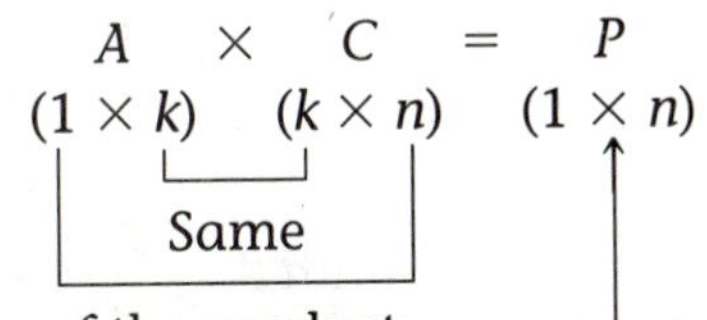

Example: We want to look at some other combinations of pizzas and salads before making our final decision about how many to order. Use matrix multiplication to calculate the totals for (1) four pizzas and three salads and (2) four pizzas and four salads at each of the three pizza houses. Be sure to label your matrices. Check the steps in your work and discuss the interpretations of each of your answers with other students in your class. Did you get (1) [$56.07 $60.86 $60.65] and (2) [$59.76 $65.20 $65.00]?

Exercises

1. Refer to matrix T in the first example of this lesson (page 111).
   a. What does matrix T represent?
   b. What is the cost of four pizzas at Sal's?
   c. Interpret T_{12} and T_{21}.

2. Nancy has a small shop in the Oldmarket where she makes and sells four different kinds of jewelry: earrings (e), pins (p), necklaces (n), and bracelets (b). She fashions each item out of either cultured pearls or jade beads. The matrix below represents Nancy's sales for May.

$$M = \begin{array}{c} \\ \text{pearl} \\ \text{jade} \end{array} \begin{array}{c} \begin{array}{cccc} \text{e} & \text{p} & \text{n} & \text{b} \end{array} \\ \begin{bmatrix} 8 & 4 & 6 & 5 \\ 20 & 10 & 12 & 9 \end{bmatrix} \end{array}.$$

   Nancy hopes to sell twice as many of each piece in June.
   a. Calculate a matrix, J, where $J = 2M$ to represent the number of each item that Nancy will sell in June if she reaches her goal.
   b. Label the rows and columns of matrix J.
   c. How many jade necklaces does Nancy expect to sell in June?
   d. Interpret J_{21} and J_{12}.

3. Matt reads on the side of his cereal box that each ounce of cereal contains the following percentages of the minimum daily requirements of:

vitamin A	25%
vitamin C	25%
vitamin D	10%

   If Matt eats 3 ounces of cereal for breakfast, what percentage of each vitamin will he get? Show the matrices and matrix operation in your calculation. Label your matrices.

4. The regents at a state university recently announced a 7% raise in tuition rates per semester hour. The current rates per semester hour are shown in the table on the next page.

	Undergraduate	Graduate
Resident	\$ 53.50	\$ 71.00
Nonresident	\$145.50	\$175.00

a. Write and label a matrix that represents this information.
b. Find a new matrix that represents the tuition rates per semester hour after the 7% raise goes into effect. Label your matrix.
c. Find a matrix that represents the actual dollar increase for each of the categories. Label your matrix.

5. As you saw in the second example in this lesson (page 112), if Q is a (1×5) row matrix and P is a (5×1) column matrix, then the product Q times P is a single-value matrix with dimensions (1×1). Suppose you multiply the column matrix P times the row matrix Q.
   a. What will be the dimensions of the product P times Q? Justify your answer using a schematic diagram similar to the ones following the examples in this lesson.
   b. Multiply matrix P times matrix Q. (Refer to the second example in this lesson for the values in these matrices.)
   c. Does the product P times Q have a meaningful interpretation of this situation? Explain your answer.

6. Susan's credit union has investments in three states—Massachusetts, Nebraska, and California. The deposits in each state are divided between consumer loans and bonds. The amount of money (in thousands of dollars) invested in each category is displayed in the table below.

	Mass.	Neb.	Cal.
Loans	230	440	680
Bonds	780	860	940

The current yields on these investments are 6.5% for consumer loans and 7.2% for bonds. Use matrix multiplication to find the total earnings for each state. Label your matrices.

7. Mary is remodeling her home. She makes a trip to the lumber company to pick up ten 2×6's, four 4×6's, and two 5×5's. In 8-foot lengths, the 2×6's cost \$3.00, the 4×6's cost \$8.50, and the 5×5's cost \$9.50.
   a. Write and label a row matrix and a column matrix to represent the information in this problem.
   b. Will everyone necessarily write the same row and column matrices? Explain your answer.
   c. Perform a matrix multiplication to find the total cost of Mary's purchases.

8. Peter has \$10,000 in a 12-month CD at 7.3% (annual yield), \$17,000 in a credit union at 6.5%, and \$12,000 in bonds at 7.5%. Use vector multiplication to find Peter's earnings for a year. Label your vectors.

9. The **transpose** (A^T) of a matrix A is the matrix obtained by interchanging the rows and columns of matrix A.
   a. Describe the transpose of a row matrix and of a column matrix.
   b. Write and label the transpose (M^T) of matrix M in Exercise 2.
   c. What might be a reason for wanting to know the transpose of a matrix?

10. Refer to Exercise 2. It takes Nancy 2 hours to make a pair of earrings, 1 hour to make a pin, 2.5 hours to make a necklace, and 1.5 hours to make a bracelet.
    a. Write and label a row matrix that represents this information.
    b. Use matrix multiplication to find a matrix that represents the total number of hours that Nancy spends making each type of jewelry (cultured pearls or jade) for the month of May. (Hint: Use the transpose of matrix M that you found in the previous exercise.)
    c. Label your product matrix.
    d. Interpret each of the entries in the product.

11. Nancy expanded her jewelry business and now has shops in the Westmarket and Eastmarket plazas as well as in the Oldmarket. Her sales of cultured pearls sets for July are shown in the table below.

	Old	West	East
Earrings	10	8	12
Pins	6	5	4
Necklaces	3	2	2
Bracelets	4	3	2

Earrings sell for \$40 a pair, pins for \$35 each, necklaces for \$80, and bracelets for \$45. Use matrix multiplication to find Nancy's total sales at each location. Label your matrices.

12. During the first week of a recent fund-raiser for the math club at Happy High, Anne sold the following number of candy canes.

	Mon	Tues	Wed	Thurs	Fri
Canes	10	15	20	30	50

a. Write this information in a column matrix C. Label your matrix.

b. Find a row matrix N so that the product N times C gives the total number of candy canes that Anne sold for the week.

c. Find a row matrix A so that the product A times C gives the average number of candy canes that Anne sold each day. (Hint: By what fraction would you multiply the total number of canes to find the average?)

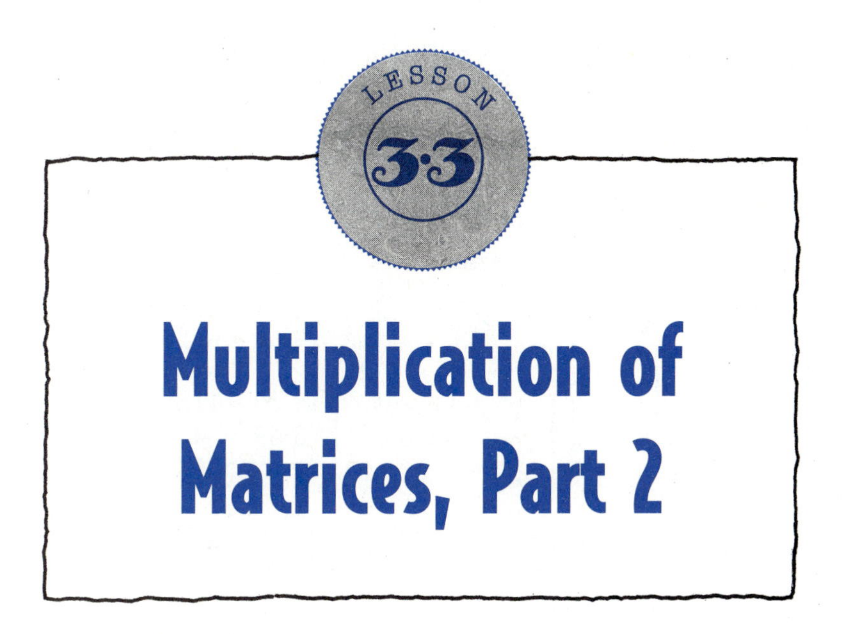

Lesson 3.3 Multiplication of Matrices, Part 2

In this lesson, we continue to explore matrix multiplication by looking at the products of multidimensional matrices. There was an example of this type of matrix multiplication in the examples of Lesson 3.2. Now, if we look at the three multiplications we did in Lesson 3.2 (pages 114–115) to compare the cost of different combinations of pizza and salad, it makes sense to combine all three options into a single matrix, *B*, and to perform a single matrix multiplication.

$$B = \begin{array}{r} \\ \text{Option 1} \\ \text{Option 2} \\ \text{Option 3} \end{array} \begin{array}{c} \begin{array}{cc} \text{No. pizzas} & \text{No. salads} \end{array} \\ \begin{bmatrix} 4 & \quad\quad\quad 3 \\ 4 & \quad\quad\quad 4 \\ 5 & \quad\quad\quad 3 \end{bmatrix} \end{array}$$

If we multiply matrix *B* times matrix *C* (see page 114), the product (call it *D*) will be a 3×3 matrix whose rows will represent the three options and whose columns will represent the three pizza houses. The components of this matrix give us the total cost for each of the three options at each of the three pizza houses.

Let's follow the steps in this multiplication. Notice that the computations are exactly the same as those for our three separate calculations in the previous lesson. We would expect, then, that row 1 of the product will represent the cost of four pizzas and three salads, that row 2 of the product will represent the cost of four pizzas and four salads, and that row 3 of the product will represent the cost of five pizzas and three salads at each of the pizza houses.

Then, matrix $B \times$ matrix $C =$

$$\begin{array}{l} \\ \text{Option 1} \\ \text{Option 2} \\ \text{Option 3} \end{array} \begin{array}{c} \text{pizzas} \quad \text{salads} \\ \begin{bmatrix} 4 & 3 \\ 4 & 4 \\ 5 & 3 \end{bmatrix} \end{array} \times \begin{array}{l} \\ \text{pizzas} \\ \text{salads} \end{array} \begin{array}{c} \text{Vin's} \quad \text{Toni's} \quad \text{Sal's} \\ \begin{bmatrix} 11.25 & 11.96 & 11.90 \\ 3.69 & 4.34 & 4.35 \end{bmatrix} \end{array}$$

or

$$\begin{bmatrix} 4 & 3 \\ 4 & 4 \\ 5 & 3 \end{bmatrix} \times \begin{bmatrix} 11.25 & 11.96 & 11.90 \\ 3.69 & 4.34 & 4.35 \end{bmatrix}$$

$$= \begin{bmatrix} 4(11.25) + 3(3.69) & 4(11.96) + 3(4.34) & 4(11.90) + 3(4.35) \\ 4(11.25) + 4(3.69) & 4(11.96) + 4(4.34) & 4(11.90) + 4(4.35) \\ 5(11.25) + 3(3.69) & 5(11.96) + 3(4.34) & 5(11.90) + 3(4.35) \end{bmatrix}$$

$$= \begin{bmatrix} 45.00 + 11.07 & 47.84 + 13.02 & 47.60 + 13.05 \\ 45.00 + 14.76 & 47.84 + 17.36 & 47.60 + 17.40 \\ 56.25 + 11.07 & 59.80 + 13.02 & 59.50 + 13.05 \end{bmatrix}$$

$$= \begin{bmatrix} 56.07 & 60.86 & 60.65 \\ 59.76 & 65.20 & 65.00 \\ 67.32 & 72.82 & 72.55 \end{bmatrix} = D.$$

When we label our product matrix for clarity's sake we have

$$D = \begin{array}{l} \\ \text{Option 1} \\ \text{Option 2} \\ \text{Option 3} \end{array} \begin{array}{c} \text{Vin's} \quad \text{Toni's} \quad \text{Sal's} \\ \begin{bmatrix} 56.07 & 60.86 & 60.65 \\ 59.76 & 65.20 & 65.00 \\ 67.32 & 72.82 & 72.55 \end{bmatrix} \end{array}$$

In this matrix, D_{11} represents the cost of four pizzas and three salads at Vin's. How would you interpret D_{23} and D_{33}?

In order for the multiplication of two matrices to be defined, the matrices must be conformable, which means that the number of columns in the first matrix must equal the number of rows in the second matrix.

Notice, also, that the order of the product matrix is the number of rows of the first matrix by the number of columns of the second. This can be shown schematically as follows:

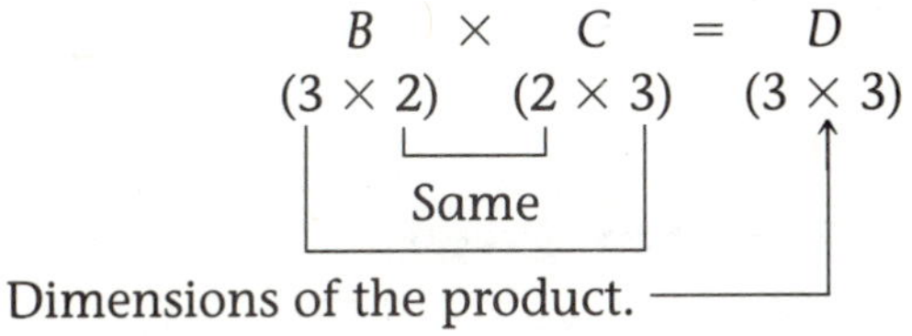

$$\underset{(3 \times 2)}{B} \times \underset{(2 \times 3)}{C} = \underset{(3 \times 3)}{D}$$

Same

Dimensions of the product.

In general, if we multiply a matrix P with m rows and k columns times a matrix Q with k rows and n columns, the product will be a matrix R with m rows and n columns.

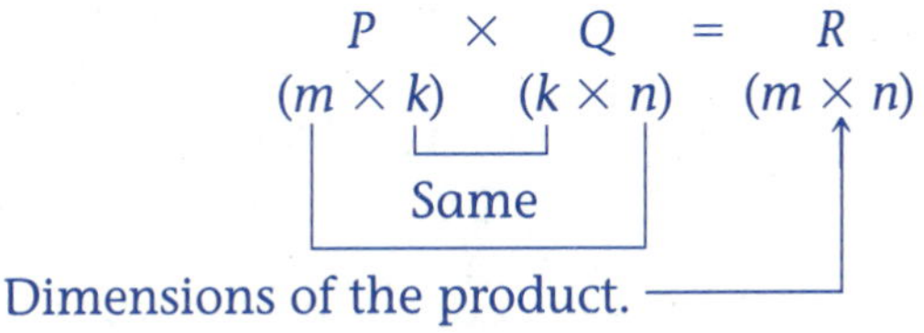

$$\underset{(m \times k)}{P} \times \underset{(k \times n)}{Q} = \underset{(m \times n)}{R}$$

Same

Dimensions of the product.

The dimensions of these matrices can also be described using the row and column labels. Matrix B classifies our data according to Options (rows) and Foods (columns). Hence we can refer to matrix B as an Options-by-Foods matrix. Likewise we can describe C as a Foods-by-Houses matrix. The product B times C, in turn, results in a matrix of dimensions Options by Houses, which is what we wanted to know. Notice, also, that when we performed the multiplication, the common label (Foods) was eliminated, leaving the product matrix with the row label of the first factor and the column label of the second factor. Schematically, the dimensions of each of the matrices involved in computing our product can be described as follows:

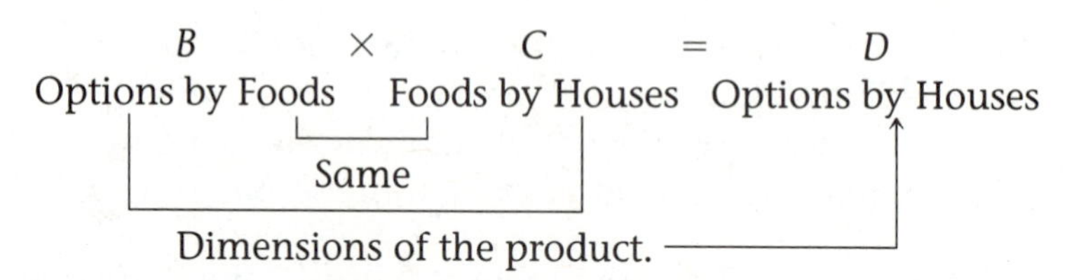

Using the row and column labels in this manner helps determine whether a matrix multiplication will result in a meaningful interpretation or whether, indeed, it will give us the results that we wish.

Exercises

1. Mike, Liz, and Kate are heirs to an estate that consists of a condominium, a customized BMW, and choice season tickets to the Nebraska Cornhusker football games, and for the purposes of fair division they have submitted the bids shown in matrix E.

$$E = \begin{array}{c} \\ \text{Mike} \\ \text{Liz} \\ \text{Kate} \end{array} \begin{array}{c} \begin{array}{ccc} \text{Condo} & \text{BMW} & \text{Tickets} \end{array} \\ \begin{bmatrix} \$185{,}000 & \$76{,}000 & \$200 \\ \$175{,}000 & \$60{,}000 & \$180 \\ \$180{,}000 & \$75{,}000 & \$250 \end{bmatrix} \end{array}$$

The awarding of the items in the estate is indicated by matrix A.

$$A = \begin{array}{c} \\ \text{Condo} \\ \text{BMW} \\ \text{Tickets} \end{array} \begin{array}{c} \begin{array}{ccc} \text{Mike} & \text{Liz} & \text{Kate} \end{array} \\ \begin{bmatrix} 1 & 0 & 0 \\ 1 & 0 & 0 \\ 0 & 0 & 1 \end{bmatrix} \end{array}$$

   a. Find the matrix product $P = EA$. Label the rows and columns of P.
   b. Write an interpretation of the entries in matrix P. (Refer to Exercise 6 in Lesson 2.2 on page 56.)

2. Emma and Ken go out to eat at Sammy's Drive In. Ken has a small cheeseburger, a baked potato with sour cream, and a shake. Emma orders a Sammy's special, medium fries, and a shake. The

approximate number of calories, grams of fat, and milligrams of cholesterol in each of these foods is represented by the table below.

	Calories	Fat (g)	Cholesterol (mg)
Cheeseburger	450	40	50
Sammy's special	570	48	90
Potato/sour cream	500	45	25
French fries	300	30	0
Shake	400	22	50

a. Write a matrix Q that describes Emma's and Ken's choices, with the columns representing the foods. Label the rows and columns of this matrix.

Industries such as the fast-food business have changed dramatically since the advent of the computer in using matrices as a natural tool for storing and manipulating data.

b. Write a matrix C that represents the information in the table on the previous page, with the rows representing the foods. Label the rows and columns of this matrix.
c. What are the dimensions of matrix Q and of matrix C?
d. What are the dimensions of the product Q times C? By using a schematic diagram, show why your answer is correct.
e. The dimensions of matrix Q could be described as Persons by Foods. Describe the dimensions of matrices C and Q times C in a similar manner. Justify your answer for matrix Q times C with a schematic diagram.
f. Multiply matrix Q times matrix C to get a matrix R. Label the rows and columns of matrix R.
g. Interpret R_{12}, R_{21}, and R_{23}.

3. a. What must be true about the dimensions of matrixes A and B if the product $C = AB$ is defined?
b. If the products AB and BA both are defined, what must be true about the dimensions of matrices A and B? Why?
c. If AB and BA both are defined, it does not necessarily follow that $AB = BA$ (i.e., in general, matrix multiplication is not commutative). Using 2×2 matrices, find examples where $AB = BA$ and where AB is not equal to BA.

4. Let A be any 3×3 matrix and let

$$I = \begin{bmatrix} 1 & 0 & 0 \\ 0 & 1 & 0 \\ 0 & 0 & 1 \end{bmatrix}.$$

Show that $IA = AI = A$.

The matrix I is called an **identity matrix.** An identity matrix is any matrix in which each of the entries along the main diagonal are ones and all other entries are zeros. Identity matrices act in the same way for matrix products as the number 1 does for number products.

5. Given the matrices A, B, and C on the next page, do you think that $A(BC) = (AB)C$? Test your guess by computing the products $A(BC)$ and $(AB)C$.

$$A = \begin{bmatrix} 1 & 0 & 1 \\ 0 & 1 & 1 \end{bmatrix} \quad B = \begin{bmatrix} 3 & 1 \\ 2 & 2 \\ -1 & 1 \end{bmatrix} \quad C = \begin{bmatrix} 1 & -1 & 0 \\ 2 & 1 & 1 \end{bmatrix}.$$

6. Find two (2×2) matrices, A and B, to demonstrate that $(A + B)(A - B)$ is not necessarily equal to $A^2 - B^2$.

7. In algebra, two numbers whose product is 1 (the identity element for multiplication) are called inverses of each other. For example, 5 and $\frac{1}{5}$ (or 5^{-1}) are inverses of each other, since $5(\frac{1}{5}) = (\frac{1}{5})\,5 = 1$.

   Similarly, if A and B are two square matrices such that $AB = BA = I$, then A and B are called **inverses** of each other. The inverse of A is denoted as A^{-1}.

   Verify that the matrices A and B below are inverses of each other, by computing AB and BA.

$$A = \begin{bmatrix} 2 & 3 \\ 1 & 2 \end{bmatrix} \quad B = \begin{bmatrix} 2 & -3 \\ -1 & 2 \end{bmatrix}.$$

8. Carefully plot the points $A(0,0)$, $B(6,2)$, $C(8,6)$, and $D(2,4)$ on graph paper. Connect the points forming a polygon $ABCD$. This polygon can be represented with a matrix P, as follows:

$$\begin{matrix} & A & B & C & D \end{matrix}$$
$$P = \begin{bmatrix} 0 & 6 & 8 & 2 \\ 0 & 2 & 6 & 4 \end{bmatrix}.$$

   a. Multiply the matrix representing polygon $ABCD$ by the matrix

$$T_1 = \begin{bmatrix} -1 & 0 \\ 0 & 1 \end{bmatrix}.$$

   b. Plot and label the four points represented in your new matrix as A', B', C', and D'. Connect the points to form polygon $A'B'C'D'$.
   c. Describe the relationship between polygon $A'B'C'D'$ and polygon ABCD.
   d. Multiply the matrix representing polygon $A'B'C'D'$ by the matrix

$$T_2 = \begin{bmatrix} 1 & 0 \\ 0 & -1 \end{bmatrix}.$$

e. Plot and label the four points represented in your new matrix as A'', B'', C'', and D''. Describe the relationship between polygon $A''B''C''D''$ and polygon $A'B'C'D'$.

f. Multiply T_2T_1 to get a new matrix, R. Multiply R times the matrix P, which represents the original polygon, $ABCD$, and plot the resulting points. What effect does R have on $ABCD$? Do the following to test your answer: Use a blank sheet of unlined paper and trace both your axes and polygon $ABCD$. Leave your copy on top of the original polygon, and place the point of your pencil on the origin. Now, holding the original paper in place, rotate the top sheet until your copy of $ABCD$ rests on top of polygon $A''B''C''D''$. Describe what happens to polygon $ABCD$.

g. Find a matrix, T_3, that reflects polygon $A''B''C''D''$ about the y-axis into quadrant IV of your graph.

h. Find a matrix, T_4, that rotates polygon $A'B'C'D'$ about the origin into quadrant IV. How does T_4 relate to T_2 and T_3?

For Exercises 9, 10, and 11, you will need either a calculator or access to computer software that performs matrix operations.

9. The matrix A below is called an *upper triangular matrix*.

$$A = \begin{bmatrix} 1 & 1 & 1 \\ 0 & 1 & 1 \\ 0 & 0 & 1 \end{bmatrix}$$

a. Calculate A^2, A^3, and A^4.

b. Make a conjecture about the form of A^k.

c. Test your conjecture by computing additional powers of A.

10. The Fancy Bag manufacturing company that makes and markets fine leather bags has three factories—one in New York, one in Nebraska, and one in California. One of the bags they make comes in three styles—handbag, standard shoulder bag, and roomy shoulder bag. The production of each bag requires three kinds of work—cutting the leather, stitching the bag, and finishing the bag.

Matrix T gives the time (in hours) of each type of work required to make each type of bag.

$$T = \begin{array}{l} \\ \text{Handbag} \\ \text{Standard} \\ \text{Roomy} \end{array} \begin{array}{c} \begin{array}{ccc} \text{Cutting} & \text{Stitching} & \text{Finishing} \end{array} \\ \begin{bmatrix} 0.4 & 0.6 & 0.4 \\ 0.5 & 0.8 & 0.5 \\ 0.6 & 1.0 & 0.6 \end{bmatrix} \end{array}$$

Matrix P gives daily production capacity at each of the factories.

$$P = \begin{array}{l} \\ \text{New York} \\ \text{Nebraska} \\ \text{California} \end{array} \begin{array}{c} \begin{array}{ccc} \text{Handbag} & \text{Standard} & \text{Roomy} \end{array} \\ \begin{bmatrix} 10 & 15 & 20 \\ 25 & 15 & 12 \\ 20 & 12 & 10 \end{bmatrix} \end{array}$$

Matrix W provides the hourly wages of the different workers at each factory.

$$W = \begin{array}{l} \\ \text{New York} \\ \text{Nebraska} \\ \text{California} \end{array} \begin{array}{c} \begin{array}{ccc} \text{Cutting} & \text{Stitching} & \text{Finishing} \end{array} \\ \begin{bmatrix} 7.50 & 8.50 & 9.00 \\ 7.00 & 8.00 & 8.50 \\ 8.40 & 9.60 & 10.10 \end{bmatrix} \end{array}$$

Matrix D contains the total number of orders received at each factory for the months of May and June.

$$D = \begin{array}{l} \\ \text{Handbag} \\ \text{Standard} \\ \text{Roomy} \end{array} \begin{array}{c} \begin{array}{cc} \text{May} & \text{June} \end{array} \\ \begin{bmatrix} 600 & 800 \\ 800 & 1{,}000 \\ 400 & 600 \end{bmatrix} \end{array}$$

a. Matrix T can be described as a Bag-by-Work matrix. Describe matrices P, W, and D in a similar manner.

Use the matrices above (or their transposes) to compute the following. Label the rows and columns of the matrix in each answer. Hint: The label dimensions from part a will help you decide what your matrix products should look like.

b. The number of hours of each type of work needed each month to fill all orders.

c. The production cost per bag at each factory.

d. The cost of filling all May orders at the Nebraska factory. (Hint: In this example the answer, a single value, is the product of a row vector and a column vector.)
e. The daily number of hours of each type of work needed at each factory if production levels are at capacity.

11. (For those students who have studied trigonometry.)
a. Plot the polygon $ABCD$ represented in Exercise 8.
b. Multiply the matrix P by the following transformation matrix:

$$T_1 = \begin{bmatrix} \cos 30° & -\sin 30° \\ \sin 30° & \cos 30° \end{bmatrix}$$

c. Plot the resulting polygon and label it $A'B'C'D'$. How does polygon $A'B'C'D'$ relate to polygon $ABCD$? Try repeating the transformation using 180° to test your answer.
d. Write a matrix that rotates a polygon through 60°. Does this transformation matrix have the same effect as applying T_1 twice? Test your answer.
e. Find a matrix that rotates polygon $ABCD$ through 90° and another that rotates the polygon through −90°. Find the product of these two transformation matrices. What is the relationship between these two matrices? Test your answer by finding the product of the matrices that rotates the polygon through 60° and −60°.

Population Growth: The Leslie Model, Part 1

Age-specific population growth is a topic that is of great concern to people in fields as diverse as urban planning and wildlife management. Urban policymakers are interested in knowing how many people there will be in various age groups after certain periods of time have elapsed. Those in wildlife management worry about maintaining animal populations at levels that can be supported in their natural habitats without damage to the environment.

If the age distribution of a population at a certain date is known, along with birth and survival rates for age-specific groups, a model can be created to determine the age distributions of the survivors and descendants of the original population at successive intervals of time. The problem that we use to illustrate this model was posed in 1945 by P. H. Leslie of the Bureau of Animal Population at Oxford University in Oxford, England.

In this problem, we examine the growth rate of a population of a species of small brown rats, *Rattus norvegicus.* The life span of these rodents is 15 to 18 months. They have their first litter at approximately 3 months and continue to reproduce every 3 months until they reach

TABLE 1 ▸ Birth and Survival Rates

Age (months)	Birthrate	Survival Rate
0–3	0	0.6
3–6	0.3	0.9
6–9	0.8	0.9
9–12	0.7	0.8
12–15	0.4	0.6
15–18	0	0

the age of 15 months. Birthrates and age-specific survival rates for 3-month periods are summarized in Table 1. In order to simplify the situation as much as possible, we assume that birthrates and survival rates remain constant over time, and we consider only the female population.

The actual number of female births in a particular age group can be found by multiplying the birthrate by the number of females currently in the age group. The survival rate is the probability that a rodent will survive and move into the next age group.

The original female rodent population is 42 animals, with the age distribution shown in the table below.

Age (months)	0–3	3–6	6–9	9–12	12–15	15–18
Number	15	9	13	5	0	0

In examining this model, one question that might be asked is how many rodents there will be after 3 months have passed. Also of interest might be the age distribution of this new group. To answer these questions, it is necessary to find the number of new babies introduced into the population and the number of rats that survive in each group and move up to the next age group.

The number of new births after 3 months *(one cycle)* can be found by mutiplying the number of female rodents in each age group times the corresponding birthrates and then finding the sum:

$$15(0) + 9(0.3) + 13(0.8) + 5(0.7) + 0(0.4) + 0(0)$$
$$= 0 + 2.7 + 10.4 + 3.5 + 0 + 0 = 16.6.$$

The number of female rodents in the 0–3 age group after 3 months is about 17.

The number of rodents that survive in each age group and move up to the next can be found as follows (SR is survival rate):

Age	No.	SR	Number moving up to the next age group
0–3	15	0.6	(15)(0.6) = 9.0 move up to the 3–6 age group.
3–6	9	0.9	(9)(0.9) = 8.1 move up to the 6–9 age group.
6–9	13	0.9	(13)(0.9) = 11.7 move up to the 9–12 age group.
9–12	5	0.8	(5)(0.8) = 4.0 move up to the 12–15 age group.
12–15	0	0.6	(0)(0.6) = 0 move up to the 15–18 age group.
15–18	0	0	No rodent lives beyond 18 months.

Hence, after 3 months the female population will have grown from 42 to approximately 50, with the distribution shown in Table 2.

To calculate the female population of rodents after 6 months (two cycles), this process can be repeated using the numbers from Table 2.

Exercises

1. Using Table 2 (the distribution of the rat population after 3 months),
   a. Calculate the number of newborn rats (aged 0 to 3) after 6 months (two cycles). (It is suggested that the number of rats not be rounded to the nearest integer when the values are to be used for further analysis, even though we know we can't have a fraction of a rodent. Rounding off can mean a significant difference in calculations over time.)

TABLE 2 ▸ End of 3 Months

Age	0–3	3–6	6–9	9–12	12–15	15–18
Number	16.6	9.0	8.1	11.7	4.0	0

b. Calculate the number of rats that survive in each age group after 6 months and move up to the next age group.
c. After 6 months how many rodents will be in the population, and what will be the distribution of that number?
d. Using your population distribution from part c, calculate the number of rats and the approximate number in each age group after 9 months (three cycles). Continue this process to find the number of rodents after 12 months (four cycles).
e. Compare the original number of rats to the numbers after 3, 6, 9, and 12 months. What do you observe?
f. What do you think might happen to this population if we extended the calculations to 15, 18, 21, . . . months?

2. A species of deer has the following birth and survival rates:

Age (years)	Birthrate	Survival Rate
0–2	0	0.6
2–4	0.8	0.8
4–6	1.7	0.9
6–8	1.7	0.9
8–10	0.8	0.7
10–12	0.4	0

a. Given that the initial population for this species is 148 deer with the following distribution,

Age (years)	0–2	2–4	4–6	6–8	8–10	10–12
Number	50	30	24	24	12	8

find the number of newborn deer after 2 years (one cycle).
b. Arrange the initial population distribution in a row matrix and the birthrates in a column matrix. Multiply the row matrix times the column matrix. Interpret this result.
c. Calculate the number of deer that survive in each age group after 2 years and move up to the next age group.

d. Explore the possibility of multiplying the initial population distribution in a row matrix times some column matrix to find the number of deer after 2 years that move from
   i. The 0–2 group to the 2–4 group. (Hint: The column matrix that you use needs to contain several zeros in order to produce the desired product.)
   ii. The 2–4 to the 4–6 group.
   iii. The 4–6 group to the 6–8 group.
   iv. The 6–8 group to the 8–10 group.
   v. The 8–10 group to the 10–12 group.

3. Using the birth and survival rate information for *Rattus norvegicus* from the lesson (Table 1), find the population total and distribution after 3 months (one cycle) for the following initial populations:
   a. [35 0 0 0 0 0].
   b. [5 5 5 5 5 5].

4. Using the birth and survival rate information for the deer population in Exercise 2, find the population total and distribution after 2 years (one cycle), 4 years (two cycles), 6 years (three cycles), 8 years (four cycles), and 10 years (five cycles) if the initial population is [25 0 0 0 0 0].

Population Growth: The Leslie Model, Part 2

In our beginning exploration of the Leslie model for population growth, it was possible to use an initial population distribution, birthrates, and survival rates to predict population figures at future times. Looking two, three, or even four cycles into the future is not impossible, but the arithmetic soon becomes cumbersome. What do the wildlife manager and the urban planner do if they want to look ten, twenty, or even more cycles into the future?

In Lesson 3.4, Exercise 2 on page 134 used the actual model that P. H. Leslie devised. The use of matrices seems to hold the key, and with the aid of computer software or a calculator, looking ahead many cycles is not difficult. In fact, some fascinating results are produced.

Let's return to our rodent model. If the original population distribution (P_0) and a matrix that we call L are multiplied, the population distribution at the end of cycle 1 (P_1) can be calculated.

$$P_0L = [15 \quad 9 \quad 13 \quad 5 \quad 0 \quad 0] \times \begin{bmatrix} 0 & 0.6 & 0 & 0 & 0 & 0 \\ 0.3 & 0 & 0.9 & 0 & 0 & 0 \\ 0.8 & 0 & 0 & 0.9 & 0 & 0 \\ 0.7 & 0 & 0 & 0 & 0.8 & 0 \\ 0.4 & 0 & 0 & 0 & 0 & 0.6 \\ 0 & 0 & 0 & 0 & 0 & 0 \end{bmatrix}$$

$$= [15(0) + 9(0.3) + 13(0.8) + 5(0.7) + 0(0.4)$$
$$+ 0(0) \quad 15(0.6) \quad 9(0.9) \quad 13(0.9) \quad 5(0.8) \quad 0(0.6) \quad 0(0)]$$
$$= [16.6 \quad 9.0 \quad 8.1 \quad 11.7 \quad 4.0 \quad 0 \quad 0] = P_1.$$

The matrix L (called the **Leslie matrix**) is formed by augmenting or joining the column vector containing the birthrates of each age group and a series of column vectors that contain a survival rate as one entry and zeroes everywhere else. Notice that the survival rates (of which there are $n - 1$, since no animal survives beyond the 15–18 age group) lie along the **superdiagonal** that is immediately above the main diagonal of the matrix.

When the matrix L is multiplied by a population distribution P_k, a new population distribution P_{k+1} results. To find population distributions at the end of other cycles, the process can be continued.

$$P_1 = P_0L$$
$$P_2 = P_1L = (P_0L)L = P_0(LL) = P_0L^2.$$

In general, $P_k = P_0L^k$.

Using this formula to find the population distribution for the rodents after 24 months (eight cycles) and the total population of the rodents, we have

$$P^8 = P_0L^8 = [15 \quad 9 \quad 13 \quad 5 \quad 0 \quad 0] \begin{bmatrix} 0 & 0.6 & 0 & 0 & 0 & 0 \\ 0.3 & 0 & 0.9 & 0 & 0 & 0 \\ 0.8 & 0 & 0 & 0.9 & 0 & 0 \\ 0.7 & 0 & 0 & 0 & 0.8 & 0 \\ 0.4 & 0 & 0 & 0 & 0 & 0.6 \\ 0 & 0 & 0 & 0 & 0 & 0 \end{bmatrix}^8$$

$$= [21.03 \quad 12.28 \quad 10.90 \quad 9.46 \quad 7.01 \quad 4.27].$$

Total population $= 21.03 + 12.28 + 10.90 + 9.46 + 7.01 + 4.27 = 64.95$, or approximately 65 rodents.

Exercises

Note: For the following exercises, you will need to have access to either a calculator or computer software that performs matrix operations.

1. Using the original population distribution, [15 9 13 5 0 0], and the Leslie matrix from the *Rattus norvegicus* example,
   a. Find the population distribution after 15 months (five cycles).
   b. Find the total population after 15 months. (Hint: Multiply P_0L^5 times a column matrix consisting of 6 ones.)
   c. Find the population distribution and the total population after 21 months.

2. Suppose the *Rattus norvegicus* start dying off from overcrowding when the total female population for a colony reaches 250. Find how long it will take for this to happen when the initial population distribution is
   a. [18 9 7 0 0 0].
   b. [35 0 0 0 0 0].
   c. [5 5 5 5 5 5].
   d. [25 15 10 11 7 13].

3. Complete the table for the given cycles of *Rattus norvegicus* using the original population distribution of [15 9 13 5 0 0].

Cycle	Total Population	Growth Rate
original	42	
1	49.4	17.6%
2	56.08	13.5%
3	57.40	2.4%
4		
5		
6		

4. Using the population totals from Exercise 3,
   a. Find the growth rates from the original population to the end of year 1, year 1 to year 2, year 2 to 3, 3 to 4, 4 to 5, and 5 to 6.
   b. What do you observe about these rates?
   c. Calculate the total populations for P_{19}, P_{20}, and P_{21}. What is the growth rate between these successive years?

5. One characteristic of the Leslie model is that growth does stabilize at a rate called the **long-term growth rate** of the population. As you observed in Exercise 4, the growth rate of *Rattus norvegicus* converges to about 3.04%. This means that for a large enough k, the total population in cycle k will equal about 1.0304 times the total population in the previous cycle.
   a. Find the long-term growth rate of the total population for each of the initial population distributions in Exercise 2.
   b. How does the initial population distribution seem to affect the long-term growth rate?

NEWS CLIP

Md. May Open Season on Bears

Wildlife Officials Ponder Renewing Hunts to Curb Population

Associated Press

CUMBERLAND, Md.—Wildlife managers may propose the state's first bear-hunting season in 38 years because the swelling population of the animals is raising public concern.

"It's one of the options to manage the population," said Joshua L. Sandt, forest wildlife program manager for the Maryland Department of Natural Resources. "Bears are neat animals to have around, but we want to keep them at a population that is compatible with landowners. They can be a nuisance."

The bear population, which is about 200 in Garrett, Allegany and Washington counties, began increasing about eight years ago when officials in Pennsylvania moved about 7,000 of them from the northern part of the state to forests near its southern border.

6. Consider once again our deer species from Exercise 2 in Lesson 3.4 on page 134. The birth and survival rates are listed below.

Age (years)	Birthrate	Survival Rate
0–2	0	0.6
2–4	0.8	0.8
4–6	1.7	0.9
6–8	1.7	0.9
8–10	0.8	0.7
10–12	0.4	0

a. Construct the Leslie matrix for this animal.
b. Given that $P_0 = [50 \quad 30 \quad 24 \quad 24 \quad 12 \quad 8]$, find the long-term growth rate.
c. Suppose the natural range for this animal can sustain a herd that contains a maximum of 1,250 females. How long before this herd size is reached?
d. Once the long-term growth rate of the deer population is reached, how might the population of the herd be kept constant?

7. In his study of the application of matrices to population growth, H. P. Leslie (1945) was particularly interested in a special case in which the birthrate vector has only one nonzero element. The following example falls into this special case. Suppose there is a certain kind of bug that lives at most 3 weeks and reproduces only in the third week of life. Fifty percent of the bugs born in 1 week

survive into the second week, and 70% of the bugs who survive into their second week also survive into their third week. On the average, six new bugs are produced for each bug that survives into its third week. A group of five 3-week-old female bugs decide to make their home in a storage box in your basement.

a. Construct the Leslie matrix for this bug.
b. What is P_0?
c. How long before there will be at least 1,000 female bugs living in your basement?

8. The problem in Exercise 7 is an example of a population that grows in waves. We wonder whether the population growth for this population will stabilize in any way over the long run. To explore this question, make a table of the population distributions P_{22} through P_{30}.
   a. Examine the population change from one cycle to the next. Can you find a pattern in the population growth?
   b. Examine the population change from P_{22} to P_{25}, P_{23} to P_{26}, P_{24} to P_{27}, P_{25} to P_{28}, P_{26} to P_{29}, and P_{27} to P_{30}. Are you surprised at the results? Why?

9. Change the initial population in the bug problem to $P_0 = [4\ 4\ 4]$ and repeat the instructions in Exercise 8, looking at the total population growth for each cycle.

10. Examine the changes in successive age groups from P_{22} to P_{25}, P_{23} to P_{26}, P_{24} to P_{27}, P_{25} to P_{28}, P_{26} to P_{29}, and P_{27} to P_{30}. Make a conjecture based on your results.

11. Using mathematical induction, prove that $P_k = P_0 L^k$ for any original population P_0 and Leslie matrix L.

1. Can a matrix with dimensions 3 by 5 be added to a matrix with dimensions 5 by 3? Explain your answer.

2. The math club is planning a Saturday practice session for an upcoming math contest. For lunch they ordered 35 Mexican special combination lunches, 6 large bags of corn chips, 6 containers of salsa, and 12 six-packs of assorted cold drinks.
   a. Write this information in a row matrix L. Label your matrix.
   b. Interpret L_2 and L_4.
   c. The math club pays \$4.50 per lunch, \$1.97 per bag of corn chips, \$2.10 for each container of salsa, and \$2.89 for each six-pack of cold drinks. Use the multiplication of a row and column vector to find the total cost to the math club. Label your vectors.

3. A youth fellowship group is planning a spring retreat. They have contacted three lodges in the vicinity to inquire about rates. They found that Crystal Lodge charges \$13.00 per person per day for lodging, \$20.00 per day for food, and \$5.00 per person for use of the recreational facilities. Springs Lodge charges \$12.50 for lodging, \$19.50 for meals, and \$7.50 for use of the recreational facilities. Bear Lodge charges \$20.00 per night for lodging, \$18.00 a day for meals, and no extra charge for using the recreational facilities. Beaver Lodge charges a flat rate of \$40.00 a day for lodging (meals included) and no additional fee for use of the recreational facilities.

a. Display this information in a matrix, C. Label the rows and columns.
b. State the values of C_{22} and C_{43}.
c. Interpret C_{13} and C_{31}.

4. Mrs. Jones has been bothered by flies, spiders, and a variety of beetles on her summer porch. She has been shopping for a vacuum-powered insect disposal system. She found one at Z-Mart and another model at Base Hardware. The Z-Mart system cost $39.50; disposal cartridges were 6 for $24.50; and storage cases were $8.50. At Base Hardware the system cost $49.90; cartridges were 6 for $29.95; and cases were $12.50.
   a. Write and label a matrix showing the prices for the three parts at the two stores.
   b. Mrs. Jones decided to wait and see whether the prices for insect disposal systems would be reduced during the upcoming end-of-summer sales. When she went back during the sales, the Z-Mart prices were reduced by 10% and the Base Hardware prices were reduced by 20%. Construct a matrix showing the sale prices for each of the three parts at the two stores.
   c. Use matrix subtraction to compute how much Mrs. Jones could save for each part at the two stores.
   d. Suppose Mrs. Jones is interested in purchasing the vacuum-powered insect disposal systems for herself and three of her friends. Use multiplication of a matrix by a scalar to show how much she would pay for four of each part of the system at the two stores at the sale prices.

5. An artist fashions plates and bowls from small pieces of colored woods. He currently has orders for five 10-inch plates, three large bowls, and seven small bowls. Each plate requires 100 pieces of ebony, 800 pieces of walnut, 600 pieces of rosewood, and 400 pieces of maple. It takes 200 ebony pieces, 1,200 walnut pieces, 1,000 rosewood pieces, and 800 pieces of maple to make a large bowl. A small bowl takes 50 pieces of ebony, 500 walnut pieces, 450 rosewood pieces, and 400 pieces of maple.
   a. Write a row matrix showing the current orders for this artist's work.
   b. Construct a matrix showing the number of pieces of wood used in an individual plate or bowl.

c. Use matrix multiplication to compute the number of pieces of each type of wood that the artist will need for the plates and bowls that are on order.

d. Suppose it takes the artist 3 weeks to fashion a plate, 4 weeks to make a large bowl, and 2 weeks to complete a small bowl. Use matrix multiplication to show how long it will take the artist to fill all his orders for plates and bowls.

6. Ed has money invested in three sports complexes in San Diego. His return (annual) from a \$50,000 investment in a tennis club is 8.2%. He receives 6.5% from a \$100,000 investment in a golf club and 7.5% on a \$75,000 investment in a soccer club. Use vector multiplication to find Ed's income from his investments for 1 year. Label your vectors.

7. Three music classes at Central High are selling candy as a fund-raiser. The number of each kind of candy sold so far by each of the three classes is shown in the table below.

	Jazz Band	Symphonic Band	Orchestra
Chocolate delights	300	220	250
Chocolate overdose	240	330	400
Chocolate chewies	150	200	180
Sour balls	175	150	160

The profit for each type of candy is as follows: sour balls, 30 cents; chocolate overdose, 50 cents; chocolate delights, 25 cents; and chocolate chewies, 35 cents. Use matrix multiplication to compute the profit made by each class on its candy sales.

8. Write the transpose (A^T) of matrix A where

$$A = \begin{bmatrix} 4 & 2 & 6 \\ 5 & 1 & 3 \end{bmatrix}.$$

9. The dimensions of matrices P, Q, R, and S are 3×2, 3×3, 4×3, and 2×3, respectively. If matrix multiplication is possible, find the dimensions of the following matrix products?

a. QP.
b. RQ.
c. QS.
d. RPS.

10. Let matrix

$$M = \begin{bmatrix} 1 & 1 \\ 1 & 1 \end{bmatrix}.$$

a. Calculate M^2, M^3, and M^4.
b. Predict the components of M^5 and check your prediction.
c. Generalize to M^n where n is a natural number.
d. Repeat parts a, b, and c for the matrix

$$M = \begin{bmatrix} 1 & 0 \\ 2 & 3 \end{bmatrix}.$$

11. Complete the following statement: If a square matrix, A, has an inverse, A^{-1}, then the product AA^{-1} = the ——— matrix I, where I is a ——— .

12. Which of the following matrices are inverses of each other? Explain your answers.

a. $\begin{bmatrix} -1 & 3 \\ 2 & -5 \end{bmatrix}$ and $\begin{bmatrix} 5 & 3 \\ 2 & 1 \end{bmatrix}$

b. $\begin{bmatrix} 1 & 0 \\ 0 & 1 \end{bmatrix}$ and $\begin{bmatrix} 1 & 0 \\ 0 & 1 \end{bmatrix}$

c. $\begin{bmatrix} 2 & 1 & 0 \\ 3 & 2 & 1 \end{bmatrix}$ and $\begin{bmatrix} 1 & -1 \\ -1 & 2 \\ -1 & 2 \end{bmatrix}$

13. The student body at Central High is planning to hire a band for the senior prom. Their choices are bands *A*, *B*, and *C*. They survey the sophomore, junior, and senior classes and find that the following percentages of students (regardless of sex) prefer the bands as follows:

$$\begin{array}{c} \\ A \\ B \\ C \end{array}\begin{array}{c} \begin{array}{ccc} \text{10th} & \text{11th} & \text{12th} \end{array} \\ \begin{bmatrix} 20\% & 35\% & 40\% \\ 30\% & 30\% & 25\% \\ 50\% & 35\% & 35\% \end{bmatrix} \end{array}.$$

The student body population by class and sex is as follows:

$$\begin{array}{c} \\ \text{10th} \\ \text{11th} \\ \text{12th} \end{array}\begin{array}{c} \begin{array}{cc} \text{Male} & \text{Female} \end{array} \\ \begin{bmatrix} 235 & 225 \\ 205 & 215 \\ 175 & 190 \end{bmatrix} \end{array}.$$

Use matrix multiplication to find

a. The number of males and females who prefer each band.
b. The total number of students who prefer each band.

14. The characteristics of the female population of a herd of small mammals are shown in the following table:

Age Groups (months)

Rates	0–4	4–8	8–12	12–16	16–20	20–24
Birth	0	0.5	1.1	0.9	0.4	0
Survival	0.6	0.8	0.9	0.8	0.6	0

Suppose the initial female population for the herd is

$$[22 \quad 22 \quad 18 \quad 20 \quad 7 \quad 2].$$

a. What is the expected life span of this mammal?
b. Construct the Leslie matrix for this population.
c. Determine the long-range growth rate for the herd.
d. Suppose this mammal starts dying off from overcrowding when the total female population for the herd reaches 520. How long will it take for this to happen?

Bibliography

Aksamit, D. L., J. V. Mitchell, and B. J. Pozehl. 1987. "Relationships between PPST and ACT Scores and Their Implications for Basic Skills Testing of Prospective Testing." *Journal of Teacher Education,* November–December, pp. 48–52.

Cozzens, M. B., and R. D. Porter. 1987. *Mathematics and Its Applications.* Lexington, MA: Heath.

Kemeny, J. G., J. N. Snell, and G. L. Thompson. 1957. *Finite Mathematics.* Englewood Cliffs, NJ: Prentice-Hall.

Leslie, P. H. 1945. "On the Uses of Matrices in Certain Population Mathematics." *Biometrika* 33: 183–212.

Lincoln Journal Star, June 9, 1991, p. 2E, and June 16, 1991, p. 4D.

Maurer, S. B., and A. Ralston. 1991. *Discrete Algorithmic Mathematics.* Reading, MA: Addison-Wesley.

North Carolina School of Science and Mathematics. 1988. *New Topics for Secondary School Mathematics.* Reston, VA: National Council of Teachers of Mathematics.

Ross, K. A., and C. R. B. Wright. 1985. *Discrete Mathematics.* Englewood Cliffs, NJ: Prentice-Hall.

Tuchinsky, Philip M. 1986. *Matrix Multiplication and DC Ladder Circuits.* Lexington, MA: COMAP.

U.S. Department of Education. 1991. *Youth Indicators 1991: Trends in the Well-being of American Youth.* Washington, DC: U.S. Government Printing Office.

UNISYS
DATA NETWORK
UNISYS

Graphs and Their Applications

When the boundaries or names of countries change, cartographers have to be prepared to provide the public with new maps. For years, map makers and mathematicians alike have wondered about the number of colors it takes to color a map.

What is the minimum number of colors needed to color any map? How do you optimally color a map? What do map coloring and scheduling meeting times for your school organizations have in common? Whether you're a cartographer who must find the number of colors needed to color a map, a businessperson who must determine whether a project can be completed on time, or a planner who wants to know the most efficient way to route a city's garbage trucks, you'll find the answer in an area of mathematics known as graph theory.

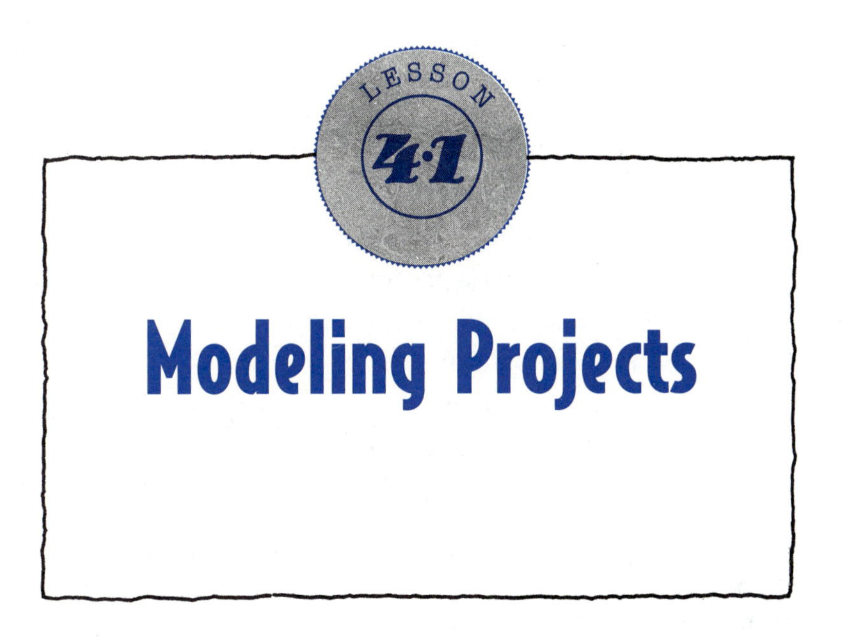

Modeling Projects

How does a building contractor organize all of the jobs needed to complete a project? How do your parents manage to get all parts of a Thanksgiving dinner done at the same time? Many people believe that planning is a simple activity. After all, everybody does it. Planning your day-to-day activities seems to be second nature. What most people fail to realize is that for people in the business world who must plan and work on extensive projects, this haphazard manner of planning is not the most efficient way to complete a job. A more scientific, organized method must be used.

Explore This

The Central High yearbook staff has only 16 days left before the deadline for completing their book. They are running behind schedule and still have several tasks left to finish. The tasks and times that it takes to do each task are listed in the table at the top of the next page.

Task	Time
Start	0
A. Buy film	1 day
B. Load cameras	1
C. Take photos of clubs	3
D. Take sports photos	2
E. Take photos of teachers	1
F. Develop film	2
G. Design layout	5
H. Print and mail pages	3

Will it be possible for the yearbook to be completed on time if the tasks have to be done one after the other? If some tasks can be done at the same time, could the deadline be met?

As you may have noticed, some of the yearbook staff jobs can be done at the same time, yet many of them cannot be started until others have been completed. Assuming the following prerequisites, how soon can the project be completed?

Task	Time	Prerequisite Task
Start	0	—
A. Buy film	1	None
B. Load cameras	1	*A*
C. Take photos of clubs	3	*B*
D. Take sports photos	2	*C*
E. Take photos of teachers	1	*B*
F. Develop film	2	*D, E*
G. Design layout	5	*D, E*
H. Print and mail pages	3	*G, F*

Drawing a diagram or **graph** of this information makes it easier to see the relationships among the tasks. In the graph in Figure 4.1, tasks are represented by points, or **vertices,** and the arrows, or **edges,** indicate which tasks must be finished before a new task can begin. Each edge also shows the number of days it takes to complete the preceding task. Note that tasks with the same prerequisites are aligned vertically. Although this is not necessary, it makes the graph easier to follow.

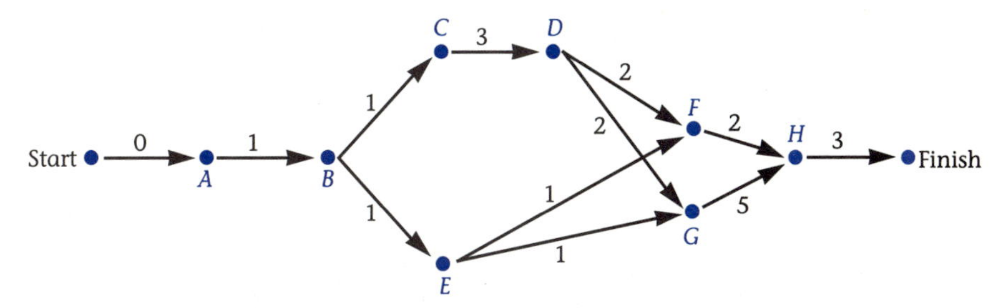

Figure 4.1 Diagram of the order of tasks that are necessary to complete the yearbook.

Exercises

In Exercises 1 through 4, draw a graph and label the vertices for each task table.

1.

Task	Time	Prerequisites
Start	0	
A	5	None
B	6	*A*
C	4	*A*
D	4	*B*
E	8	*B*, *C*
F	4	*C*
G	10	*D*, *E*, *F*
Finish		

2.

Task	Time	Prerequisites
Start	0	
A	1	None
B	2	None
C	3	*A*, *B*
D	5	*B*
E	5	*C*
F	5	*C*, *D*
G	4	*D*, *E*
H	4	*E*, *F*
Finish		

3.

Task	Time	Prerequisites
Start	0	
A	5	None
B	7	*A*
C	4	*A*
D	3	*B*
E	7	*B*, *C*
F	5	*C*
G	8	*D*, *E*, *F*
Finish		

4.

Task	Time	Prerequisites
Start	0	
A	5	None
B	8	*A*, *D*
C	9	*B*, *I*
D	7	None
E	8	*B*
F	12	*I*
G	4	*C*, *E*, *F*
H	9	None
I	5	*D*, *H*
Finish		

5. In order to get a family dinner completed in the shortest amount of time, Mrs. Shu listed the following tasks:

Task	Time	Prerequisite Task
Start	0	—
A. Wash hands		
B. Defrost hamburger		
C. Shape meat into patties		
D. Cook hamburgers		
E. Peel and slice potatoes		
F. Fry potatoes		
G. Make salad		
H. Set table		
I. Serve food		

a. Construct reasonable time estimates in minutes for each of these tasks, and label the prerequisites.
b. Construct a graph of the table.
c. What is the least amount of time needed to prepare dinner?

6. Your best friend, Matt, has always been very disorganized. He is now trying to get ready to leave for college and desperately needs your help.
   a. Develop a table of at least six activities that will need to be completed to get Matt on his way. Give the times and prerequisites of these activities.
   b. Construct a corresponding graph.
   c. What is the least amount of time it will take to get Matt off to school?

7. Consider the following graph:

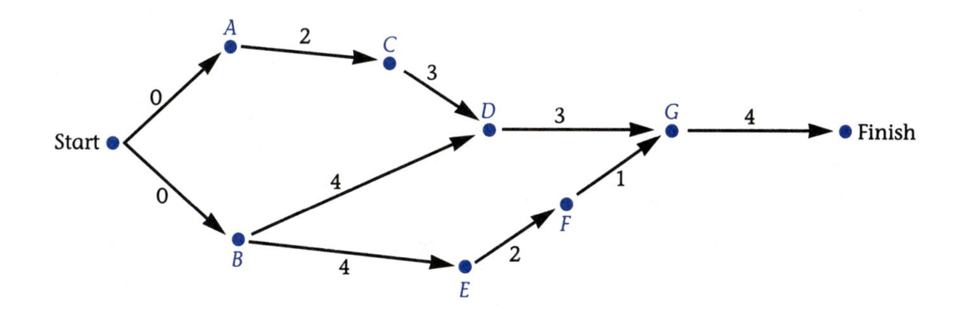

Complete the task table below for this graph.

Task	Time	Prerequisite Task
Start	0	—
A		
B		
C		
D		
E		
F		
G		
Finish		

Critical Paths

Finding the shortest time to complete a project is relatively easy if only a few activities make up the project. But as the tasks increase in number, the problem becomes more difficult to solve by inspection alone.

In the 1950s the U.S. government was faced with the need to complete very complex systems such as the U.S. Navy Polaris Submarine project. In order to do this efficiently, a method was developed called PERT (Program Evaluation and Review Technique) in which those tasks that were critical to the earliest completion of the project were targeted. This path of targeted tasks from the start to the finish of the project became known as a **critical path.**

Recall the graph that represented the Central High yearbook project. How might one go about finding the critical path for this project? To do this, an **earliest-start time** (EST) for each task must be found. The EST is the earliest that an activity can begin if all the activities preceding it begin as early as possible.

To calculate the EST for each task, begin at the start and label each vertex with the smallest possible time that will be needed before the

task can begin. The label for C in Figure 4.2 is found by adding the EST of B to the one day that it takes to complete task $B(1 + 1 = 2)$. In the case of task G, it cannot be completed until both predecessors, D and E, have been completed. Hence, G cannot begin until seven days have passed.

In the case of the yearbook staff, the earliest time in which the project can be completed is 15 days. The time that it takes to complete all of the tasks in the project corresponds to the total time for the longest path from start to finish. A path with this longest time is the desired critical path. In Figure 4.2, the critical path is Start–*ABCDGH*–Finish.

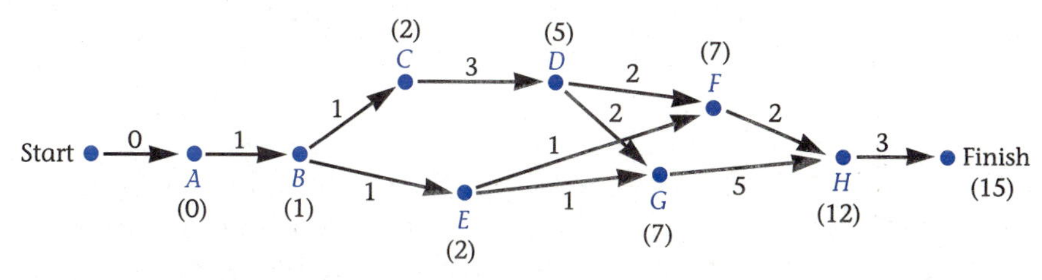

Figure 4.2 Yearbook diagram showing the earliest-start time for each task.

Example

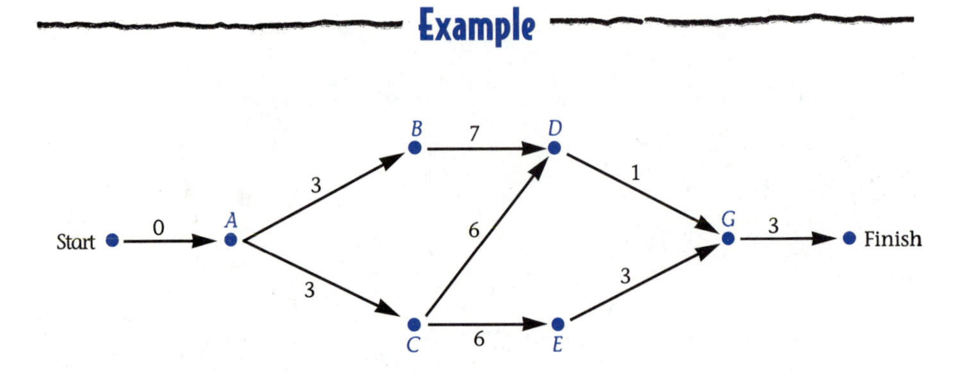

1. Copy the graph and label the vertices with the EST for each task, and determine the earliest completion time for the project. The times are in minutes.

2. Find the critical path.

The solutions are as follows:

1.

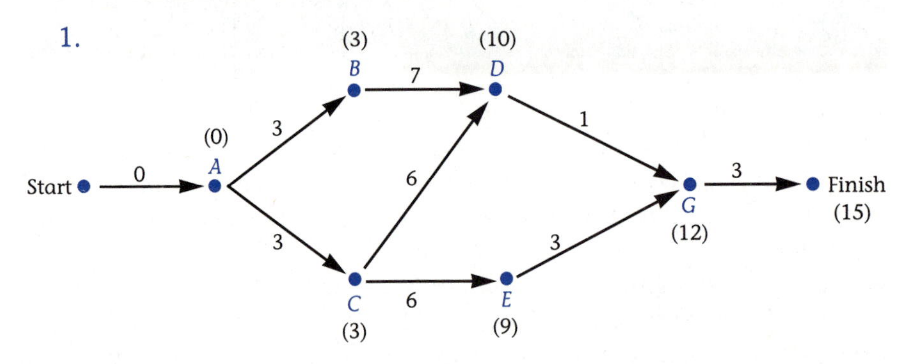

The earliest time in which the project can be completed is 15 minutes.

2. Since the critical path is the longest path from the start to finish, the critical path is Start–*ACEG*–Finish.

If it is desirable to cut the completion time of a project, it can be done by shortening the length of the critical path once it is found. In the preceding example, the completion time of the project could be cut to 14 minutes if the time to complete task *E* could be cut to 2 minutes.

Welding represents just one element of a much larger project. The efficient management of projects like the construction of a building requires the use of critical path analysis.

Exercises

1.

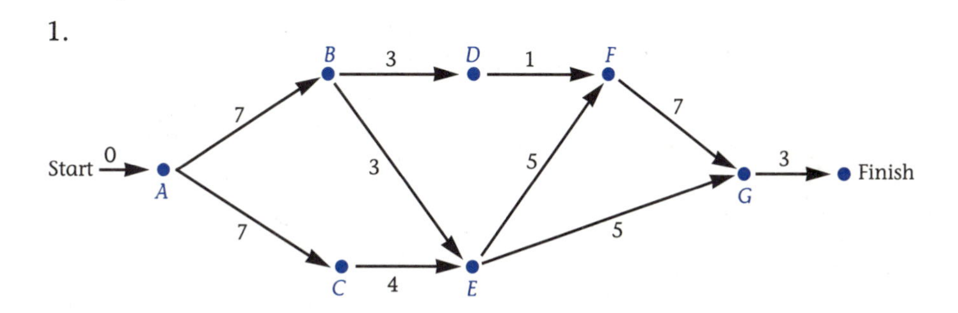

Complete the following:

Vertex	Earliest-Start Time
A	0
B	7
C	
D	
E	
F	
G	
Minimum project time =	
Critical path(s) =	

In Exercises 2 and 3, list the vertices of the graphs and give their earliest-start time, as in Exercise 1. Determine the minimum project time and all of the critical paths.

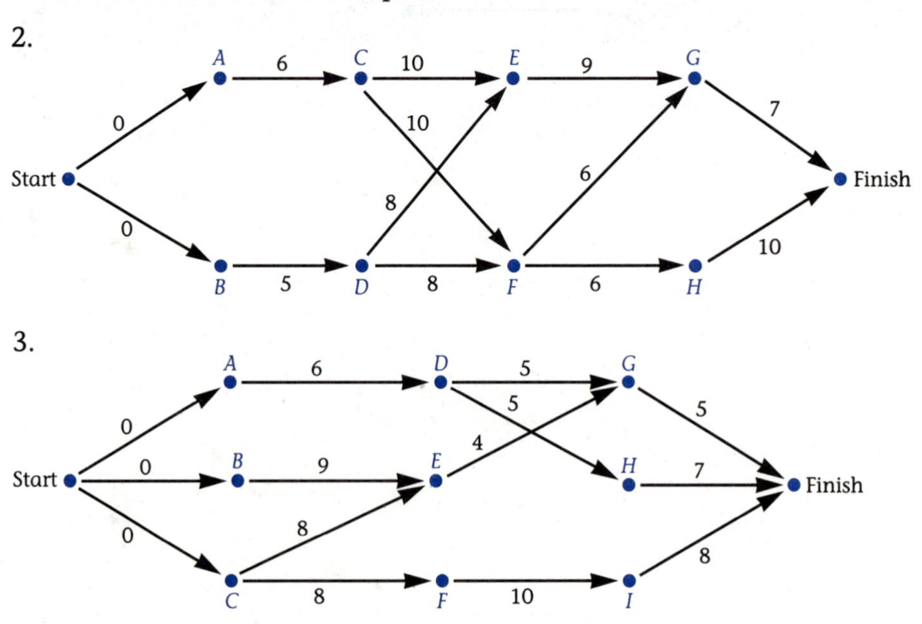

4. From the table below, construct a graph to represent the information, and label the vertices with their earliest-start time. Determine the minimum project time and the critical path.

Task	Time	Prerequisites
Start	0	
A	2	None
B	4	None
C	3	*A*, *B*
D	1	*A*, *B*
E	5	*C*, *D*
F	6	*C*, *D*
G	7	*E*, *F*

5.

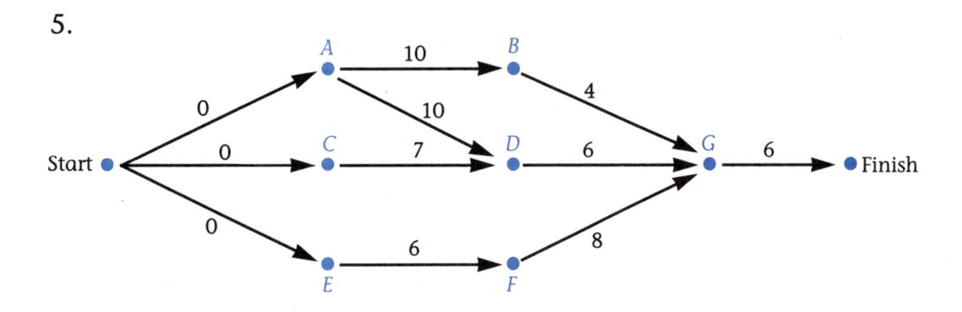

a. Copy the graph, and label the vertices with the earliest-start time.
b. How quickly can the project be completed?
c. Determine the critical path.
d. What will happen to the minimum project time if task *A*'s time can be reduced to 9 days? To 8 days?
e. Will the project time continue to be affected by reducing the time of task *A*? Why or why not?

6. Construct a graph with three critical paths.

7. Determine the minimum project time and the critical path.

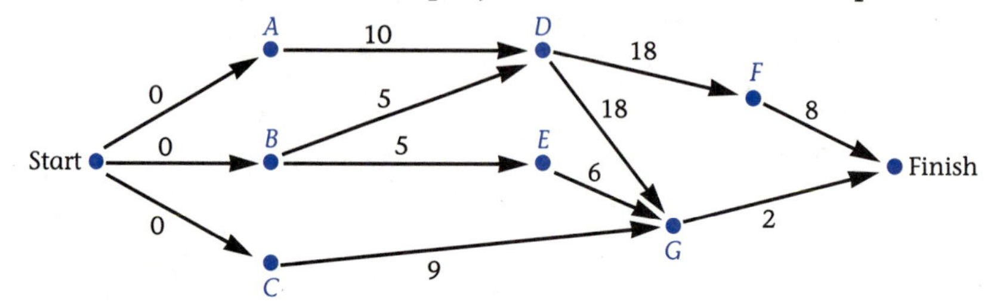

8. In the graph below, the ESTs for the vertices are labeled and the critical path is marked.

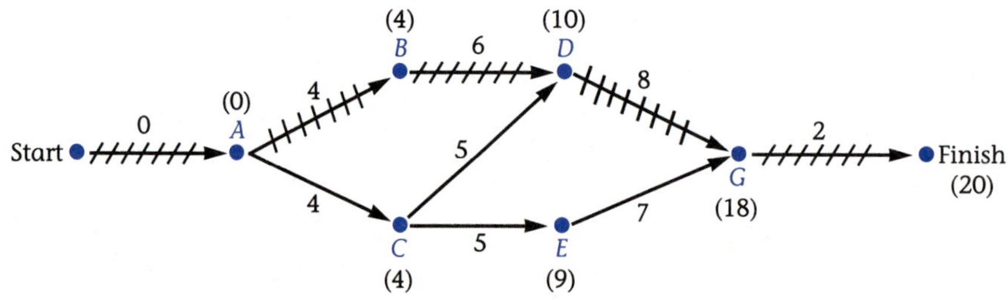

a. Task *E* can begin as early as day 9. If it begins on day 9, when will it be completed? If it begins on day 10? On day 11? What will happen if it begins on day 12

b. To complete task *E* by day 18, the day on which task *G* is to begin, what is the latest day on which *E* can begin?

If an activity is not on the critical path, it is possible for it to start later than its earliest-start time. The latest that a task can begin without delaying the project's minimum completion time is known as the **latest-start time** (LST) for the task.

c. To find the LST for vertex *C*, the times of the two vertices (*D* and *E*) need to be considered. Since vertex *D* is on the critical path, the latest it can start is on day 10. For *D* to begin on time, what is the latest day on which *C* can begin? In part b, we found that the latest *E* can start is on day 11. In that case, what is the latest *C* can begin? From this information, what is the latest (LST) that *C* can begin without delaying *either* task *D* or *E*?

9. To find the LST for each task, it is necessary to begin with the Finish and work through the graph in reverse order to the Start. Write an algorithm to find the LSTs for the tasks in the graph. Test your algorithm on the graph in Exercise 1.

The Vocabulary and Representations of Graphs

Graphs have many applications in addition to critical path analysis. They are frequently used in social science, computer science, chemistry, biology, transportation, and communications. In the following sections several of these applications are examined.

Recall that a **graph** is a set of points called **vertices** and a set of line segments called **edges.** Often graphs are used to model situations in which the vertices represent objects, and edges are drawn between the vertices on the basis of a particular relationship between the objects. The important characteristics of a graph will remain unchanged if the edges are curved.

Explore This

Case 1

Suppose the vertices of Figure 4.3 represent the starting five players on a high school basketball team, and the edges denote friendships. This graph indicates that player *C* is friends with all of the other players and

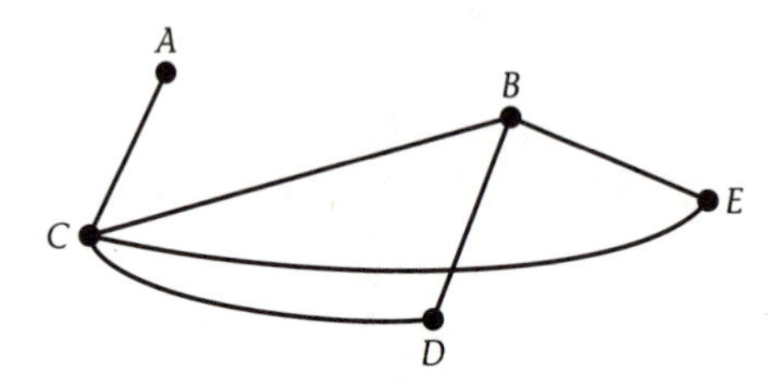

Figure 4.3 Graph showing five vertices and five edges.

that *E* has only two friends, *C* and *B*. Note that edge *CE* and edge *DB* intersect in this graph but that their intersection does not create a new vertex.

1. Which player has only one friend?
2. How many friends does *E* have? Who are they?
3. Redraw the graph so that *A* has no friends.

Consider the following solutions:

1. *A*.
2. Two, *C* and B.
3. The graph shown in Figure 4.4.

The graph in Figure 4.4 is not connected. A graph is **connected** if there is a path between each pair of vertices. In this graph, there is no path from *A* to any of the other vertices.

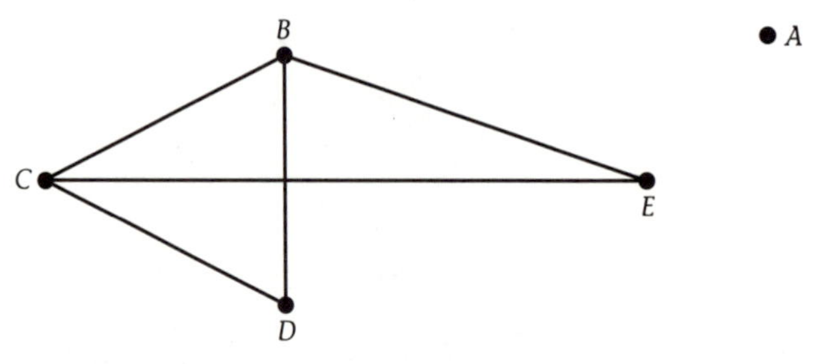

Figure 4.4 Graph that shows *A* with no friends.

Case 2

Now let the vertices in Figure 4.3 represent rooms in your school. The vertices are connected if there are direct hallways between two rooms. Then, according to the graph in Figure 4.3, a student can get from room *C* directly to any of the other four rooms.

When two vertices are connected with an edge they are said to be **adjacent.** *C* is adjacent to *A*, *B*, *D*, and *E*. Although there is no direct route from *D* to room *A*, it is possible to get from room *D* to room *A* by going through room *C*. In this case, although a path exists between *D* and *A*, they are not adjacent.

Try drawing a graph in which there is direct access from each room to every other room. Figure 4.5 shows two possible ways to represent this graph. Even though these two graphs appear to be different, they are structurally the same, and so they are considered the same graph.

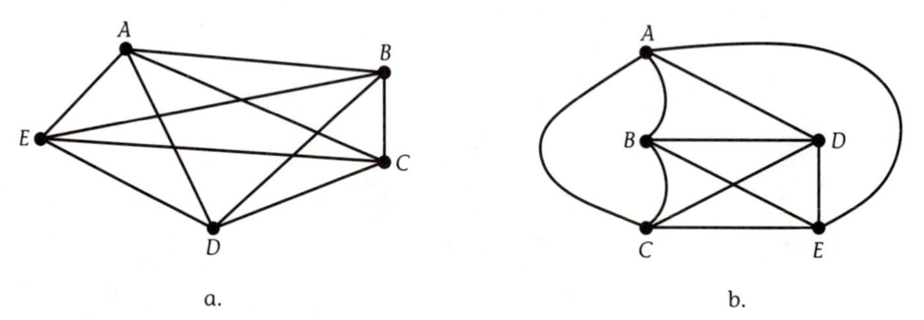

Figure 4.5 Two different representations of five rooms in a school, with each having direct access to every other room.

Graphs such as the ones in Figure 4.5, in which every pair of vertices is adjacent, are called **complete** graphs. Complete graphs are often denoted by K_N, where N is the number of vertices in the graph. Figure 4.5 depicts a K_5 graph.

There are other ways to represent graphs besides a diagram, such as the one in Figure 4.3. Another way is to list the set of vertices and the set of edges. For example, the graph in Figure 4.3 on page 166 could be described by the following:

$$\text{Vertices} = \{A, B, C, D, E\} \qquad \text{Edges} = \{AC, CB, CE, CD, BD, BE\}.$$

A third way to represent a graph is with an **adjacency matrix.** This type of representation can be used to represent the vertices and edges of the graph in a computer.

To represent Figure 4.3, a 5×5 matrix is formed in which both the rows and the columns correspond to vertices A, B, C, D, and E. If an edge exists between vertices, a 1 will appear in the corresponding position in the matrix; otherwise a 0 will appear.

$$\begin{array}{c} \\ A \\ B \\ C \\ D \\ E \end{array} \begin{array}{c} \begin{array}{ccccc} A & B & C & D & E \end{array} \\ \begin{bmatrix} 0 & 0 & 1 & 0 & 0 \\ 0 & 0 & 1 & 1 & 1 \\ 1 & 1 & 0 & 1 & 1 \\ 0 & 1 & 1 & 0 & 0 \\ 0 & 1 & 1 & 0 & 0 \end{bmatrix} \end{array}$$

The entry in row 2, column 4, is a 1, which indicates that an edge exists between vertices B and D.

Exercises

1. Mr. Butler bought six different types of fish. Some of the fish can live in the same aquarium, but others cannot. Guppies can live with Mollies; Swordtails can live with Guppies; Plecostomi can live with Mollies and Guppies; Gold Rams can live only with Plecostomi; and Piranhas cannot live with any of the other fish. Draw a graph to illustrate this.

2. Construct a graph for each of the following sets of vertices and edges. Which of the graphs are connected? Which are complete?
   a. $V = \{A, B, C, D, E\}$
      $E = \{AB, AC, AD, AE, BE\}$
   b. $V = \{M, N, O, P, Q, R, S\}$
      $E = \{MN, SR, QS, SP, OP\}$
   c. $V = \{E, F, G, J, K, M\}$
      $E = \{EF, KM, FG, JM, EG, KJ\}$
   d. $V = \{W, X, Y, Z\}$
      $E = \{WX, XZ, YZ, XY, WZ, WY\}$

3. Draw a diagram representing the graph with vertices = $\{A, B, C, D, E, F\}$ and edges = $\{AB, CD, DE, EC, EF\}$.
   a. Name two vertices that are not adjacent.
   b. F, E, C is one possible path from F to C. This path has a length of 2, since two edges were traveled to get from F to C. Name a path from F to C with a length of 3.
   c. Is this graph connected? Why or why not?
   d. Is the graph complete? Why or why not?

4. Draw a graph with 5 vertices in which vertex W is adjacent to Y; X is adjacent to Z; and V is adjacent to each of the other vertices.

5. Construct a graph for each adjacency matrix. Label the vertices $A, B, C, \ldots$.

a. $\begin{bmatrix} 0 & 1 & 0 & 0 \\ 1 & 0 & 1 & 1 \\ 0 & 1 & 0 & 1 \\ 0 & 1 & 1 & 0 \end{bmatrix}$ b. $\begin{bmatrix} 0 & 1 & 0 & 0 & 1 \\ 1 & 0 & 1 & 0 & 1 \\ 0 & 1 & 0 & 1 & 1 \\ 0 & 0 & 1 & 0 & 1 \\ 1 & 1 & 1 & 1 & 0 \end{bmatrix}$ c. $\begin{bmatrix} 0 & 1 & 0 & 0 & 0 \\ 1 & 0 & 0 & 0 & 0 \\ 0 & 0 & 0 & 1 & 1 \\ 0 & 0 & 1 & 0 & 1 \\ 0 & 0 & 1 & 1 & 0 \end{bmatrix}$

6. Determine an adjacency matrix for each of the following graphs:

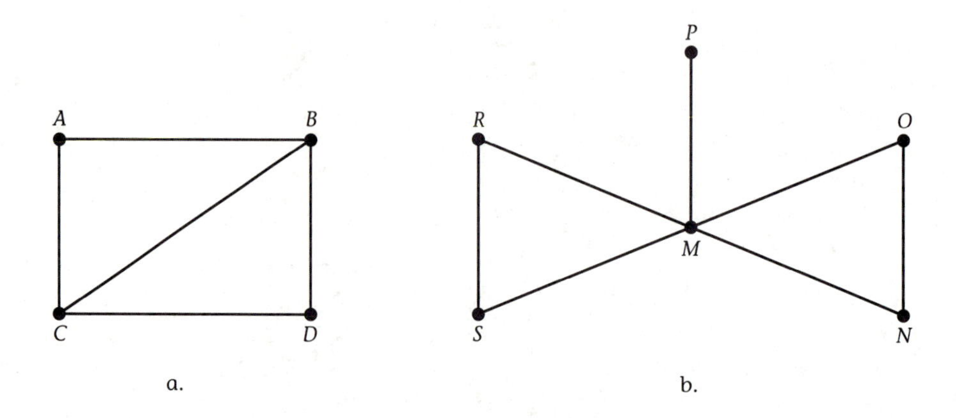

7. Give the adjacency matrix for the following graph:

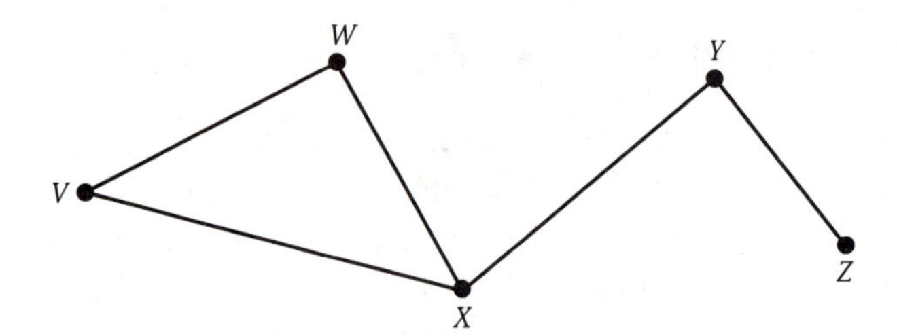

a. What do you notice about the main diagonal of the matrix?
b. A matrix may be symmetric with respect to one of its rows, columns, or diagonals. Does the matrix above possess symmetry? If so, where?
c. What would a 1 on the main diagonal indicate? What would a 2 in the second row, first column, indicate?

8. Using the graph and the adjacency matrix in Exercise 7, find the sum of each row of the matrix. What does the sum of the rows tell you about the graph?

9. The number of edges that have a specific vertex as an endpoint is known as the **degree** or **valence** of that vertex. In the graph at the top of the next page, the degree of vertex W, denoted by deg(W), is 4. Find the degree of each of the other vertices.

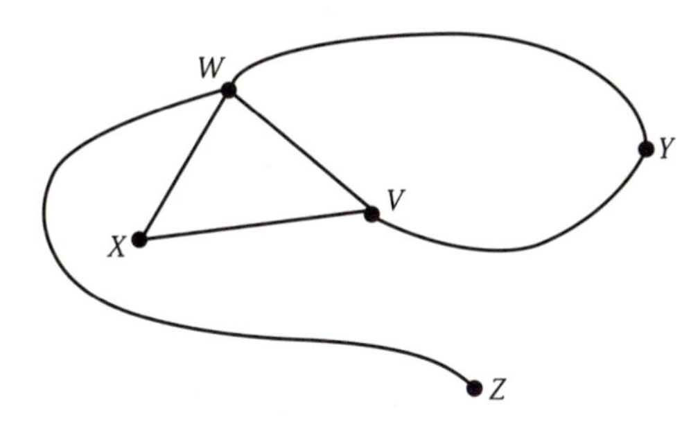

10. An edge that connects a vertex to itself is called a **loop.** If a graph contains a loop or multiple edges (more than one edge between two vertices), the graph is known as a **multigraph.**
    a. Give the adjacency matrix for the following multigraph:

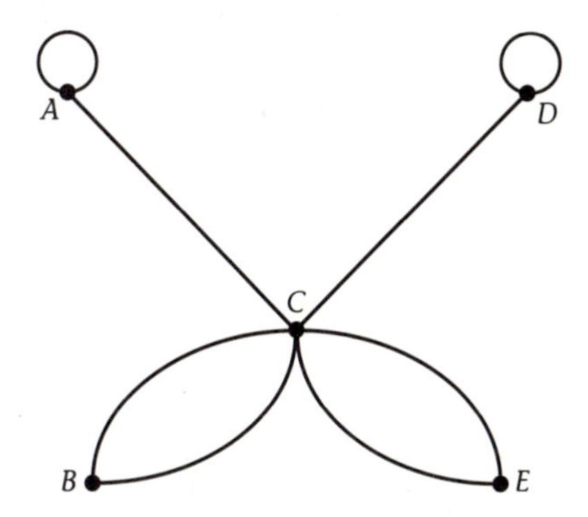

    b. What is the degree of each of the five vertices?

11. Complete the chart below for the sum of the degrees of the vertices in a complete graph.

Graph	Number of Vertices	Sum of the Degrees of All of the Vertices	Recurrence Relation
K_1	1	0	
K_2	2	2	$T_2 = T_1 + 2$
K_3	3	6	$T_3 = T_2 + 4$
K_4	—	—	———
K_5	—	—	———
K_6	—	—	———

Write a recurrence relation that expresses the relationship between the sum of the degrees of all of the vertices for K_N and the sum for K_{N-1}.

12. a. Try to construct a graph with four vertices, two of the vertices with degree 3 and two with degree 2. No loops or multiple edges are to be used.
    b. Try to construct a graph with five vertices, three of the vertices with an odd degree and two with even degree. No loops or multiple edges are to be used.
13. Describe the adjacency matrix of a complete graph.
14. Complete the table for the following complete graphs:

Graph	Number of Vertices	Number of Edges	Recurrence Relation
K_1	1	0	
K_2	2	1	$S_2 = S_1 + 1$
K_3	3	3	$S_3 = S_2 + 2$
K_4	—	—	———
K_5	—	—	———
K_6	—	—	———

Write a recurrence relation that expresses the relationship between the number of edges of K_N and the number of edges of K_{N-1}.

15. Central High School is a member of a five-team football league. Each team in the league plays exactly two games, which must be against different teams. Show that there is only one possible graph for this schedule.

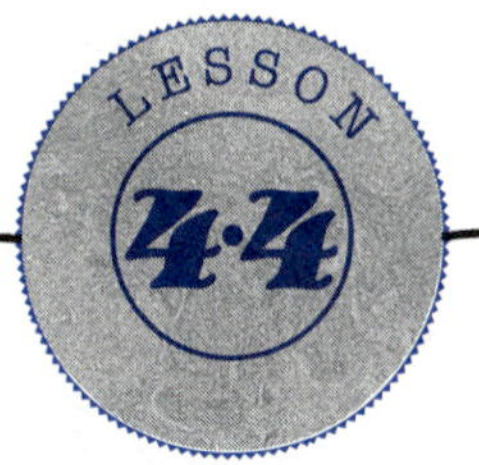

Euler Circuits and Paths

Explore This

Consider the graph in Figure 4.6. Try to draw this figure without lifting your pencil from the paper and without tracing any of the lines more than once. Is this possible?

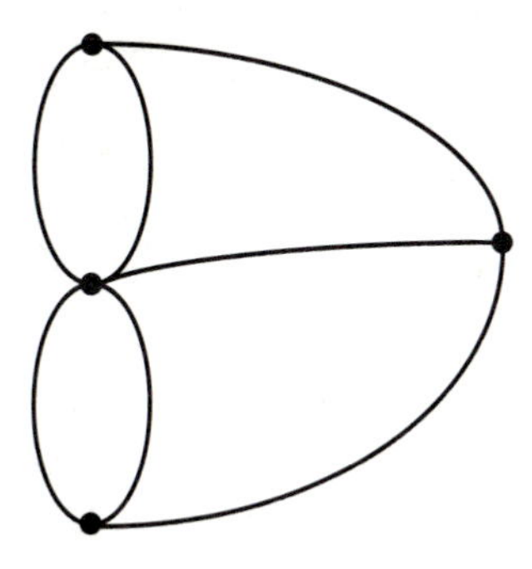

Figure 4.6 Graph.

The graph in Figure 4.6 represents an eighteenth-century problem that intrigued the famous Swiss mathematician Leonard Euler (1707–1783). The problem was one that had been posed by the residents of Königsberg, a city in what was then Prussia but is now the Russian city of Kaliningrad. In the 1700s, seven bridges connected two islands in the Pregel River to the rest of the city (see Figure 4.7). The people of Königsberg wondered whether it would be possible to walk through the city by crossing each bridge exactly once and return to the original starting point. Using a graph like the one in Figure 4.6 in which the vertices represent the landmasses of the city and the edges represent the bridges, Euler found that the desired walk through the city was not possible. In so doing, he also discovered a solution to problems of this general type.

What did Euler find? Try to traverse the graphs in Figure 4.8 without lifting your pencil or tracing the lines more than once. When can you draw the figures without retracing any edges and still end up at your starting point? When can you draw the figure without retracing and end up at a point other than the one from which you began? When can you not draw the figure without retracing?

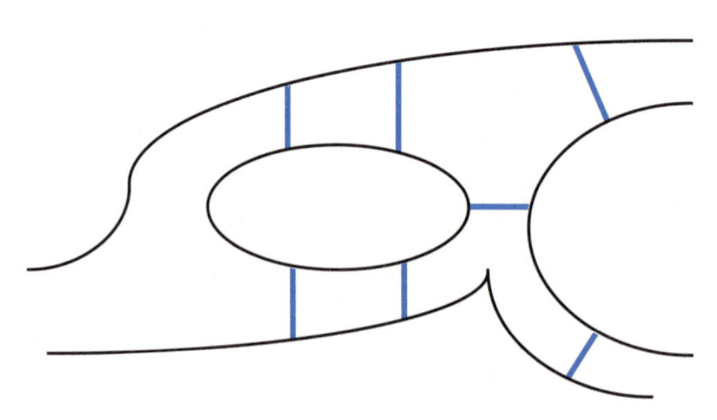

Figure 4.7 Representation of the seven bridges of Königsberg.

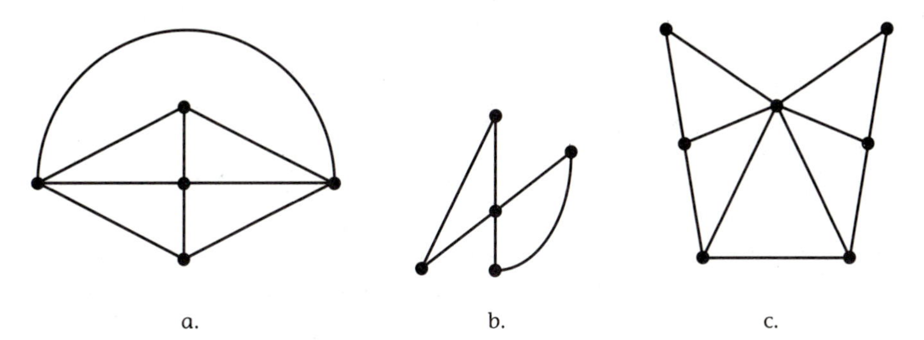

Figure 4.8 Graphs to trace.

Euler found that the key to the solution was related to the degrees of the vertices. Recalling that the degree of a vertex is the number of the edges that have the vertex as an endpoint, find the degree of each vertex of the graphs in Figure 4.8. Do you see what Euler noticed?

Euler hypothesized and later proved that in order to be able to traverse each edge of a connected graph exactly once and to end at the starting vertex, the degree of each vertex of the graph must be even, as in Figure 4.8b. In honor of Leonard Euler, a path that uses each edge of a graph exactly once and ends at the starting vertex is called an **Euler circuit.**

Euler also noticed that if a connected graph had exactly two odd vertices, it was possible to use each edge of the graph exactly once but to end at a vertex different from the starting vertex. Such a path is called an **Euler path.** Figure 4.8a is an example of a graph that has an Euler path. Figure 4.8c has four odd vertices, and so it cannot be traced without lifting the pencil. It has neither an Euler path nor an Euler circuit.

An Euler circuit in a relatively small graph can usually be found by trial and error. However, as the number of vertices and edges increases, a systematic way of finding the circuit becomes necessary. The following algorithm gives one way of finding an Euler circuit for a connected graph with all vertices of even degree.

Euler Circuit Algorithm

1. Pick any vertex, and label it *S*.
2. Construct a circuit, *C*, that begins and ends at *S*.
3. If *C* is a circuit that includes all edges of the graph, go to Step 8.
4. Choose a vertex, *V*, that is in *C* and has an edge that is not in *C*.
5. Construct a circuit *C′* that starts and ends at *V* using edges not in *C*.
6. Combine *C* and *C′* to form a new circuit. Call this new circuit *C*.
7. Go to Step 3.
8. Stop. *C* is an Euler circuit for the graph.

Example

Use the Euler circuit algorithm to find an Euler circuit for the graph below.

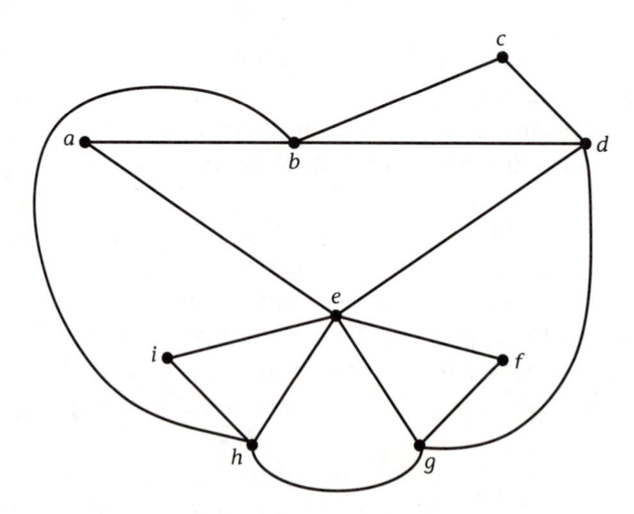

◂ Apply Step 1 of the algorithm. Choose vertex *b*, and label it *S*.

◂ Let *C* be the circuit *S*, *c*, *d*, *e*, *a*, *S*.

◂ *C* does not contain all edges of the graph, so go to Step 4 of the algorithm.

◂ Choose vertex d.

◂ Let C' be the circuit $d, g, f, e, g, h, e, i, h, b, d$.

◂ Combine C and C' by replacing vertex d in the circuit C with the circuit C'. Let C now be the circuit $S, c, d, g, f, e, g, h, e, i, h, S, d, e, a, S$.

◂ Go to Step 3 of the algorithm.

◂ C now contains all edges of the graph, so go to Step 8 of the algorithm and stop. C is an Euler circuit for the graph.

Many applications of graphs require that the edges have directions. A city with one-way streets is one such example. A graph that has directed edges, edges that can be traversed in only one direction, is known as a **digraph** (see Figure 4.9). The number of edges going into a vertex is known as the **indegree** of the vertex, and the number of edges coming out of a vertex is known as the **outdegree.**

Now look at Figure 4.9. This digraph can be described by the following set of vertices and set of ordered edges:

Vertices = $\{A, B, C, D\}$ Ordered edges = $\{AB, BA, BC, CA, DB, AD\}$.

If you follow the indicated direction of each edge, is it possible to draw this digraph and end up at the vertex where you started? That is, does the digraph have a directed Euler circuit?

Check the indegree and outdegree of each vertex. A connected digraph has an Euler circuit if the indegree and outdegree of each vertex are equal.

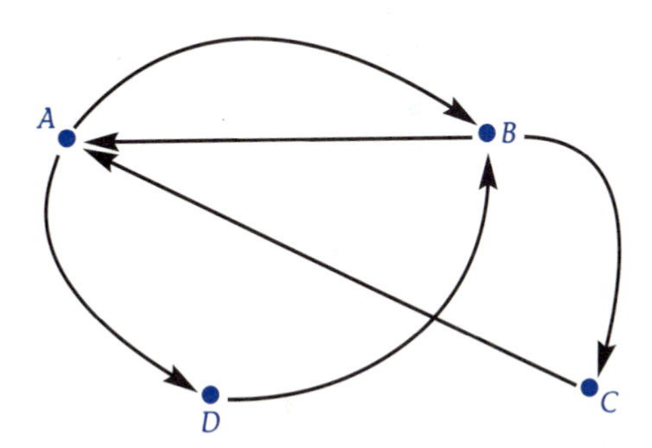

Figure 4.9 Digraph.

Exercises

1. State whether the graph has an Euler circuit, an Euler path, or neither. Explain why.

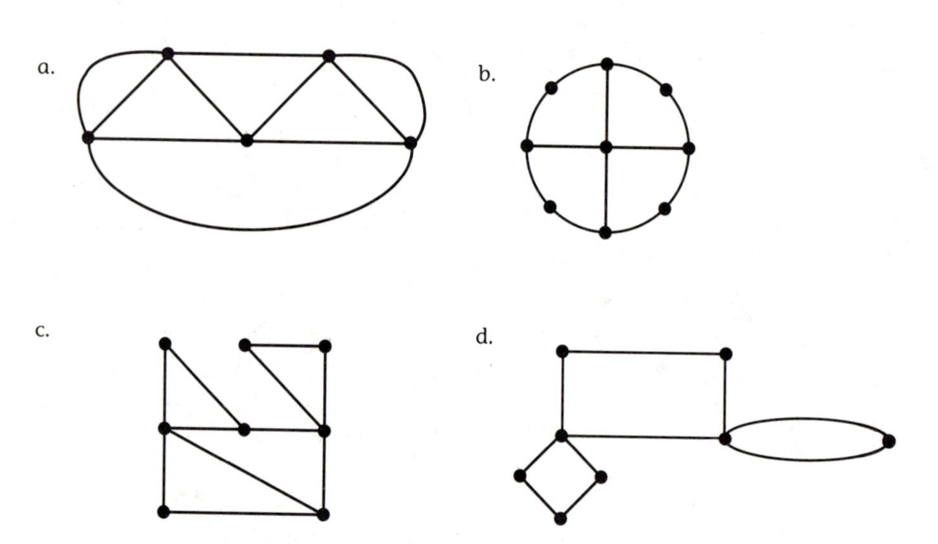

2. Sally began using the Euler circuit algorithm to find the Euler circuit for the graph below. She started at vertex d and labeled it S. The first circuit she found was S, e, f, a, b, c, S. Using Sally's start, continue the algorithm, and find the Euler circuit for the graph.

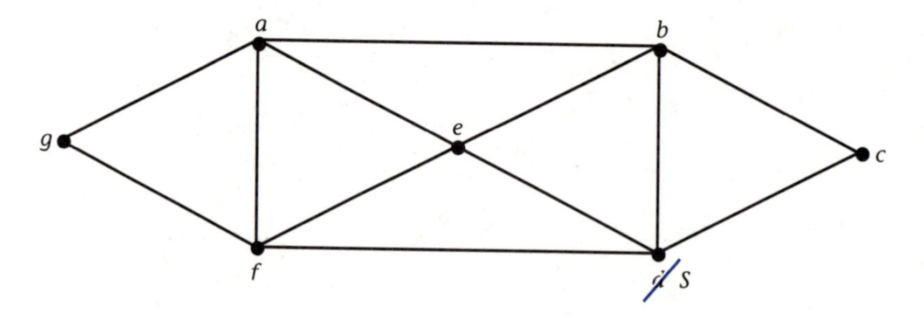

3. Use the Euler circuit algorithm to find an Euler circuit for the following graph:

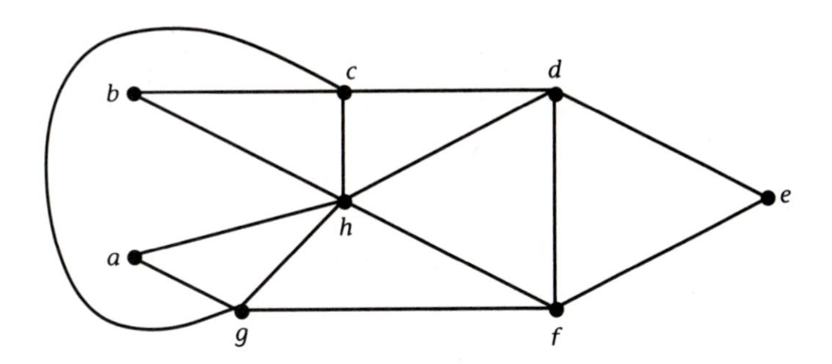

4. The Euler circuit algorithm is to be used for "connected graphs with all vertices of even degree." Why is it necessary to state that the graph must be connected? Give an example of a graph with all vertices of even degree that does not have an Euler circuit.

5. Will a complete graph with two vertices have an Euler circuit? Three vertices? Four vertices? Five vertices, or n vertices?

6. The present-day Königsberg has two more bridges than it did in Euler's time. One more bridge was added to connect the two banks of the river, A to B in the figure below. Another one was added to link the land to one of the islands, B to D. Is it now possible to make the famous walk and return to the starting point?

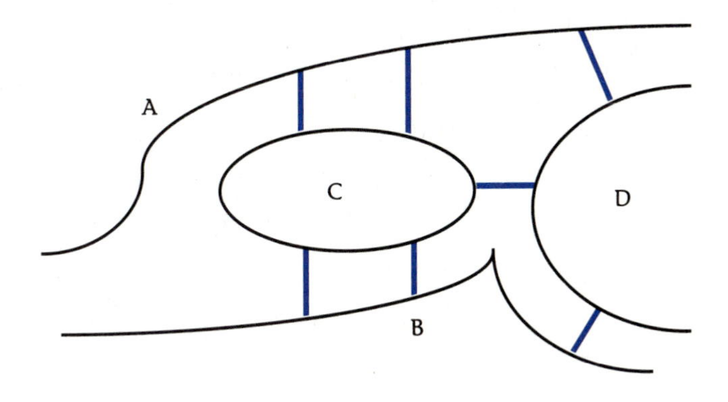

Königsberg's *original* seven bridges.

7. The streets of a city can be described by a graph in which the vertices represent the street corners and the edges represent the streets. Suppose you are the city street inspector and it is desirable to minimize time and cost by not inspecting the same street more than once.

   Is it possible to begin at the garage (G) and inspect each street only once? Will you be back at the garage at the end of inspection? Find a route that inspects all streets, repeats the least number of edges, and returns to the garage.

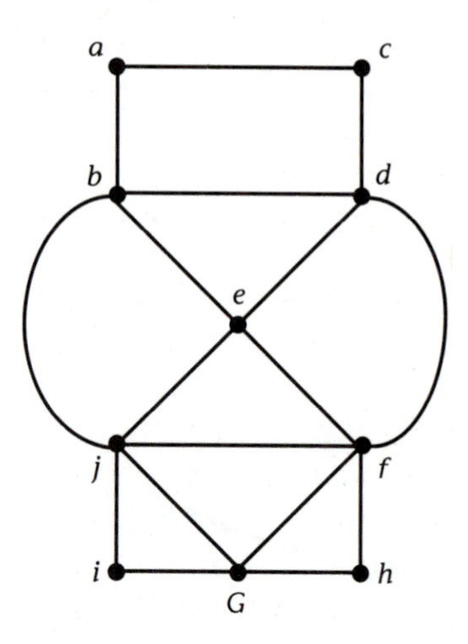

8. Construct the following digraphs:
   a. $V = \{A, B, C, D, E\}$
      $E = \{AB, CB, CE, DE, DA\}$
   b. $V = \{W, X, Y, Z\}$
      $E = \{WX, XZ, ZY, YW, XY, YX\}$

9. Determine whether the digraph has a directed Euler circuit.

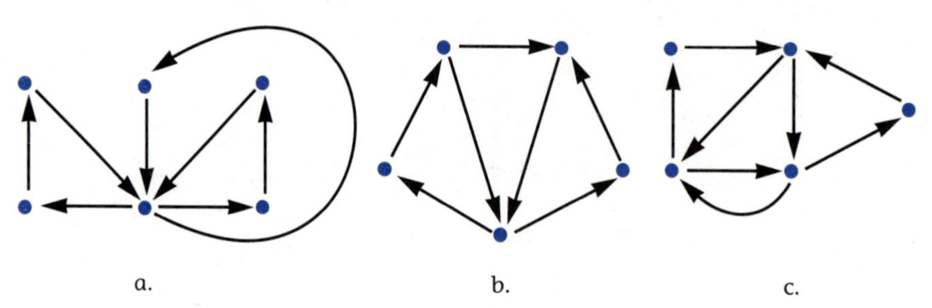

10. Does the following digraph have a directed Euler circuit? Why or why not? Does it have a directed Euler path? If it does, which vertex must be the starting vertex? When does a digraph have an Euler path?

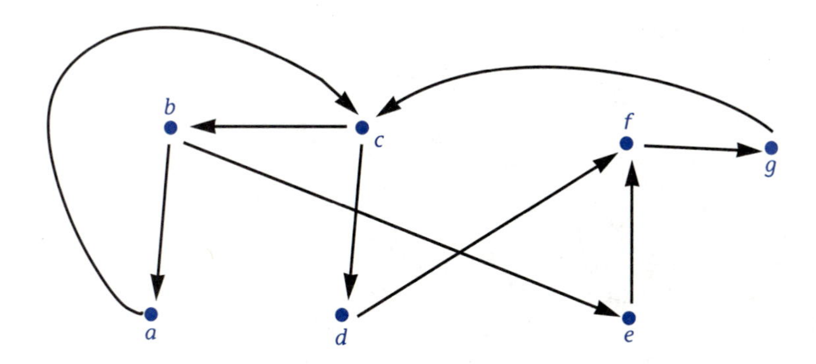

11. A digraph can be represented by an adjacency matrix. If there is a directed edge from vertex a to vertex b, then a 1 is placed in row a, column b, of the matrix; otherwise a 0 is entered. Matrix M is the adjacency matrix for the graph below:

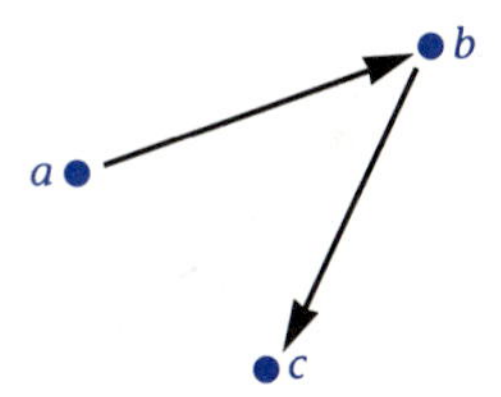

$$M = \begin{bmatrix} 0 & 1 & 0 \\ 0 & 0 & 1 \\ 0 & 0 & 0 \end{bmatrix}$$

Find the adjacency matrix for each of the following digraphs:

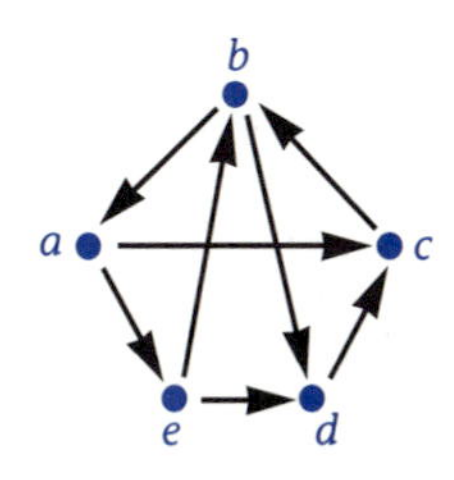

a.

b.

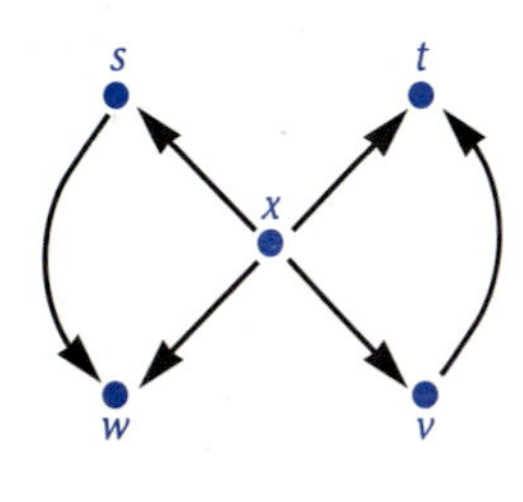

c.

12. Construct a digraph for the following adjacency matrix:

$$\begin{bmatrix} 0 & 1 & 0 & 1 & 0 \\ 1 & 0 & 1 & 0 & 1 \\ 0 & 0 & 0 & 1 & 0 \\ 0 & 1 & 1 & 0 & 0 \\ 1 & 0 & 0 & 1 & 0 \end{bmatrix}$$

a. Is there symmetry along the main diagonal of the adjacency matrix?
b. Find the sum of the numbers in the second row. What does that total indicate?
c. Find the sum of the numbers in the second column. What does that total indicate?

Hamiltonian Circuits and Paths

Explore This

Suppose once again that you are a city inspector but that this time you must inspect the fire hydrants that are located at each of the street intersections. To optimize your route, you must find a path that begins at the garage, G, visits each intersection exactly once, and returns to the garage (see Figure 4.10).

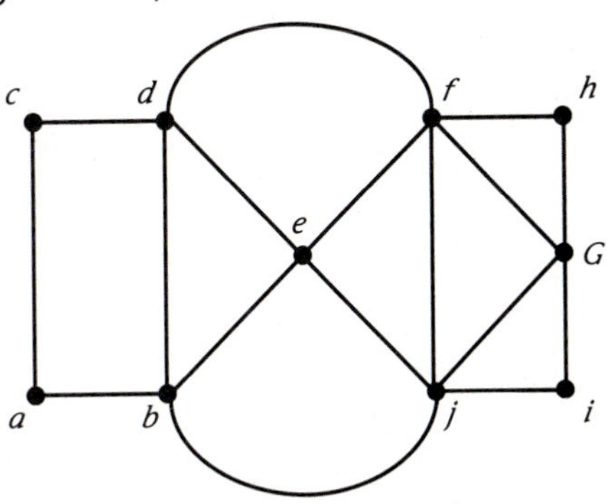

Figure 4.10 Street network.

One path that meets these criteria is *G, h, f, d, c, a, b, e, j, i, G*. Notice that it is not necessary that every edge of the graph be traversed when visiting each vertex exactly once.

In the nineteenth century an Irishman named Sir William Rowan Hamilton (1805–1865) invented a game called the Icosian game. The game consisted of a graph in which the vertices represented major cities in Europe. The object of the game was to find a path that visited each of the 20 vertices exactly once. In honor of Hamilton and his game, a path that uses each vertex of a graph exactly once is known as a **Hamiltonian path.** If the path ends at the starting vertex, it is called a **Hamiltonian circuit.**

Try to find a Hamiltonian circuit in each of the graphs in Figure 4.11.

Mathematicians continue to be intrigued by this type of problem, because a simple test for determining whether a graph has a Hamiltonian circuit has not been found. The search continues, but it now appears that a general solution may be impossible. The following theorem guarantees the existence of a Hamiltonian circuit for certain kinds of graphs.

If a connected graph has n vertices, where $n > 2$ and each vertex has degree of at least $\frac{n}{2}$, then the graph has a Hamiltonian circuit.

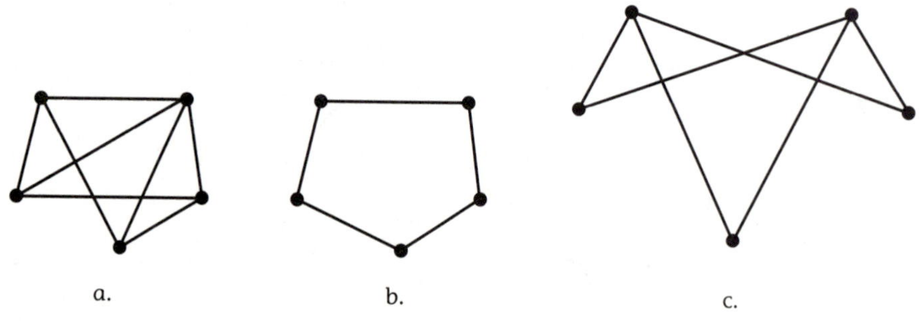

Figure 4.11 Graphs with possible Hamiltonian circuits.

Check the degree of each vertex in Figure 4.11a. Since each of the five vertices of the graph has degree of at least $\frac{5}{2}$, the graph has a Hamiltonian circuit. Unfortunately, the theorem does not tell us how to find the circuit.

If a graph has some vertices with degree less than $\frac{n}{2}$, the theorem does not apply. It may or may not have a Hamiltonian circuit. Figures 4.11b and c each has some vertices of degree less than $\frac{5}{2}$, so no conclusion can be drawn. By inspection, Figure 4.11b has a Hamiltonian circuit, but Figure 4.11c does not.

As with Euler circuits, it often is useful for the edges of the graph to have direction. Consider a competition in which each player must play every other player. To illustrate this, draw a complete graph in which the vertices represent the players, and a directed edge from vertex *A* to vertex *B* indicates that player *A* defeated player *B*. This type of digraph is known as a **tournament.** One interesting property of such a digraph is that every tournament contains a Hamiltonian path. This implies that at the end of the tournament it is possible to rank the teams in order, from winner to loser.

Example

Suppose four teams play in the school soccer round robin tournament. The results of the competition are as follows:

Game	*AB*	*AC*	*AD*	*BC*	*BD*	*CD*
Winner	*B*	*A*	*D*	*B*	*D*	*D*

Draw a digraph to represent the tournament. Find a Hamiltonian path, and then rank the participants from winner to loser.

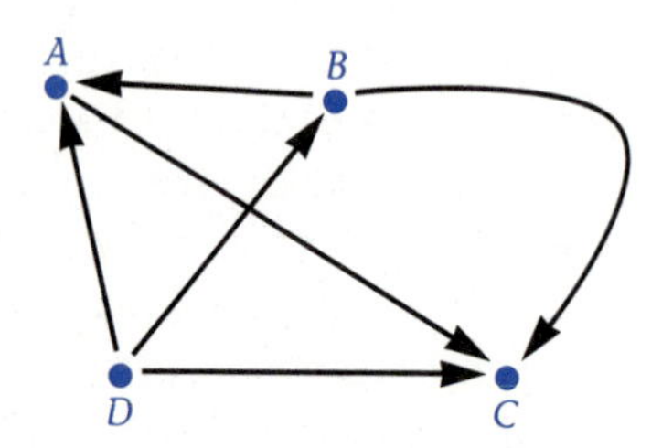

To find a solution, remember that a tournament results from a *complete* graph when direction is given to the edges. There is only one Hamiltonian path for this graph, D, B, A, C. Therefore, D finishes first; B finishes second; A is third; and C is fourth.

Exercises

1. Apply the theorem on p. 184 to the graphs below, and indicate which have Hamiltonian circuits. Explain why.

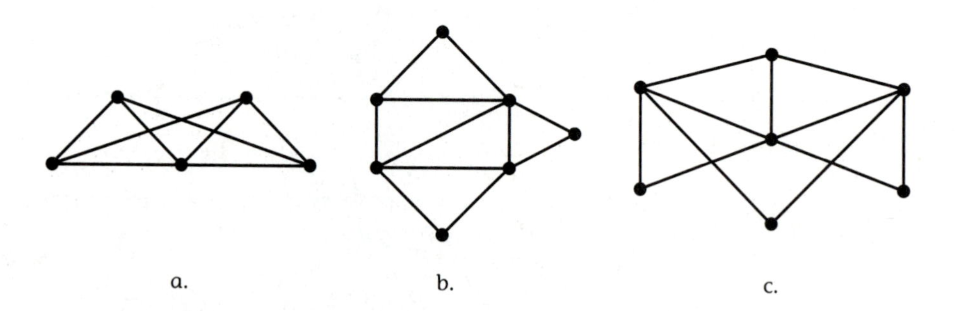

2. Give two examples of situations that could be modeled by a graph in which finding a Hamiltonian path or circuit would be of benefit.
3. a. Construct a graph that has both an Euler and a Hamiltonian circuit.
   b. Construct a graph that has neither an Euler nor a Hamiltonian circuit.

4. Hamilton's Icosian game was played on a wooden regular dodecahedron. In the planar representation of the game, find a Hamiltonian circuit for the graph. Is there only one Hamiltonian circuit for the graph? Can the circuit begin at any vertex?

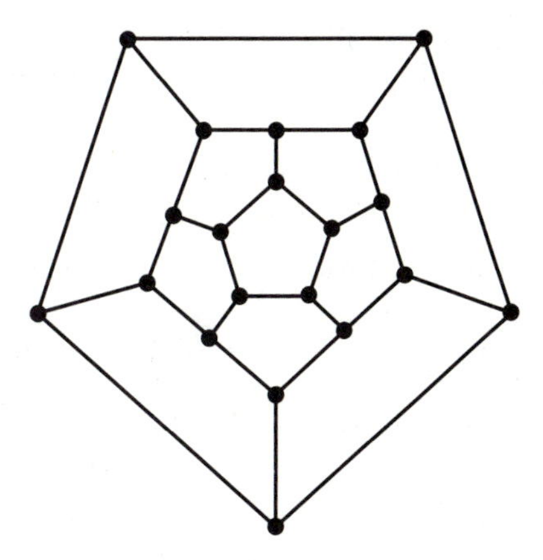

5. Draw a tournament with five players, in which player *A* beats everyone, *B* beats everyone but *A*, *C* is beaten by everyone, and *D* beats *E*.

6. Find all the directed Hamiltonian paths for the following tournaments:

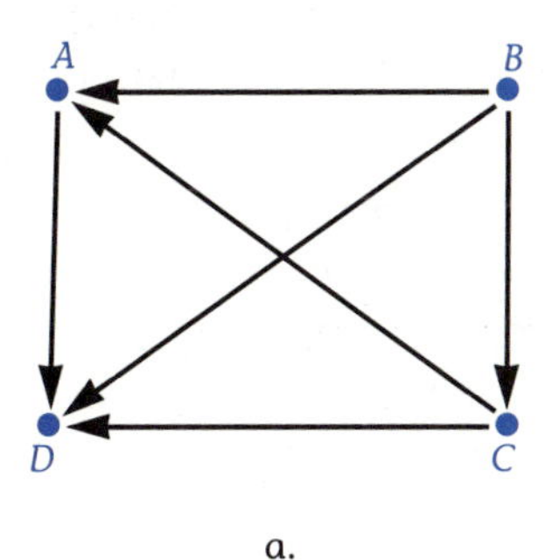

a.

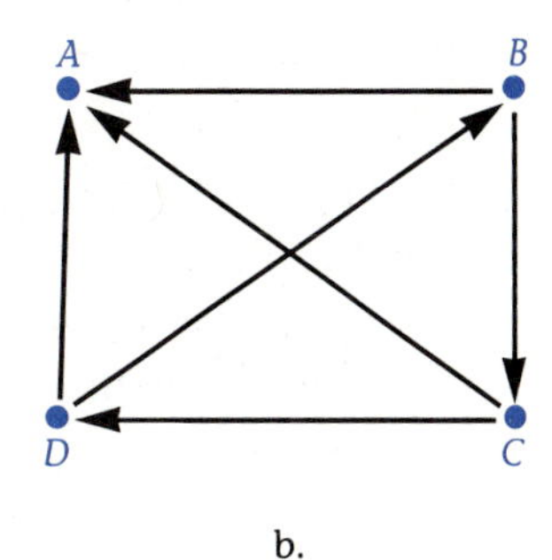

b.

7. Draw a tournament with three vertices in which
   a. One player wins all of the games it plays.
   b. Each player wins one game.
   c. Two players lose all of the games they play.

8. Draw a tournament with five vertices in which there is a three-way tie for first place.
9. When ties exist in a ranking for a tournament, is there a unique Hamiltonian path for the graph? Explain why or why not.
10. a. Write an algorithm that uses the outdegree of the vertices to find a Hamiltonian path for a tournament that has a unique path.
    b. Discuss the difficulties that arise with this algorithm when the tournament has more than one Hamiltonian path.
11. Complete the following table for a tournament:

Number of Vertices	Sum of the Outdegrees of the Vertices
1	0
2	1
3	3
4	
5	
6	

    Write a recurrence relation that expresses the relationships between S_N, the sum of the outdegrees for a tournament with n vertices, and S_{N-1}.
12. In a tournament a **transmitter** is a vertex with a positive outdegree and a zero indegree. A **receiver** is a vertex with a positive indegree and a zero outdegree. Why can a tournament have at most one transmitter and at most one receiver?
13. Consider the set of preference schedules from Lesson 1.3:

A	B	C	D
B	C	B	B
C	D	D	C
D	A	A	A
8	5	6	7

The first preference schedule could be represented by the following tournament:

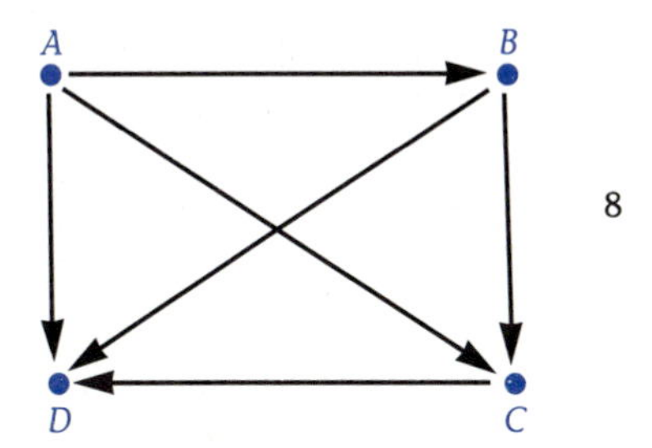

a. Construct tournaments for the other three preference schedules.
b. Construct a cumulative preference tournament that would show the overall results of the four preference schedules.
c. Is there a Condorcet winner in the election?
d. Find a Hamiltonian path for the cumulative tournament. What does this path indicate?

14. Construct an adjacency matrix for the digraph below, and call the matrix *M*.

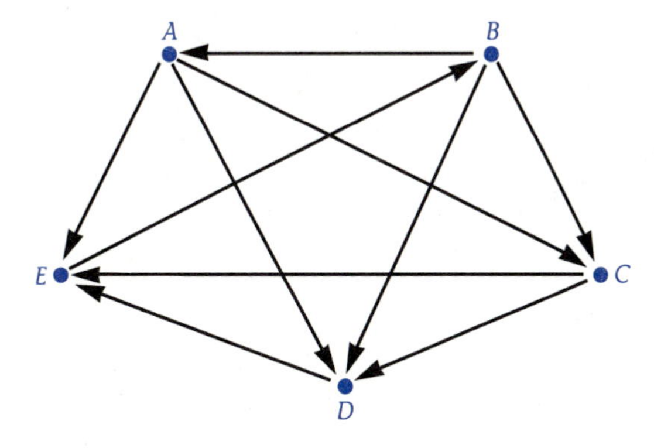

By summing the rows of *M*, it can be seen that a tie exists between *A* and *B*, each with three wins.

Square *M*. Notice that this matrix gives the number of paths of length 2 between vertices. For example, the 3 in row 2, column 5, indicates that 3 paths of length 2 exist between *B* and *E*. Those paths are *B*, *C*, *E*; *B*, *D*, *E*; and *B*, *A*, *E*. In the case of a tournament, this means that *B* has three two-stage wins or dominances over *E*. Add *M* and M^2, and use the sum to determine the total number of ways that *A* and *B* can dominate in one and two stages. Who might now be considered the winner? What would M^3 indicate? Find M^3 and see if you are correct.

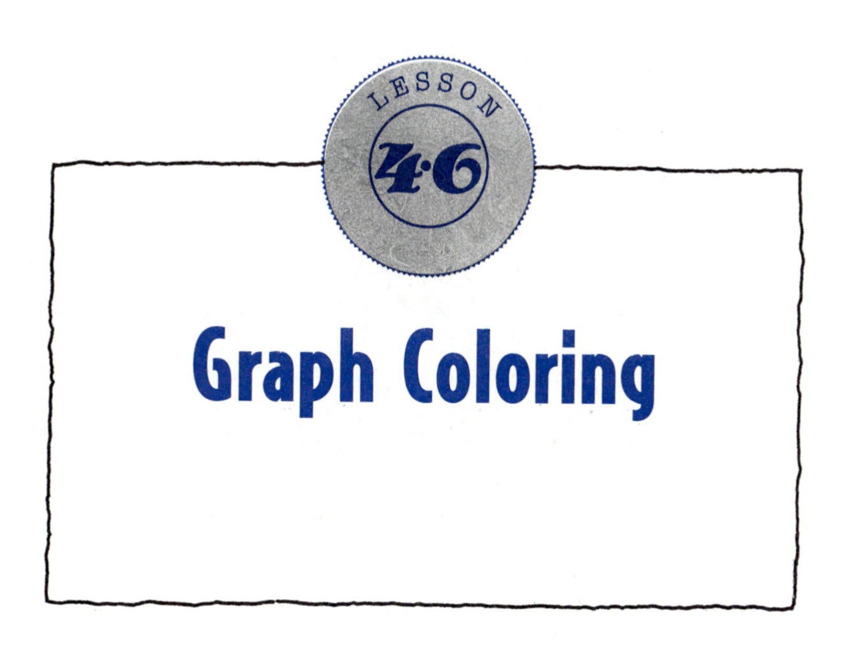

Graph Coloring

Explore This

Below is a table of organizations at Central High School and the students who hold offices in these clubs.

	Math Club	Honor Club	Science Club	Art Club	Pep Club	Spanish Club
Matt	X	X	X	—	—	—
Marty	X	—	—	X	X	—
Kim	—	X	—	—	—	X
Lois	X	—	X	—	—	—
Dot	X	—	—	—	X	—

Each club at Central High wants to meet once a week. Since several students hold offices in more than one organization, it is necessary to arrange the meeting days so that no students will be scheduled for more than one meeting on the same day. Is it possible to create such a schedule? What is the minimum number of days needed?

One possible solution to the problem is to use five days for the scheduling. The Math and Spanish clubs could meet on Monday, and the remaining clubs could meet on the other four days. If the problem is to schedule the meetings in the fewest number of days, then this solution is not optimal. It is possible to create a schedule using only three days.

Finding such a schedule by trial and error is not too difficult in this case, but a mathematical model would be helpful for more complicated problems. Construct a graph in which the vertices represent clubs at Central High School and the edges indicate that the clubs share an officer and so cannot meet on the same night.

Now begin labeling the vertices of the graph in Figure 4.12 with the days of the week. Adjacent vertices must have different labels, since this is where the conflicts occur. One way of assigning days is to begin with the Math Club, and label it with Monday. Since no one belongs to both the Math Club and the Spanish Club, also label the Spanish Club with Monday. Label the Honor Club with Tuesday. The Pep or Art Club, but not both, can receive a Tuesday label. The other is labeled with Wednesday. The Science Club can also receive a Wednesday label. The resulting schedule is an optimal solution to the problem, but notice that it is not unique.

Problems of this type are called *coloring* problems because historically the labels placed on the vertices of the graphs were called colors.

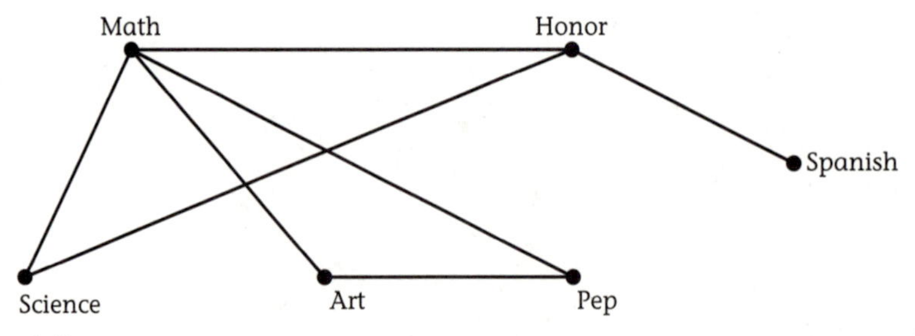

Figure 4.12 Graph showing the organizations at Central High that share an officer.

The process of labeling the graph is referred to as **coloring the graph,** and the minimum number of labels or colors that can be used is known as the **chromatic number** of the graph. The chromatic number for the graph in Figure 4.12 is three.

Problems of this type first attracted interest in the nineteenth century when mathematicians such as Augustus de Morgan, William Rowan Hamilton, and Arthur Cayley became interested in a problem known as the *four-color conjecture.* The problem stated that any map that could be drawn on the surface of a sphere could be colored with, at most, four colors.

For over 100 years, this problem intrigued mathematicians. During that time many claimed to have proved the conjecture but flaws were always found in the proofs. It wasn't until 1976 that Kenneth Appel and Wolfgang Haken of the University of Illinois actually solved the famous problem, and the four-color conjecture became the four-color theorem. The problem was proved in a way very different from earlier attempts. Appel and Haken used a high-speed computer in their verification. When the proof was finally complete, they had used over 1,200 hours of time on three different computers.

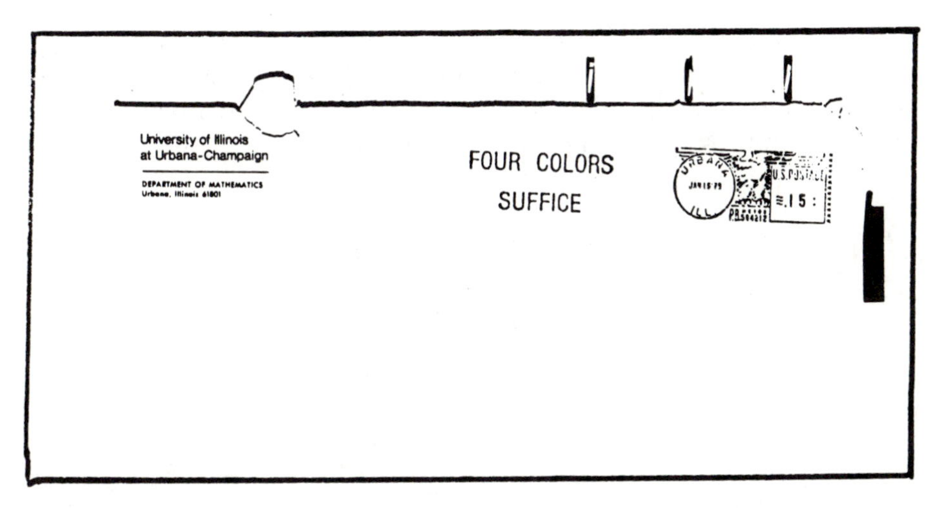

Postage meter stamp used by the University of Illinois to commemorate the proof of the four-color theorem.

One way to approach the problem of coloring a map is to represent each region of the map with a vertex of a graph. Two vertices are connected by an edge if the regions they represent have a common border. Coloring the map is then the same process as coloring the vertices of a graph so that adjacent vertices have different colors.

Example

Color the following map using four or fewer colors:

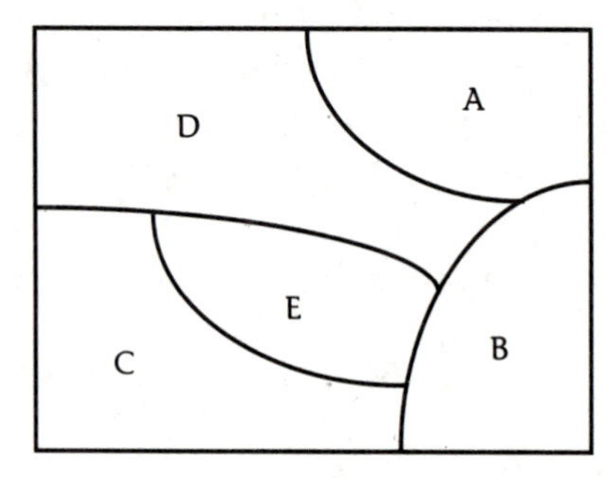

To find a solution, represent the map with a graph in which each vertex represents a region of the map, and draw edges between vertices if the regions on the map have a common border. Then label the graph with a minimum number of colors.

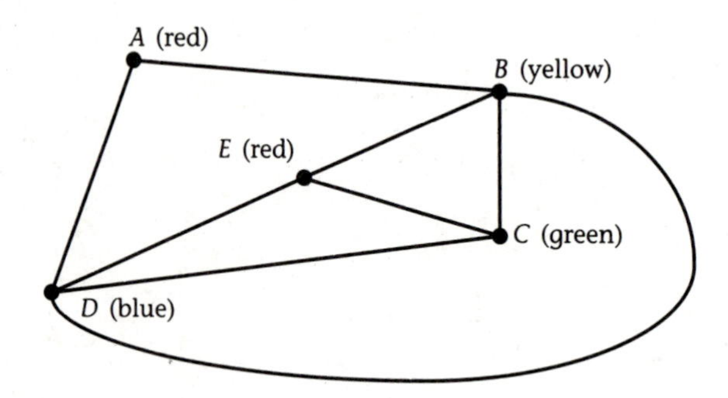

Four colors are necessary to color this map.

Exercises

1. Find the chromatic number for each of the graphs below.

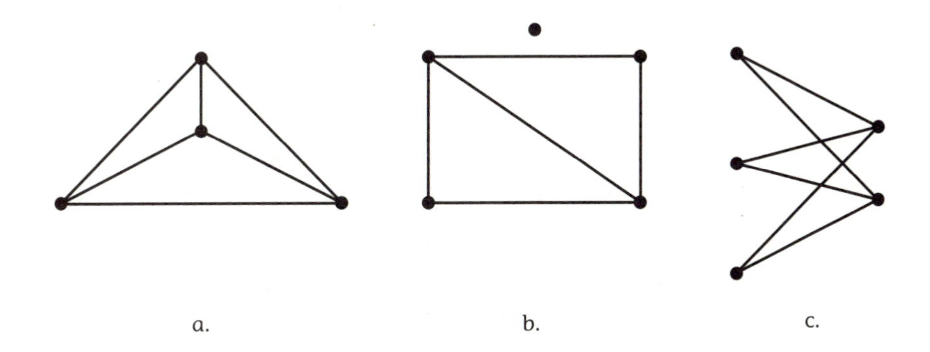

2. a. Draw a graph that has four vertices and a chromatic number of three.
   b. Draw a graph that has four vertices and a chromatic number of one.

As the number of vertices in a graph increases, a systematic method of labeling the vertices becomes necessary. One way to do this is to create a coloring algorithm.

3. One way to begin the coloring process is first to color the vertices with the most conflict. How can the vertices be ranked from those with the most conflict to those with the least conflict?
4. After having colored the vertex with the most conflict, which other vertices can receive that same color?
5. Which vertex would then get the second color? Which other vertices could get that same second color?
6. When would the coloring process be complete?
7. Refer back to Exercises 3 through 6, and create an algorithm that colors a graph.

8. Use the algorithm that was developed in Exercise 7 to color the graph below. What is the chromatic number of the graph? Did your algorithm find the proper chromatic number?

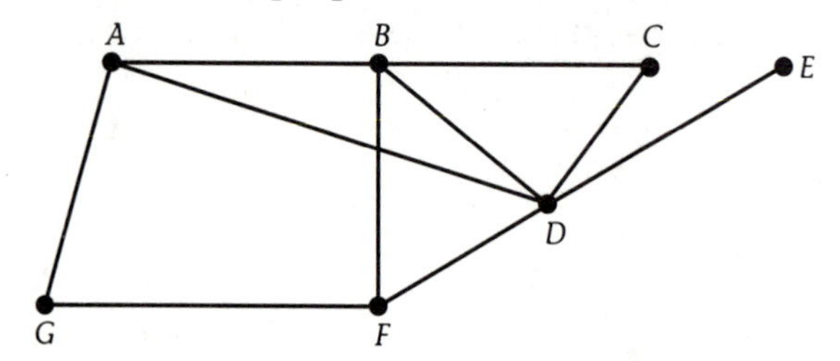

9. What is the chromatic number of K_2? K_3? K_4? K_N?

10. A **cycle** is a path that begins and ends at the same vertex and does not use any edge or vertex more than once.
    a. If a cycle has an even number of vertices, what is its chromatic number?
    b. What is the chromatic number of a cycle with an odd number of vertices?

11. Mrs. Suzuki is planning to take her history class to the art museum. Below is a graph showing those students who are not compatible. Assuming that the seating capacity of the cars is not a problem, what is the minimum number of cars necessary to take the students to the museum?

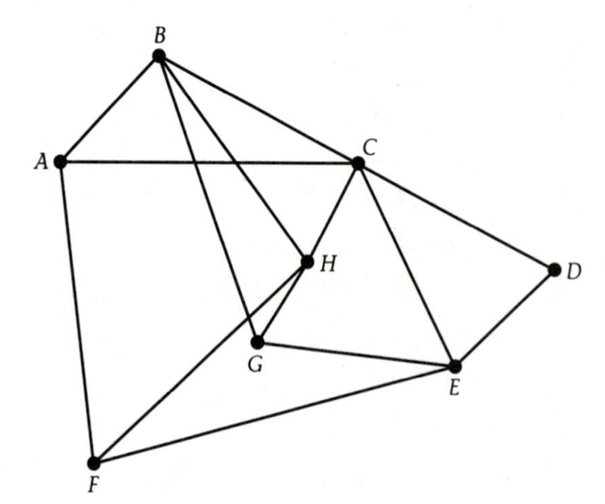

12. Refer to Exercise 1, Lesson 4.3, page 169. What is the minimum number of fish tanks needed to house the fish?

13. Below is a list of chemicals and the chemicals with which each cannot be stored.

Chemicals	Cannot Be Stored With
1	2, 5, 7
2	1, 3, 5
3	2, 4
4	3, 7
5	1, 2, 6, 7
6	5
7	1, 4, 5

How many different storage facilities are necessary in order to keep all seven chemicals?

14. Color the following map using only three colors.

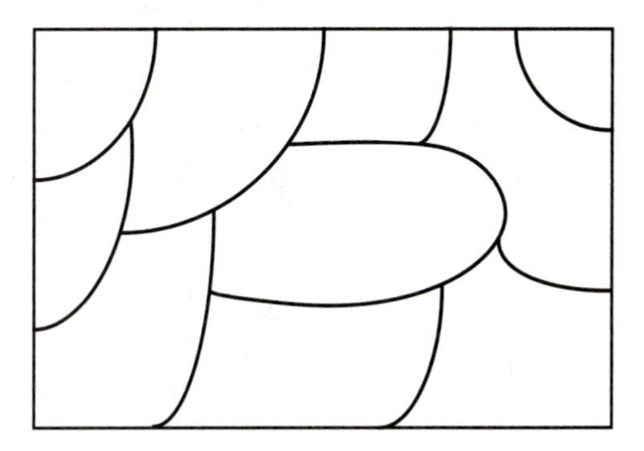

15. Draw graphs to represent the maps below. Color the graphs. What is the minimum number of colors needed to color each map?

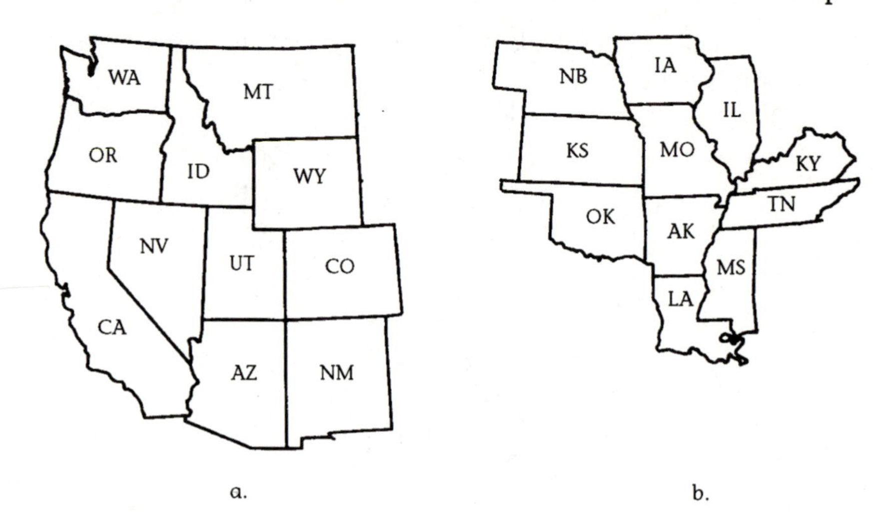

1. Draw a graph for the task table below.

Task	Time	Prerequisites
Start	0	—
A	2	None
B	4	*A*
C	4	*A*
D	3	*B*
E	2	*C*
F	1	*C*
G	2	*D*
H	5	*D,E,F*
I	3	*G,H*
Finish		

2. Complete the task table on the next page for the graph below.

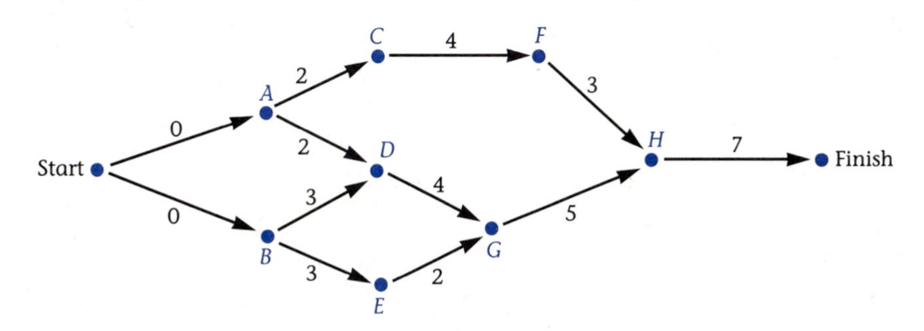

Task	Time	Prerequisites
Start	0	—
A		
B		
C		
D		
E		
F		
G		
H		
Finish		

3. a. List the vertices of the graph below, and give their earliest-start time.
   b. Determine the minimum project time.

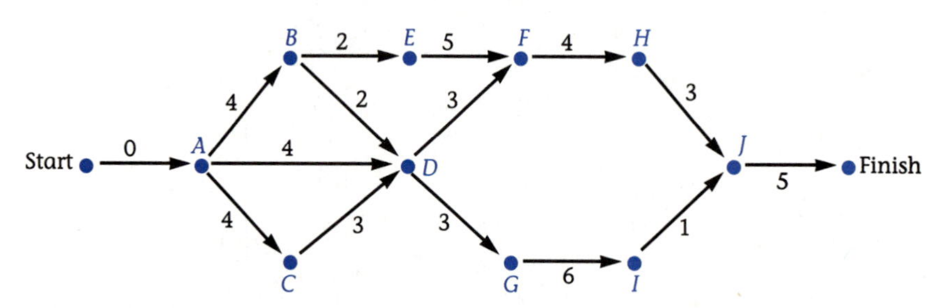

4. Use your graph from Exercise 1.
   a. Recopy it and label the vertices with their EST.
   b. Determine the critical path and the minimum project time.

5.

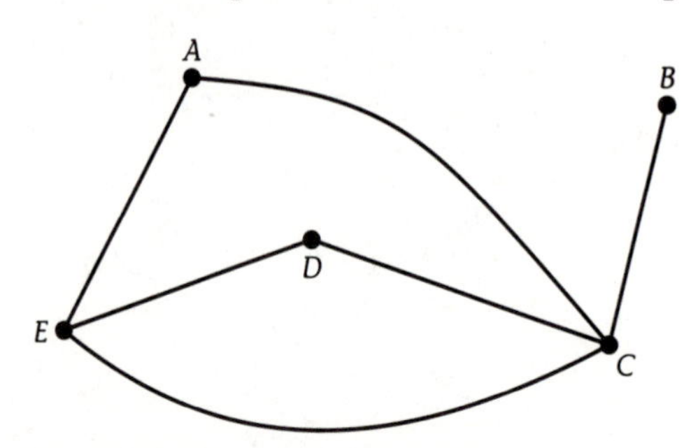

a. Is this graph connected? Why or why not?
b. Is this graph complete? Why or why not?
c. Name two vertices that are adjacent to vertex E.
d. Name a path from B to E of length 3.
e. What is the degree of vertex C?
f. Determine an adjacency matrix for the graph.

6. Tell whether the following graphs have an Euler circuit, an Euler path, or neither. Explain why.

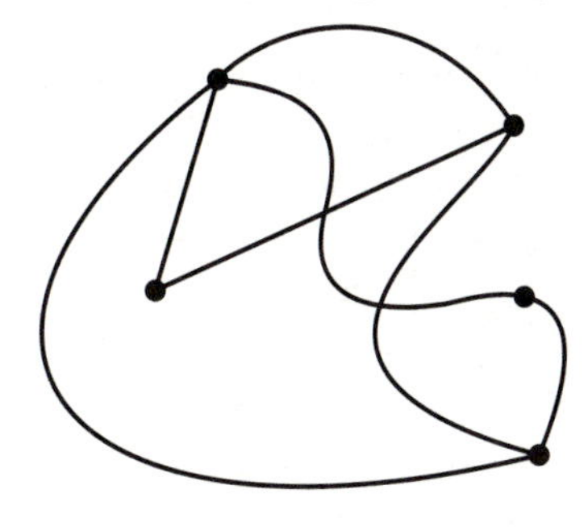

a.

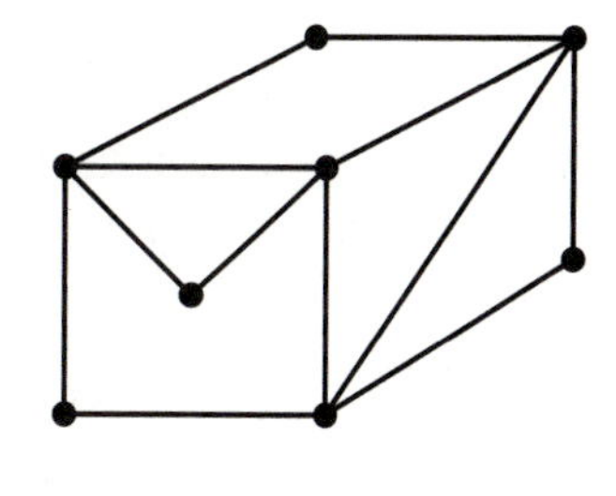

b.

7. Construct a graph for each of the following:

a. $V = \{A, B, C, D, E\}$
$E = \{AE, AB, CD, BC, DE\}$

b.
$$\begin{array}{c} \\ A \\ B \\ C \\ D \end{array}\begin{array}{c} \begin{array}{cccc} A & B & C & D \end{array} \\ \begin{bmatrix} 0 & 0 & 1 & 1 \\ 0 & 0 & 1 & 1 \\ 1 & 1 & 0 & 1 \\ 1 & 1 & 1 & 0 \end{bmatrix} \end{array}$$

8. Below is a multigraph that represents the downtown area of a small city. The local post office has decided that the mail drop boxes, which are located at the intersections of each street, must be monitored twice daily.

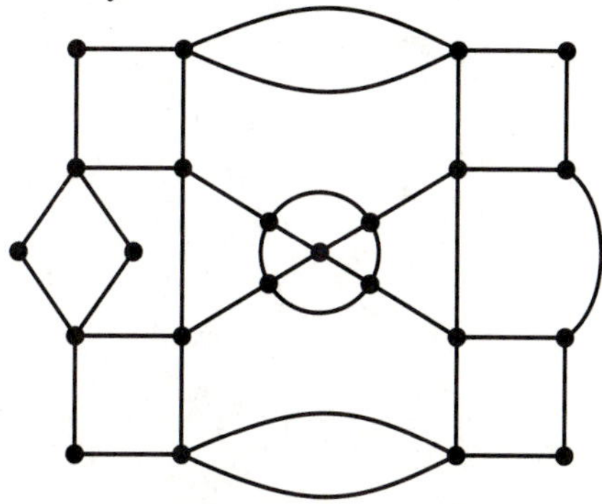

a. Is it possible to find a circuit that begins and ends at the same intersection and visits each drop box exactly once?
b. If not, is there a path that begins at one drop box, visits each drop box exactly once, and ends at a different drop box?
c. If either route exists, copy the figure and darken the edges of your proposed route.

9. Use the graph in Exercise 8.
   a. Is it possible for the local street inspector to begin at an intersection and inspect each street exactly once?
   b. Is it possible for the inspector to finish her route at the same intersection from which she began? Why or why not?

10.

	Students						
Exam	1	2	3	4	5	6	7
(M) Math	X		X		X		X
(A) Art		X		X		X	
(S) Science	X	X					X
(H) History			X			X	
(F) French				X	X		
(R) Reading	X	X		X	X		X

In scheduling the final exam for summer school at Central High, six different tests must be scheduled. The table shows the exams that are needed for seven different students.

a. Draw a graph that illustrates which exams have students in common with other exams.
b. What is the *minimum* number of time slots needed to schedule the six exams?

11.

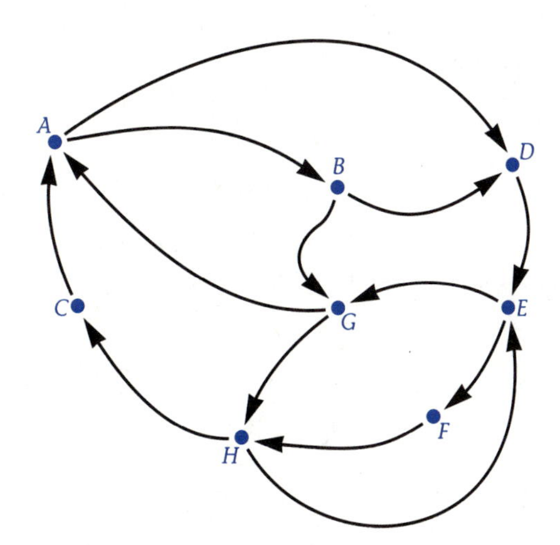

Does the above digraph have

a. A directed Euler circuit? Why or why not? If yes, list the circuit.
b. A directed Euler path? Why or why not? If yes, list the path.

12.

a. Represent the map below with a graph.
b. Color the graph.
c. What is the minimum number of colors needed to color the map?

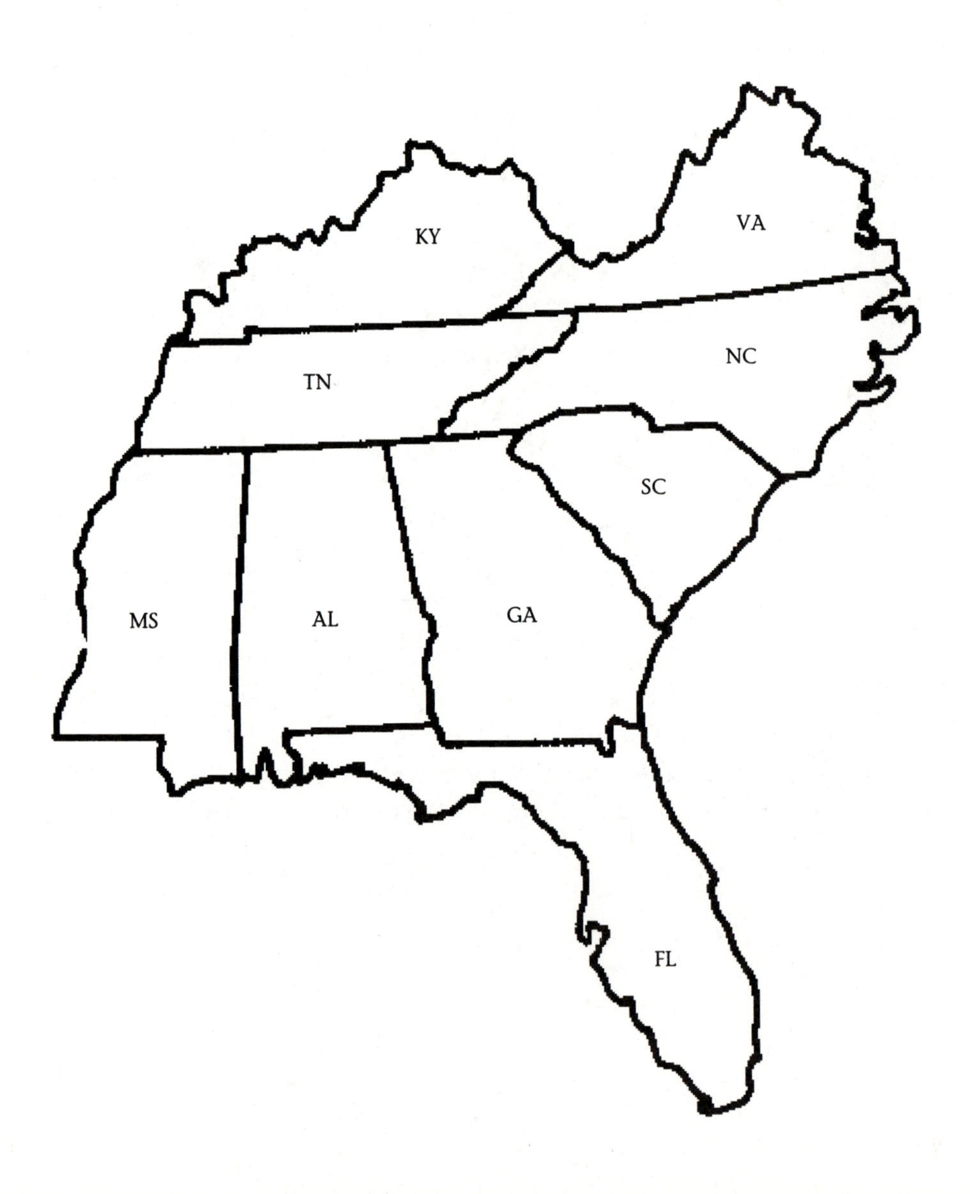

Bibliography

Biggs, N. L., E. K. Lloyd, and R. J. Wilson. 1976. *Graph Theory 1736–1936*. Oxford: Clarendon Press.

Busch, D. 1991. *The New Critical Path Method.* Chicago: Probus.

Chavey, Darrah. 1992. *Drawing Pictures with One Line.* (HistoMAP Module #21). Lexington, MA: COMAP.

COMAP. 1993. *For All Practical Purposes: Introduction to Contemporary Mathematics.* 3rd ed. New York: Freeman.

Copes, W., C. Sloyer, R. Stark, and W. Sacco. 1987. *Graph Theory: Euler's Rich Legacy*. Providence, RI: Janson Publications.

Cozzens, Margaret B., and R. Porter. 1987. *Mathematics and Its Applications.* Lexington, MA: Heath.

Cozzens, Margaret B., and R. Porter. 1987. *Problem Solving Using Graphs.* (HiMAP Module #6). Lexington, MA: COMAP.

Dossey, John, A. Otto, L. Spence, and C. Vanden Eynden. 1992. *Discrete Mathematics.* 2nd ed. Glenview, IL: Scott, Foresman.

Francis, Richard L. 1989. *The Mathematicians' Coloring Book.* (HiMAP Module #10). Lexington, MA: COMAP.

Kenney, Margaret J., ed. 1991. "Discrete Mathematics across the Curriculum, K-12." *1991 Yearbook of the National Council of Teachers of Mathematics.* Reston, VA: NCTM.

Malkevitch, J., and W. Meyer. 1974. *Graphs, Models, and Finite Mathematics.* Englewood Cliffs, NJ: Prentice-Hall.

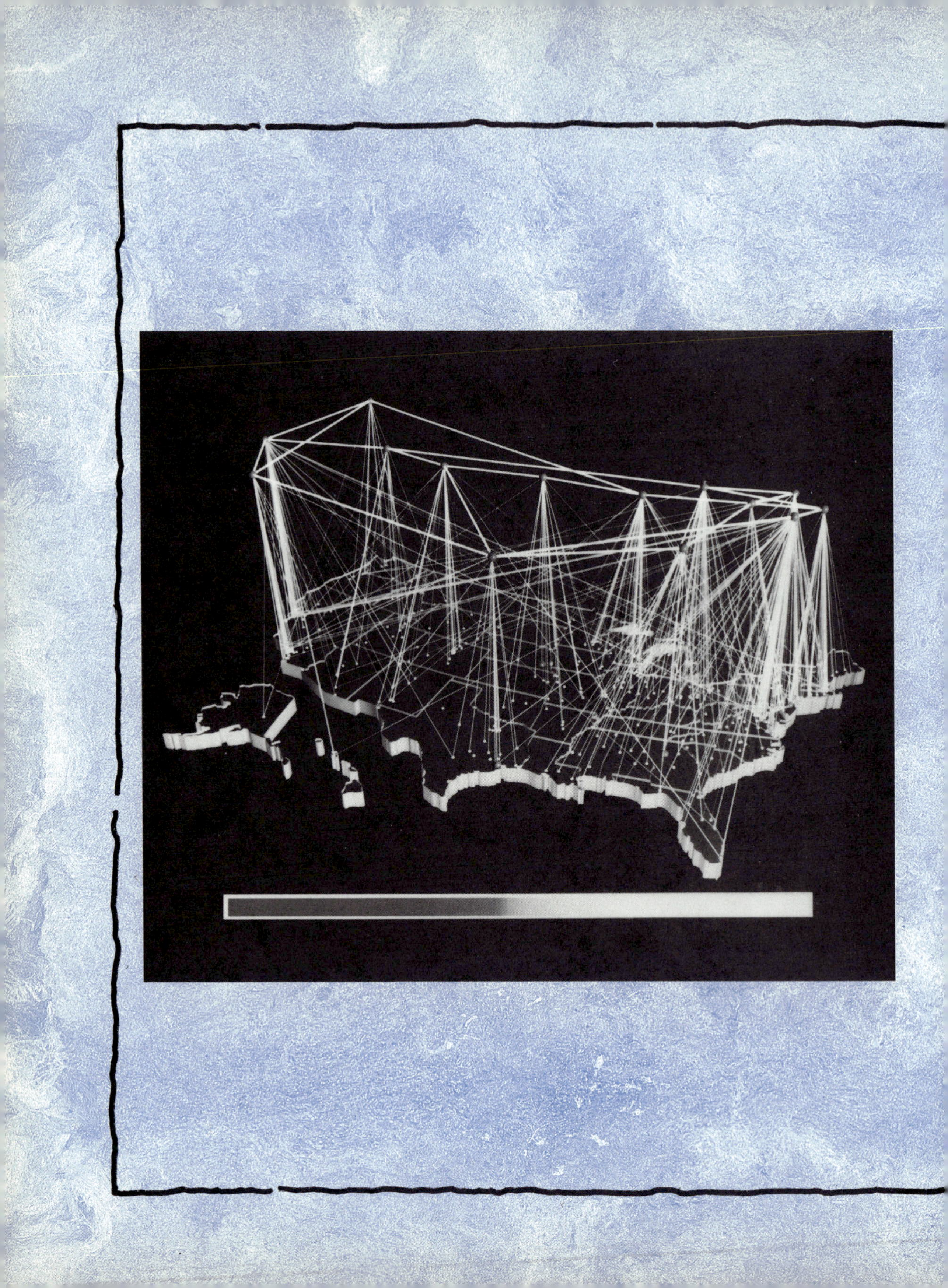

More Graphs, Subgraphs, and Trees

Because of modern technology, elaborate communication networks span the country and most of the earth. These networks allow instant transmission of information between almost any two locations. They affect many aspects of our lives, including the way we work, the way we learn, and the way in which we are entertained.

How does one construct a communications network that links several locations together, but does so at the lowest possible cost? How is the most efficient route between two locations in a network found? Can the methods used to find the best route between points in a communication network also be used to plan the best route for an automobile or plane trip? The mathematics of graph theory plays an important role in solving these and many other problems that are important in our ever-changing world.

Planarity and Coloring

In Lesson 4.6, we solved problems by modeling them with graphs and then coloring the graphs. The four-color theorem states that any map that can be drawn on the surface of a sphere can be colored with four colors or fewer. If this is true, then why does it take more than four colors to color some graphs?

Explore This

Try to redraw the graphs so that their edges intersect only at the vertices. (Think of the edges as rubber bands.)

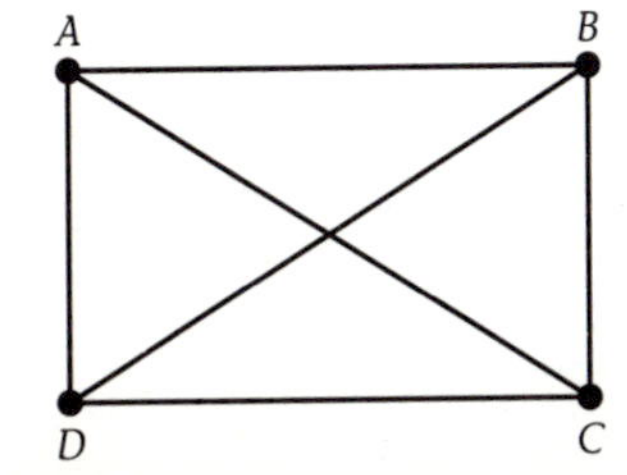

Figure 5.1 K_4 graph.

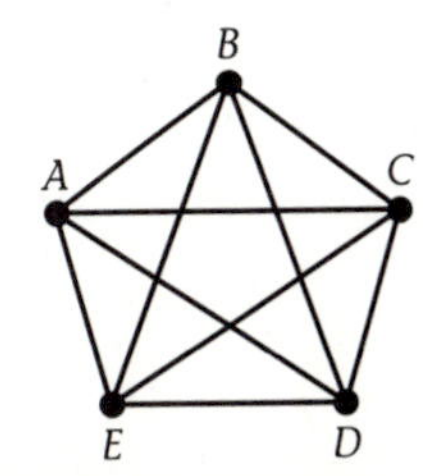

Figure 5.2 K_5 graph.

It is relatively easy to redraw Figure 5.1 so that the edges do not cross (see Figure 5.3), but no matter how hard you try, at least two edges of Figure 5.2 will always intersect (see Figure 5.4).

A graph that can be drawn so that no two edges intersect except at a vertex is called a **planar graph.** Figure 5.1 shows a planar graph, and Figure 5.2 shows a graph that is not planar. When regions of a map are represented by vertices of a graph and edges are drawn between vertices when boundaries exist between regions, the resulting graph is planar. In other words, graphs that come from a map in a plane or on a sphere are always planar. Hence, the four-color theorem can be stated in a different way:

> Every planar graph has a chromatic number that is less than or equal to four.

Why do some graphs require more than four colors? Planarity is the key. If a graph is not planar, we do not know how many colors it will take to color it.

One type of graph that is not planar, a K_5, is shown in Figure 5.2. One other nonplanar graph about which many problems (see Exercise 18 on page 215) have been written is shown in Figure 5.5. Try to draw it

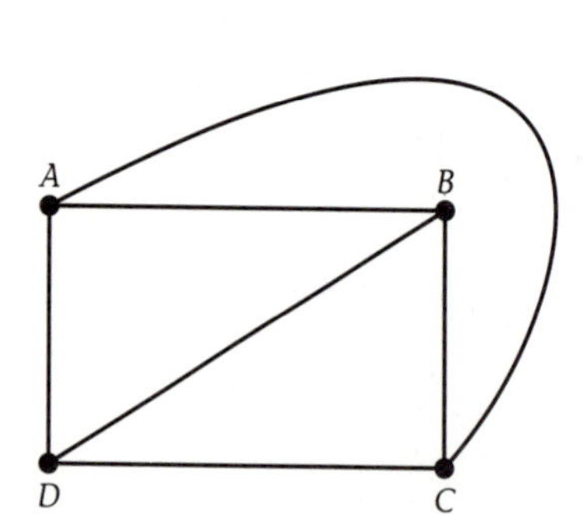

Figure 5.3 Graph in Figure 5.1 redrawn with edges not intersecting.

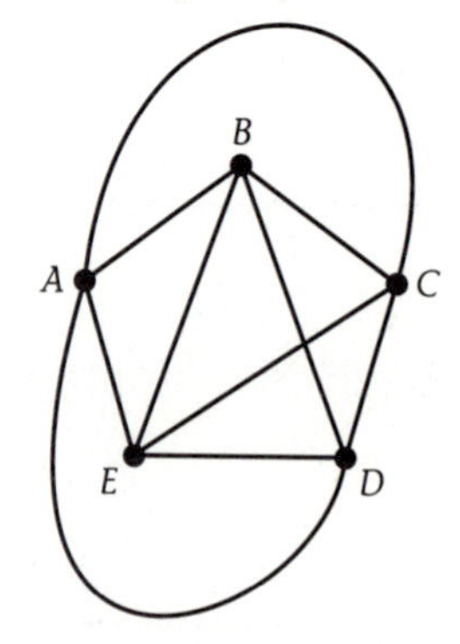

Figure 5.4 An attempt to redraw Figure 5.2 with edges not intersecting.

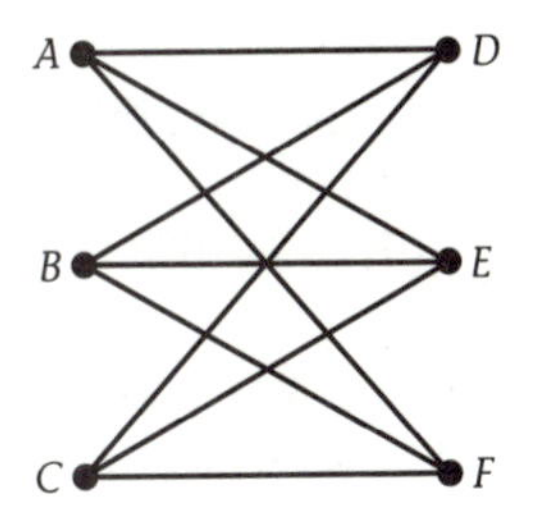

Figure 5.5 $K_{3,3}$ graph.

without the edges crossing. Once again you will discover that this is not possible.

The graph in Figure 5.5 has interesting characteristics other than not being planar. It is an example of a **bipartite** graph. A graph is bipartite if its vertices can be divided into two distinct sets so that each edge of the graph has one vertex in each set. A bipartite graph is said to be complete if it contains all possible edges between the pairs of vertices in the two distinct sets. Complete bipartite graphs can be denoted by $K_{M,N}$, where M and N are the number of vertices in the two distinct sets. Figure 5.5 is a $K_{3,3}$ graph.

Figure 5.6 is a complete bipartite graph, $K_{3,2}$, since its vertices can be separated into two distinct sets ($\{A, B, C\}$ and $\{X, Y\}$; every edge has one vertex in each set; and all possible edges from one set to the other are drawn. Determining whether or not a given graph is planar can be difficult. It would be time-consuming, if not impossible, to redraw a very large graph so that none of the edges crossed.

In 1930 Kazimierz Kuratowski, a Polish mathematician, proved a theorem that provides a resolution to this problem of determining the planarity of a graph. He identified these two important graphs, K_5 and

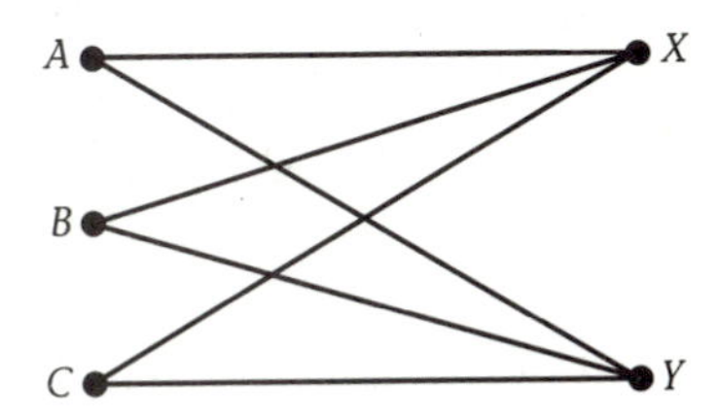

Figure 5.6 $K_{3,2}$ graph.

$K_{3,3}$, and noted that any graph that has a K_5 or $K_{3,3}$ subgraph is not planar. A graph G' is said to be a **subgraph** of graph G if all of the vertices and edges of G' are contained in G. Kuratowski also noted that the lack of a K_5 or $K_{3,3}$ does not guarantee planarity (see Exercises 20 and 21 on page 216).

In actual practice, approximately 99% of nonplanar graphs of modest size are nonplanar because of a $K_{3,3}$ subgraph rather than a K_5 subgraph.

For example, is the graph below planar?

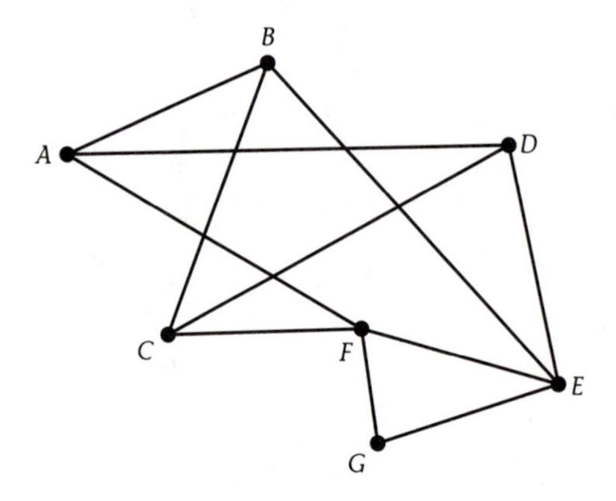

Upon close observation of vertices, A, B, C, D, E, and F, a $K_{3,3}$ subgraph can be found. Therefore, the graph is nonplanar.

Exercises

In Exercises 1 through 3 decide whether the graph is planar or nonplanar. If the graph is planar, redraw it without edge crossings.

1. 2. 3.

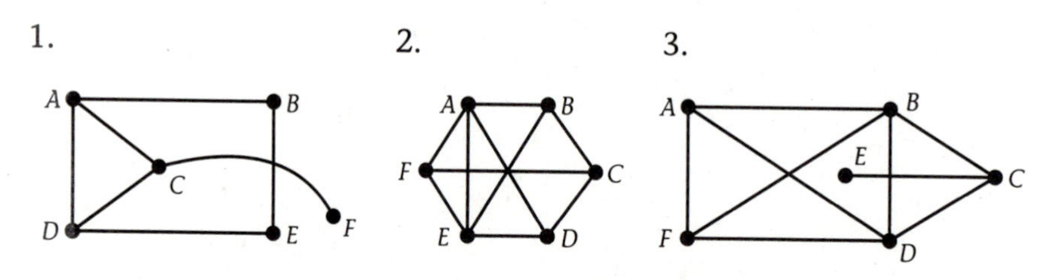

4. The graph below is planar. Draw it without edge crossings.

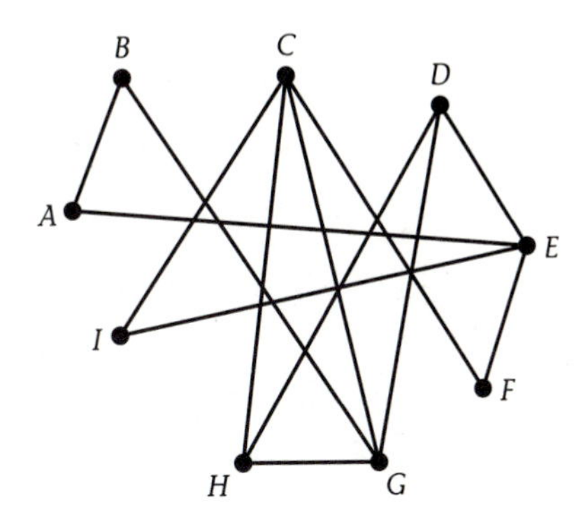

5. How do you know that it is not necessary to check the graph in Exercise 4 for a K_5 subgraph?

6. Devise a systematic method of searching a graph for a K_5 subgraph. Describe your method in a short paragraph, and try it on the graph below.

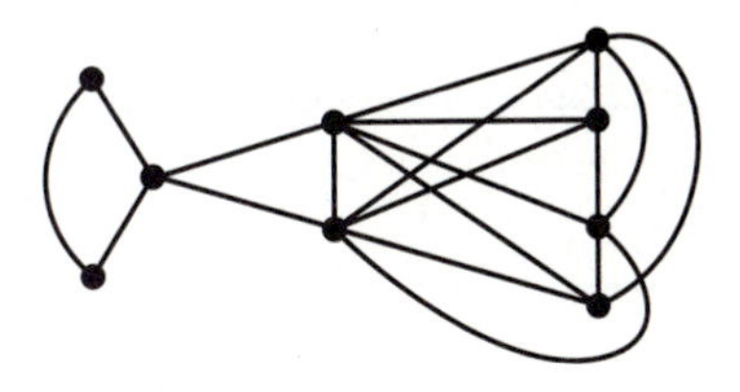

7. The **complement** of a graph G is customarily denoted by $\overline{G}$. $\overline{G}$ has the same vertices as G but its edges are those not in G. The edges of G and $\overline{G}$ along with vertices from either set would make a complete graph. Draw the complement of the graph below.

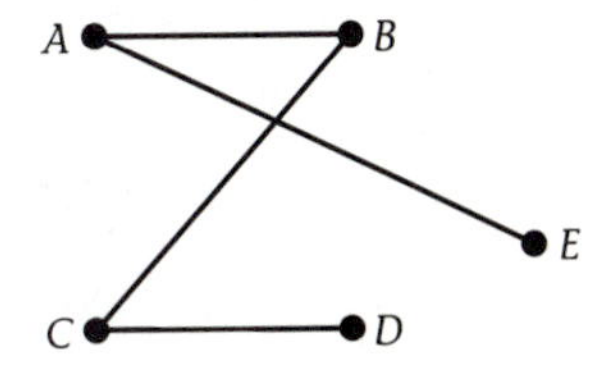

8. Every planar graph with nine vertices has a nonplanar complement. Verify this statement for one case by drawing a planar graph with nine vertices and then drawing its complement. For your case, is the complement nonplanar?

9. The concept of planarity is extremely important to printing circuit boards for the electronics industry. Why?

10. Construct the following bipartite graphs:
    a. $K_{2,3}$ b. $K_{2,4}$

11. For each of the following bipartite graphs, list the two distinct sets into which the vertices can be divided:

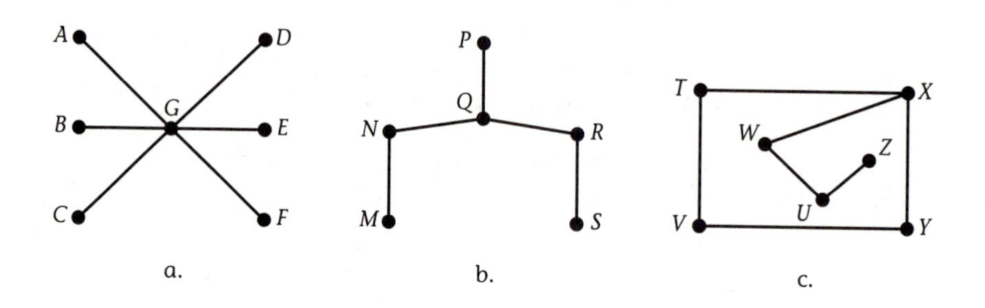

12. State whether the graphs below are bipartite.

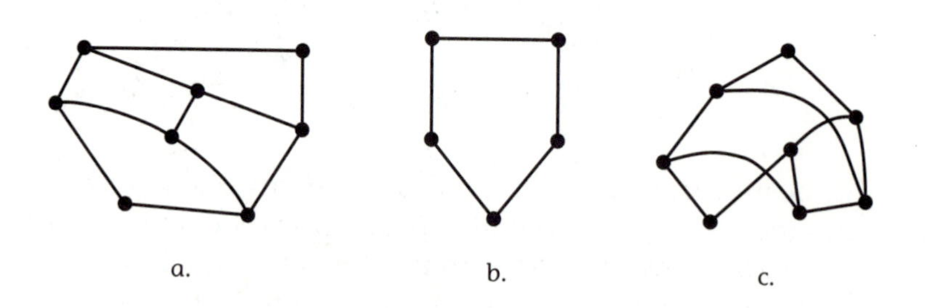

13. Devise a method of telling whether a graph is bipartite. Write a short algorithm for your method.

14. How many edges are in a $K_{2,3}$ graph? A $K_{4,3}$? A $K_{M,N}$?

15. When does a bipartite graph, $K_{M,N}$, have an Euler circuit?

16. At Ms. Johnson's party, six men and five women walk into the dining room. If each man shakes hands with each woman, how many handshakes will occur? Represent this situation with a graph. What kind of a graph is it?

17. Construct a situation that can be represented by a bipartite graph that is not complete.

18. Consider the puzzle known as the *Wells and Houses problem* or the *Utilities problem.*

    Three houses and three wells are built on a piece of land in an arid country. Because it seldom rains, the wells often run dry, and so each house must have access to each well. Unfortunately, the occupants of the three houses dislike one another and want to construct paths to the wells so that no two paths cross.

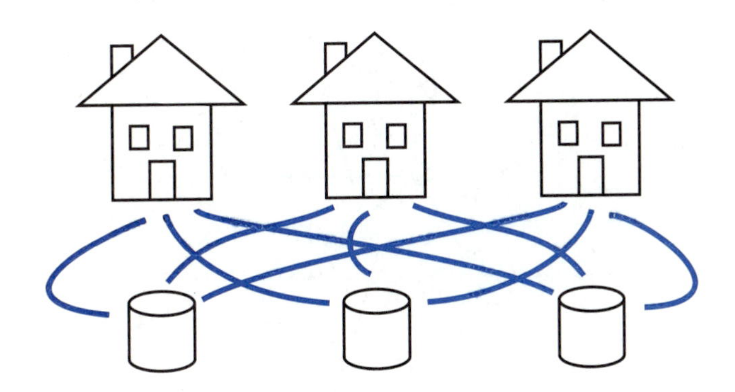

    Draw a graph to illustrate this problem. Is it possible to satisfy the feuding families? Why or why not?

19.

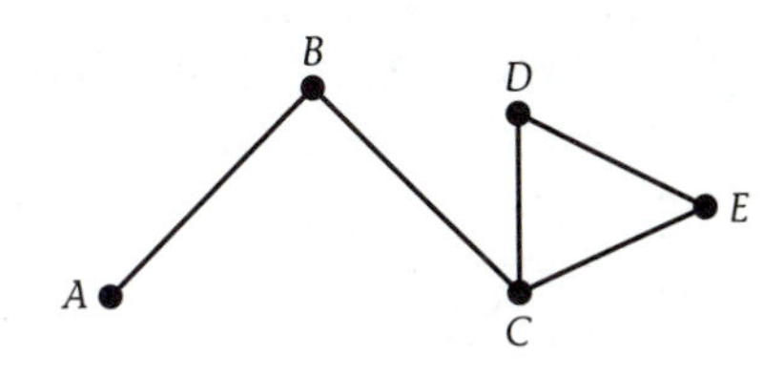

The above graph is a subgraph of which graph(s) at the top of the next page? Why?

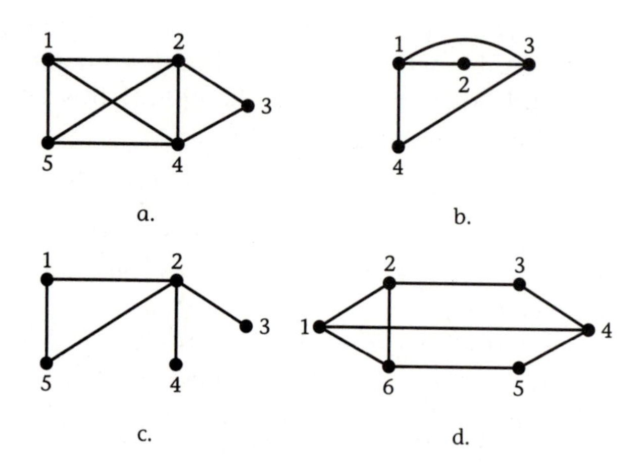

20. Is the graph below planar? Does it contain K_5 or $K_{3,3}$ as a subgraph?

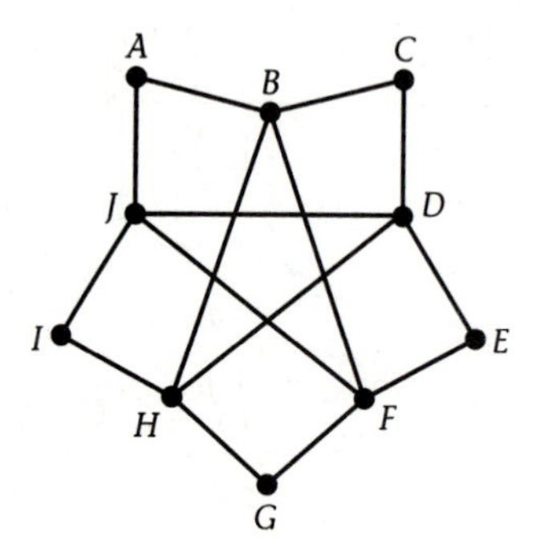

21. For many years mathematicians thought that all nonplanar graphs had K_5 or $K_{3,3}$ as subgraphs. Then nonplanar graphs such as the one in Exercise 20 were found. The graph in Exercise 20 is said to be an **extension** of a K_5 because it was formed by adding a vertex or vertices to the edges of the K_5 graph. Extensions of K_5 and $K_{3,3}$ are nonplanar.

    Redraw the graph in Exercise 20 so that is shows that it is an extension of a K_5.

The Traveling Salesperson Problem

Explore This

Suppose you are a salesperson who lives in St. Louis. Once a week you have to travel to Minneapolis, Chicago, and New Orleans and return home to St. Louis.

The graph in Figure 5.7 represents the trips that are available to you. The edges of the graph are labeled with the cost of each possible trip. For instance, the cost of making a trip from Chicago to New Or-

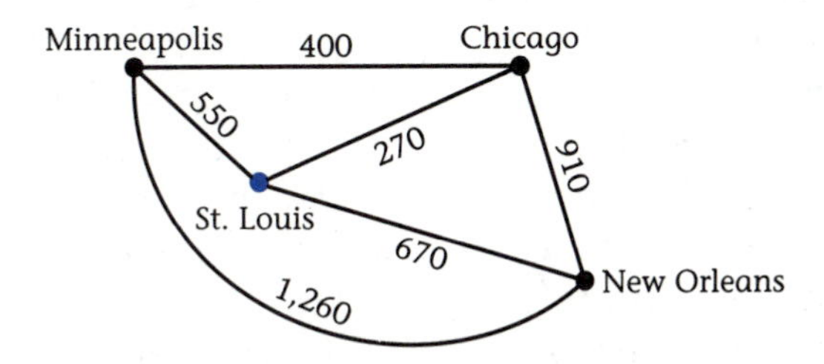

Figure 5.7 Graph of five cities with the costs of traveling between them.

leans is $910. When each edge of a graph is assigned a number (weight), the graph is called a **weighted graph.**

Since you own your own business, it is important to minimize travel costs. To save money, try to find the least expensive route that begins in St. Louis, visits each of the other cities exactly once, and returns to St. Louis.

The problem of finding a Hamiltonian circuit of the minimum total weight for a graph is referred to as a **traveling salesperson problem** (TSP). The numbers associated with the edges frequently represent such things as distance, cost, or time. Problems of this type are often found in the communication and travel industries.

One way to solve the problem of finding the minimum total weight in Figure 5.7 is to list every possible circuit, along with its cost. Using a

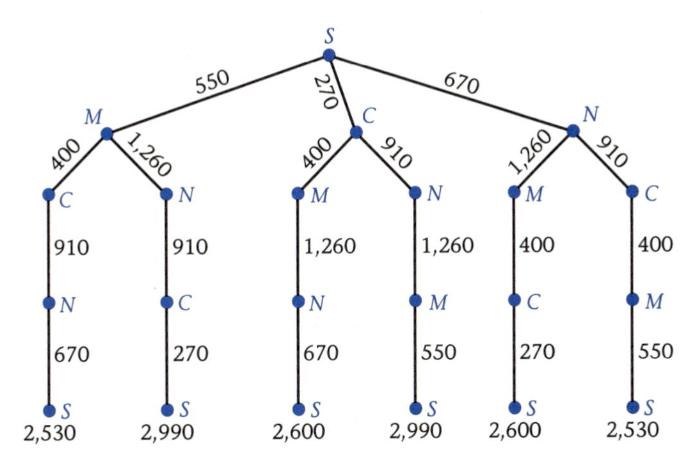

Figure 5.8 Tree diagram of every possible circuit from St. Louis to each of the other cities and back to St. Louis.

tree diagram like that in Figure 5.8 is helpful in organizing the possibilities.

Inspection of all possible routes shows that the optimal solution is the circuit *S,M,C,N,S* or the circuit in reverse order, *S,N,C,M,S*. Solving the problem for four vertices is not too difficult or time-consuming because only six possibilities need to be considered, but as the number of vertices increases so does the number of possible circuits.

As the number of vertices increases, checking all possibilities soon becomes impractical, if not impossible. Even with the help of a computer that can do computations at the rate of 1 billion per second, it would take more than 19 million years to compute the weights of every circuit for a graph with 25 vertices!

Is it possible that a faster, more efficient algorithm exists? You might try, for example, beginning at a vertex, looking for the nearest vertex, moving to it, and so on as you complete the circuit.

To explore this method for the graph in Figure 5.7 on page 217, begin at *S*, move to the nearest neighboring vertex, then to the nearest vertex not yet visited, and return to S when all of the other cities have been visited. This method of solution to the problem is known as the **nearest-neighbor algorithm.** In this case the minimum weight is $270 + 400 + 1{,}260 + 670 = 2{,}600$. Although the solution was reached quickly, notice that it is not the best one. A method such as this one that produces a quick and reasonably close-to-optimal solution is known as a **heuristic method.**

The choice of method now becomes a trade-off. The first method guarantees the best route but is prohibitively slow. The nearest-neighbor method is quick but does not necessarily produce the optimal solution. In the case of the traveling salesperson problem, there is no known computationally efficient method of solving all problems of this type.

Even though efficient solutions for special TSP problems have been known for some time, it was thought that an efficient solution to the general problem would never be found. But then in February 1991, two mathematicians, Donald Miller and Joseph Penky, found a reasonable algorithm that will handle an additional type of traveling salesperson problem. It is now believed that additional efficient algorithms will be forthcoming. Solutions of this type are of great interest to business and industry, as they could save millions of dollars by using the algorithm and implementing its results.

NEWS CLIP

Mathematicians Find New Key to Old Puzzle

By Jerry E. Bishop
THE WALL STREET JOURNAL, February 15, 1991

Two mathematicians at Du Pont Co. and Purdue University reported a solution to a classic and previously unsolvable mathematical puzzle, the so-called traveling salesman problem.

The new method, or algorithm, for solving the problem may have such industrial uses as charting the most efficient and least costly schedule of chemical processes or scheduling steps in the overhaul of jet engines. It answers different questions from those involved in the airline and telephone routing problems solved in 1984 by a mathematician at AT&T Bell Laboratories.

Exercises

In Exercises 1 through 4,

a. Construct a tree diagram showing all possible circuits that begin at *S*, visit each vertex of the graph exactly once, and end at *S*.
b. Find the total weight of each path.
c. Identify the shortest path.
d. Use the nearest-neighbor algorithm to find the shortest path.
e. Does the nearest-neighbor algorithm give the shortest path possible?

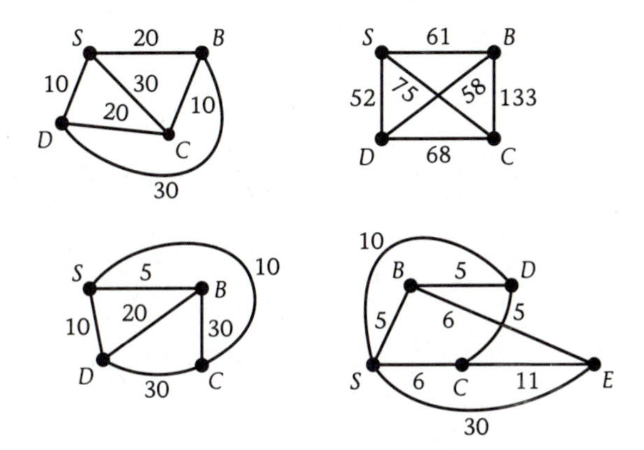

5. In a graph with 10 vertices, 9! or $9 \cdot 8 \cdot 7 \cdot 6 \cdot 5 \cdot 4 \cdot 3 \cdot 2 \cdot 1$ possible Hamiltonian circuits exist if the beginning vertex is known. Assume that a computer can perform calculations at the rate of 1 million per second. About how long will it take the computer to check all possibilities? What if the graph had 15 vertices or 14! possible circuits?

6. Give two examples of situations in which a solution to the traveling salesperson problem would be beneficial.

NEWS CLIP

VERY SMART CARS

Helping to Get There Easily

By Adam Z. Horvath *Newsday*, June 4, 1992

The next time you get lost traveling to an unfamiliar city, wouldn't it be great if you could just ask your car where you are? And then have it tell you how to get where you want to go?

That idea may sound futuristic, but in Orlando, Fla., there are 100 cars that were designed to guide people to their

destinations—and people are renting them already, at no extra charge, as part of the first large-scale experiment to test them.

Not only that, but the cars are supposed to guide their drivers out of traffic jams, and even offer information about the best restaurants and sights to visit.

The system is called "Travtek," short for "travel technology," and driving it is sort of like combining a car with a real-life video game.

As you drive, a strange-sounding voice—sort of like the computer voice in the movie "War Games"—comes from the car's dashboard. It might say, "Turn left in nine-tenths miles at East Colonial Drive." A computer screen next to the steering wheel displays moving street maps, with an arrow marking the car's location. It can also give you lists of destinations, and step-by-step directions along the way.

LESSON 5.3

Shortest Route Problems

The traveling salesperson problem asks that a Hamiltonian circuit of least total weight be found for a graph. What if you didn't need to visit every vertex in the graph and return to the starting vertex but instead needed only to find the shortest route from one vertex to another? Does an efficient method of solving this type of problem exist? The answer is yes, and the algorithm used in finding the optimal path is attributed to E. W. Dijkstra.

Edsger W. Dijkstra, born in the Netherlands in 1930, is considered one of the original theorists of modern computer science. He first published his algorithm in a German mathematics journal in 1959. The algorithm that was named after him makes it possible to find the shortest path from a given vertex of a graph to any other vertex in that graph.

The following algorithm is a modification of Dijkstra's algorithm:

Shortest Path Algorithm

1. Label the starting vertex *S* and circle it. Examine all edges that have *S* as an endpoint. Darken the edge with the shortest length and circle the vertex at the other endpoint of the darkened edge.

2. Examine all uncircled vertices that are adjacent to the circled vertices in the graph.

3. Using only circled vertices and darkened edges between the vertices that are circled, find the lengths of all paths from S to each vertex being examined. Choose the vertex and edge that yield the shortest path. Circle this vertex and darken this edge. Ties are broken arbitrarily.

4. Repeat Steps 2 and 3 until all vertices are circled. The darkened edges of the graph form the shortest routes from S to every other vertex in the graph.

Example

Use the shortest path algorithm to find the shortest path from A to$F$ in the graph.

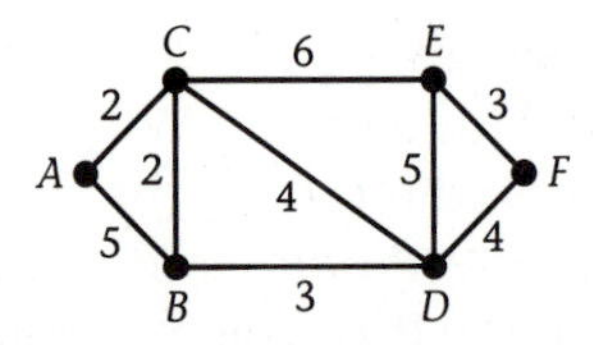

To find the solution to this problem, begin by circling vertex A and labeling it S. Examine all vertices that are adjacent to S.

	Adjacent Vertices	Path from S to Vertex	Length of Path
Adjacent to S	B	SB	5
	C	SC	2
1. Circle C, darken edge SC.			
Adjacent to S	B	SB	5
Adjacent to C	B	SCB	4
	E	SCE	8
	D	SCD	6

Continued

	Adjacent Vertices	Path from *S* to Vertex	Length of Path
2. Circle *B*, darken edge *CB*.			
Adjacent to *C*	*E*	*SCE*	8
	D	*SCD*	6
Adjacent to *B*	*D*	*SCBD*	7
3. Circle *D*, darken edge *CD*.			
Adjacent to *C*	*E*	*SCE*	8
Adjacent to *D*	*E*	*SCDE*	11
	F	*SCDF*	10
4. Circle *E*, darken edge *CE*.			
Adjacent to *E*	*F*	*SCEF*	11
Adjacent to *D*	*F*	*SCDF*	10
5. Circle *F*, darken edge *DF*.			

The shortest route from *A* to *F* is *A,C,D,F,* and the length is 10. The darkened edges also show the shortest routes from *A* to the other vertices in the graph.

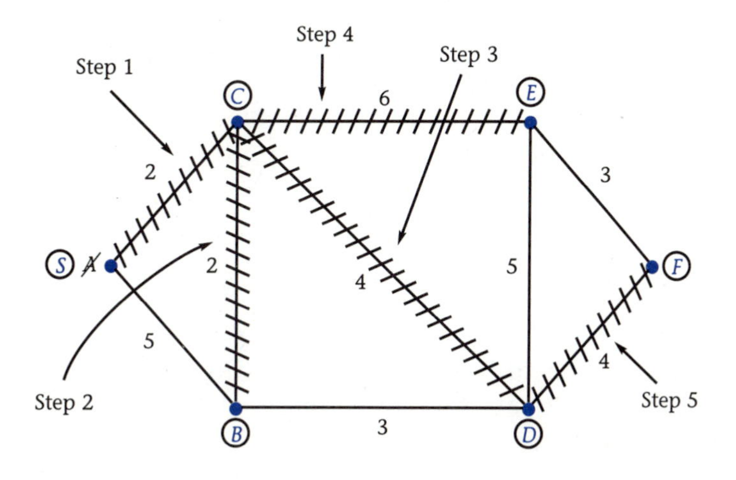

Exercises

1.

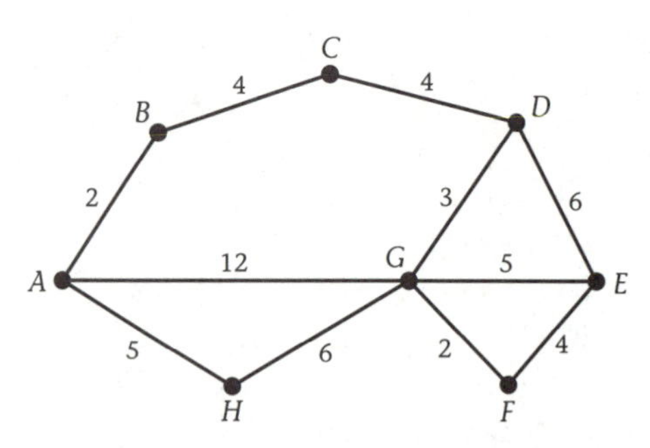

Julian began using the shortest path algorithm to find the shortest route from *A* to *E* for the graph above. The work that he was able to complete before he had to stop is shown below.

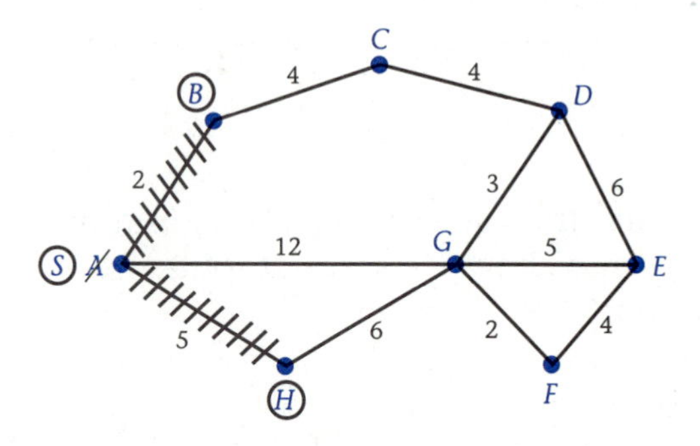

1. *SB* – ②
   SC – 5 Circle *B*, darken *SB*.
   SG – 12
2. *SBC* – 6
   SG – 12 Circle *H*, darken *SH*.
   SH – ⑤
3. *SBC* – ?
   SG – ? Circle?, darken?
   SHG – ?

Fill in the missing distances, vertex, and edge in Step 3. Then complete Julian's problem of using the shortest path algorithm to find the shortest route from *A* to *E*.

2. Use the shortest path algorithm to find the shortest route from A to F.

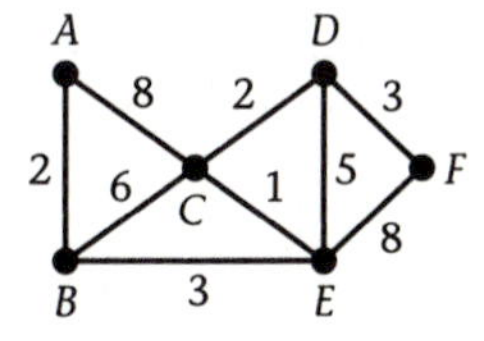

3. When might it not be necessary to repeat the procedure in the algorithm until all of the vertices are circled?

4. Use the shortest path algorithm to determine the shortest distance from S to each of the other vertices.

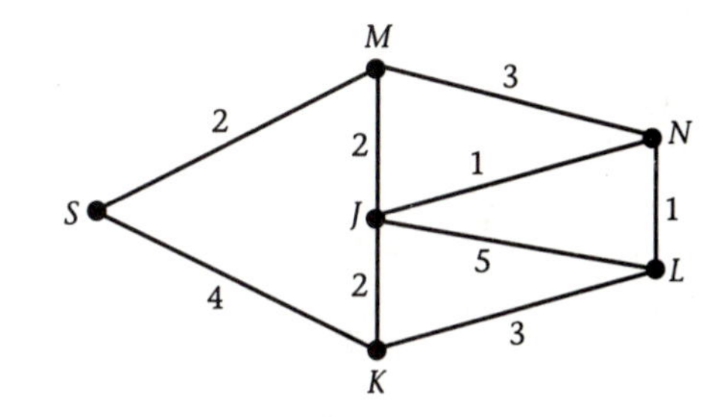

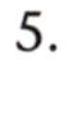
5.

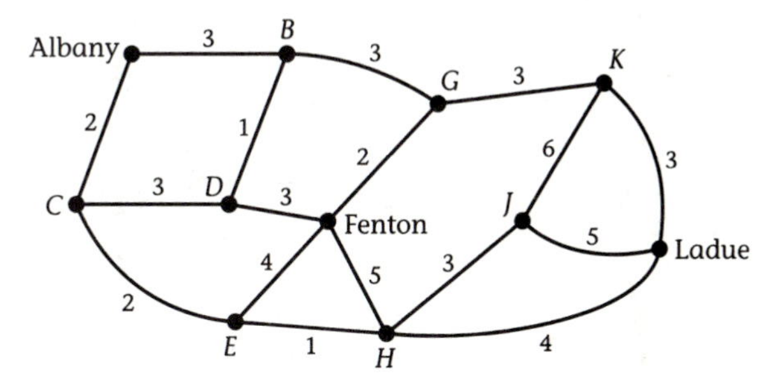

a. Use the shortest path algorithm to find the shortest route from Albany to Ladue in the graph above.
b. It is necessary to travel from Albany to Fenton to deliver a package and then to continue from there to Ladue. Find the shortest route for this trip. Explain how you arrived at the solution.

6. In the algorithm that you have been using, each time you examined all of the uncircled vertices that were adjacent to the circled ones, you had to recalculate the lengths of the paths from the starting vertex. How might the efficiency of the algorithm be increased by modifying it to avoid such recalculation?

7. The shortest path algorithm can be applied to digraphs if slight modifications are made in the algorithm. Make the appropriate changes and try the revised algorithm on the digraph below to find the shortest route from *A* to *F*.

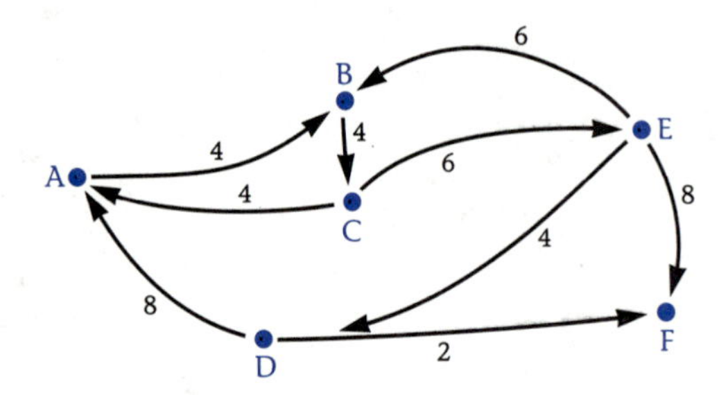

8. Mail Package, Inc., ships from certain cities in the United States to others. Below is a table of the company's shipping costs.

		To						
		Albany	Biloxi	Center	Denver	Evert	Fargo	Gale
	Albany	—	7	—	—	4	—	—
	Biloxi	—	—	—	—	—	—	6
	Center	2	—	—	—	2	—	—
From	Denver	—	—	1	—	—	—	—
	Evert	—	—	—	—	—	—	4
	Fargo	—	—	—	—	3	—	2
	Gale	1	6	—	—	—	1	—

Since a package can't be shipped directly from Denver to Biloxi, construct a digraph to represent the cost table, apply the shortest path algorithm, and find the least charge for shipping the package.

LESSON 5.4

Trees and Their Properties

In Lesson 5.2, a special type of graph called a tree was used to organize information and list all the possible routes for a traveling salesperson problem.

Tree diagrams have been used since ancient times, but it wasn't until the nineteenth century that their properties were studied in detail. In 1847, Gustav Kirchoff used trees in his study of electrical networks. Ten years later, Arthur Cayley used trees in his investigation of certain chemical compounds. Today trees are one of the most useful structures in discrete mathematics and are invaluable to computer scientists. Many of the sorts and searches that are done by computers can be modeled with trees.

Before exploring some of the properties and applications of this type of graph, it is necessary to define a tree. Recall that a **cycle** in a graph is any path that begins and ends at the same vertex and no other vertex is repeated. A **tree** is a connected graph with no cycles.

Example

Which of the graphs below are trees? Why?

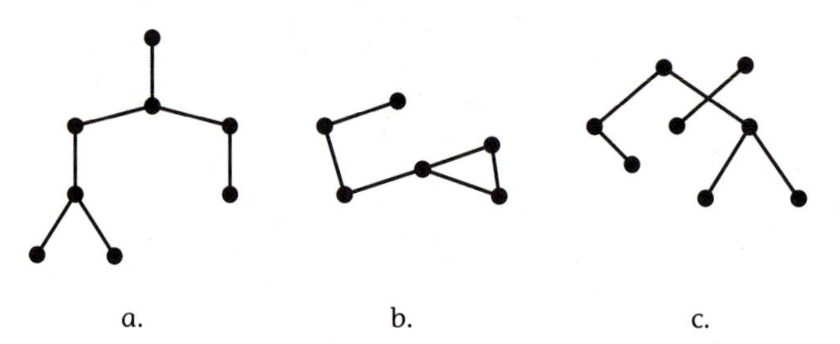

Figure 5.9 Possible trees.

To answer these questions, note that Figure 5.9a is a tree because it is a connected graph with no cycles. Figure 5.9b is not a tree because it has a cycle, and Figure 5.9c is not a tree because it is not connected.

Trees have many applications in the real world. They can be used to list and count possibilities, as was done in the traveling salesperson problems. They can also be used to model family genealogical histories (Figure 5.10), to structure decision-making processes (Figure 5.11), and to represent chemical compounds (Figure 5.12).

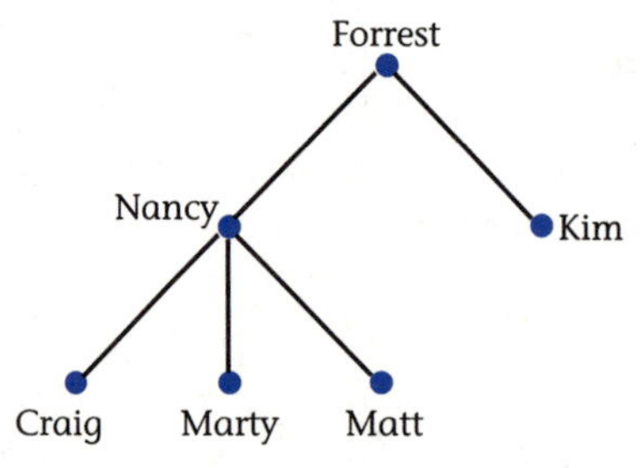

Figure 5.10 Family tree.

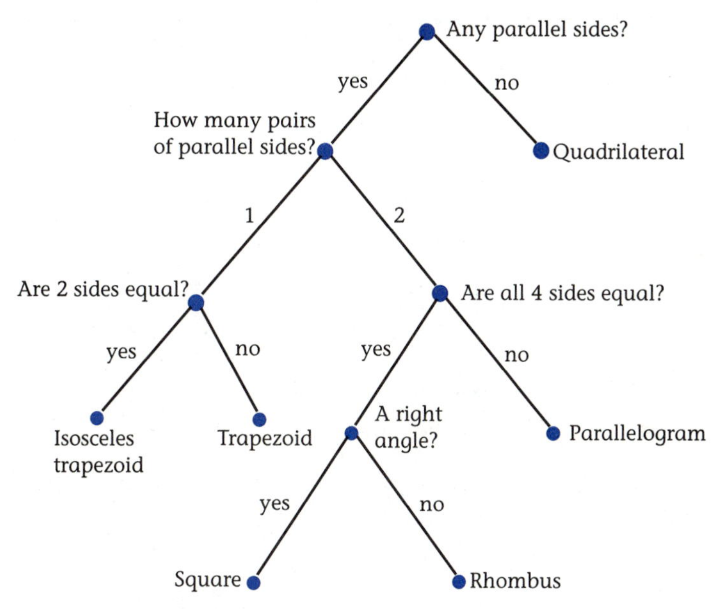

Figure 5.11 Sorting quadrilaterals.

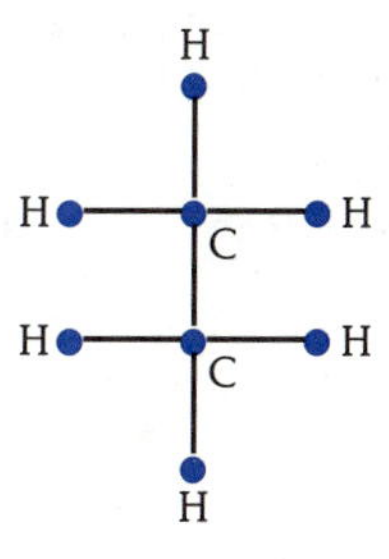

Figure 5.12 C_2H_6 (ethane).

Exercises

1. List all the cycles in the graph below.

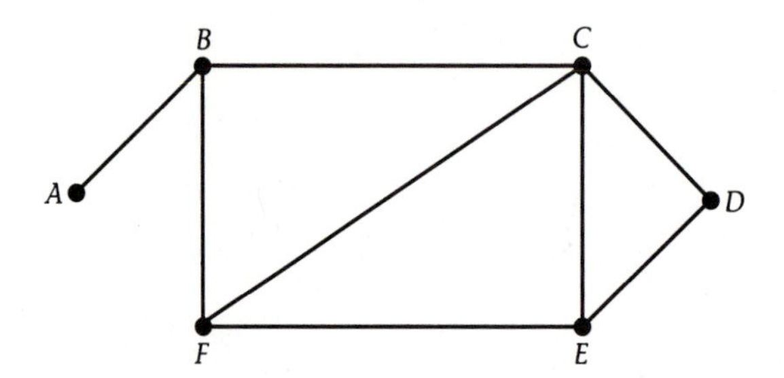

2. Determine whether the following graphs are trees. If the figure is not a tree, state why.

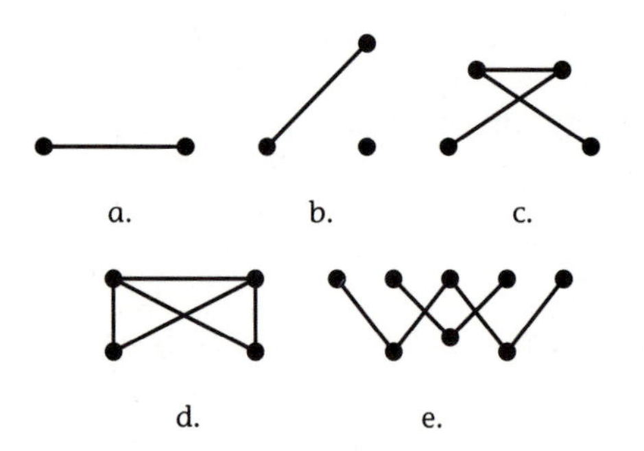

3. There is only one way to draw a tree with two vertices and only one way of drawing a tree with three vertices, but there are two distinct trees that can be formed from four vertices. Draw all of the trees that are possible for five vertices. For six vertices.

Number of vertices	Tree diagrams
2	
3	
4	 

4. Complete the table for a tree with the indicated number of vertices.

Number of Vertices	Number of Edges
1	0
2	1
3	
4	
N	

How many edges does a tree with 19 vertices have? How many vertices does a tree with 15 edges have?

5. What happens to a tree if an edge is removed from it?

6. Draw a tree with six vertices that has exactly three vertices of degree 1.

7. Complete the table for a tree with the indicated number of vertices.

Number of Vertices	Sum of the Degrees of the Vertices	Recurrence Relation
1	0	$S_1 = 0$
2	2	$S_2 = S_1 + 2$
3	4	$S_3 =$ ____
4	____	____
5	____	____
6	____	____

Write a recurrence relation that expresses the relationship between the sum of the degrees of the vertices of a tree with n vertices and the sum of the degrees of the vertices of a tree with $n - 1$ vertices.

8. Explain why the sum of the degrees of the vertices of a tree with n vertices is equal to twice the number of vertices minus 2. (Hint: Draw several different trees.)

9. Much of the terminology connected with trees is botanical in nature. For instance, a graph consisting of a set of trees is called a **forest,** and a vertex of degree 1 in a tree is called a **leaf.** Draw a forest of three trees. Circle the leaves of your graph.

10. In a hierarchical tree as in Figures 5.10 on page 230 and 5.11 on page 231, it is natural to speak of the **root** of the tree. A tree is rooted when all of the edges are directed away from the chosen root. In Figures 5.10 and 5.11, the edges are directed downward. Draw a family tree for your family, beginning with one of your grandfathers as the root of the tree. What do the leaves of your tree have in common?

11. Refer to the tree in Figure 5.11 on page 231. This graph is called a **decision tree,** and the leaves represent the final outcomes of the different decisions. Based on this tree, what is the name of a quadrilateral with two pairs of parallel sides, four sides equal, and no right angles?

12. In the following graph, find two different subgraphs that are trees.

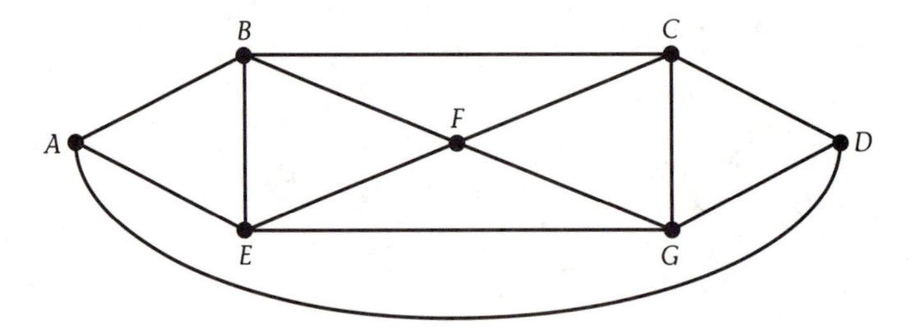

13. The children in a certain neighborhood want to communicate with one another via their very own communication network. To avoid the expense of connecting each house with every other house, a system needs to be devised that will use as few lines as possible yet allow messages to get to each person. Create such a network for the houses at the top of the following page.

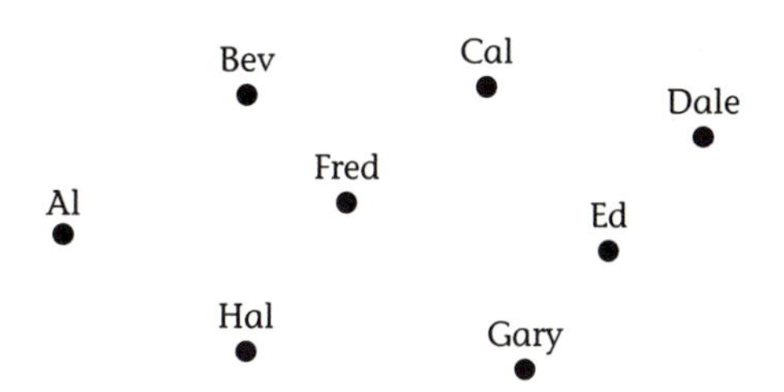

14. Clock solitaire is a card game in which the 52 cards are dealt face down into 13 piles that correspond to the 12 numbers on the face of a clock. Clock positions 11 and 12 are identified with the jack and queen, and the thirteenth or center pile is identified with the king (see Figure 5.13).

    The game is played by first turning over the top card of the king pile and putting it face up under the pile that corresponds to its value on the clock. Now turn up the top card of the pile under which you just put the card. Continue in this manner. The game is won when you have turned up all 52 cards. If a fourth king is turned up before this happens, play cannot continue, and the game is lost.

    In his book *Fundamental Algorithms* mathematician Donald E. Knuth noted two interesting things about this game. One is that the probability of winning this game is 1/13. The other is that by checking the bottom card of the 12 clock piles, you can determine whether or not you will win the game.

    To determine whether you can win a game, turn over the bottom card of each pile except for the king pile (see Figure 5.14). Draw an edge from each of the 12 cards to the clock position that

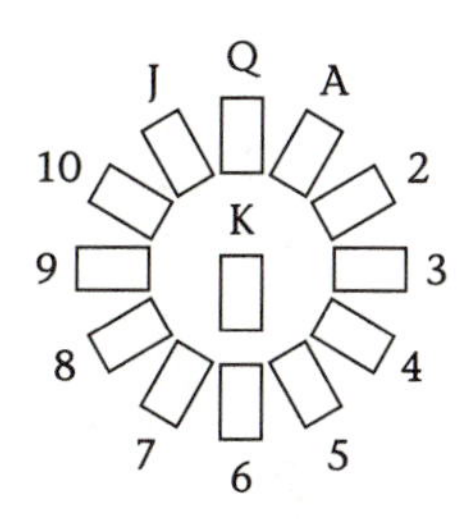

Figure 5.13 Diagram of the thirteen piles for the game of clock solitaire.

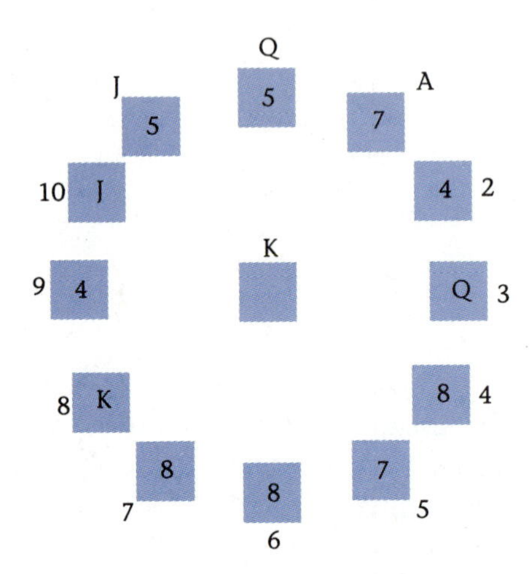

Figure 5.14 One possible configuration of the bottom cards in the game.

corresponds to the card's numeric value (see Figure 5.15). Now redraw the graph with the vertices labeled 1, 2, . . . J, Q, K. You will win if the graph is a tree that includes all 13 piles (see Figure 5.16).

Give the game a try, but before you do, record the bottom 12 cards. Predict whether you will win or lose the game by drawing a graph. Notice that only the bottom 12 cards determine your success. The arrangement of the other 40 cards makes no difference.

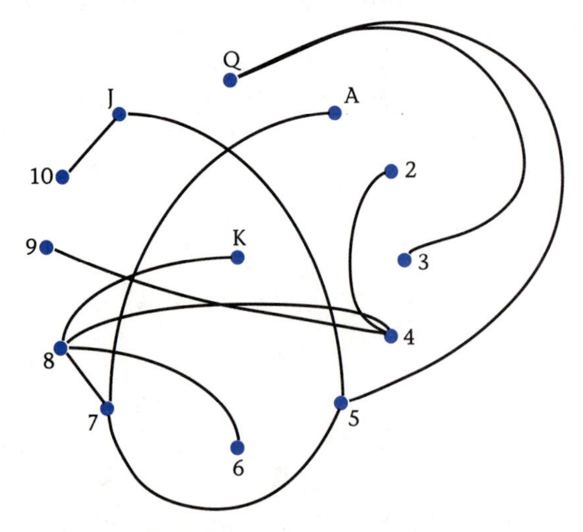

Figure 5.15 Graph showing the edges from the cards to the clock positions.

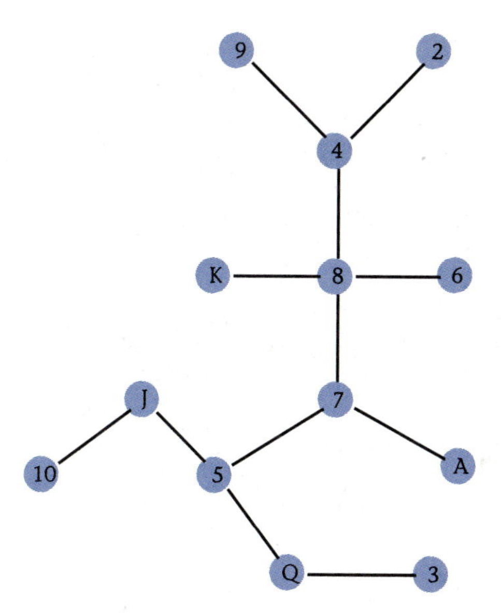

Figure 5.16 Figure showing that the graph in Figure 5.15 is a tree.

Minimum Spanning Trees

Explore This

In making earthquake preparedness plans, the St. Charles County government needs a design for repairing the county roads in case of an emergency. Figure 5.17 is a map of the towns in the county and the existing major roads between them. Devise a plan that repairs the least number of roads but keeps a route open between each pair of towns.

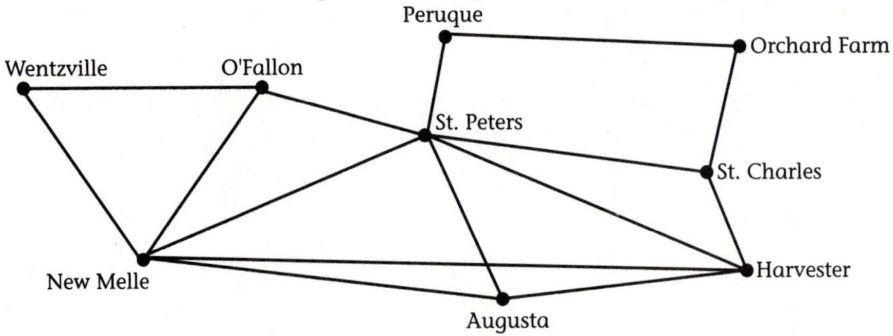

Figure 5.17 The towns in St. Charles County.

Examine your graph. Does it contain any cycles? What kind of a graph is it? If you found the minimum number of roads that contains every vertex in the map, you've found a spanning tree. A **spanning tree** of a connected graph G is a tree that is a subgraph of G and contains every vertex of G.

Compare your plan with the other plans developed in your class. Do they all contain the same edges? Do they all contain the same number of edges? It is possible for a graph to have many different spanning trees. If a graph is not connected, no spanning tree is possible.

One systematic way to find a spanning tree for a graph is to delete an edge from each cycle in the graph. Unfortunately, this is not an easy procedure for a very large graph. But there are other ways of finding a spanning tree for a graph if one exists. One method that can be easily adapted to computers is called the *breadth-first search algorithm.*

— Breadth-First Algorithm for Finding Spanning Trees —

1. Pick a starting vertex, S, and label it with a 0.

2. Find all vertices that are adjacent to S and label them with a 1.

3. For each vertex labeled with a 1, find an edge that connects it with the vertex labeled 0. Darken those edges.

4. Look for unlabeled vertices adjacent to those with the label 1 and label them 2. For each vertex labeled 2, find an edge that connects it with a vertex labeled 1. Darken that edge. If more than one edge exists, choose one arbitrarily.

5. Continue this process until there are no more unlabeled vertices adjacent to labeled ones. If not all of the vertices of the graph are labeled, then a spanning tree for the graph does not exist. If all vertices are labeled, the vertices and darkened edges are a spanning tree of the graph.

Applications of graphs often require that the edges have weights. When this is the case, it is usually desirable to find a spanning tree of minimum or maximum weight.

Return to the earthquake preparedness plan, and reconsider the problem when the distances between towns are added to the graph (see Figure 5.18). Refer back to your solution of the problem. Find the total

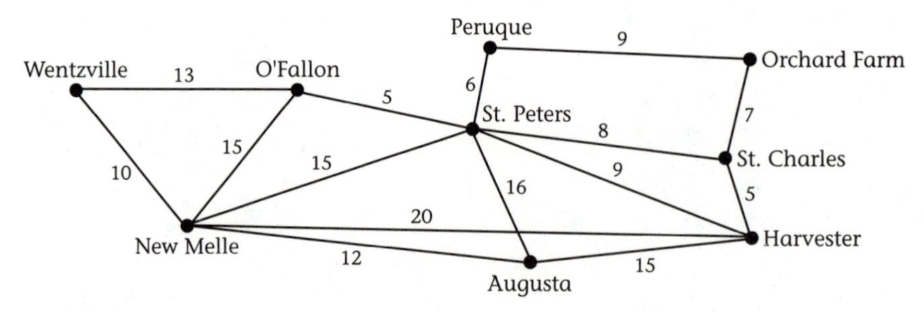

Figure 5.18 Map of St. Charles County with mileage shown.

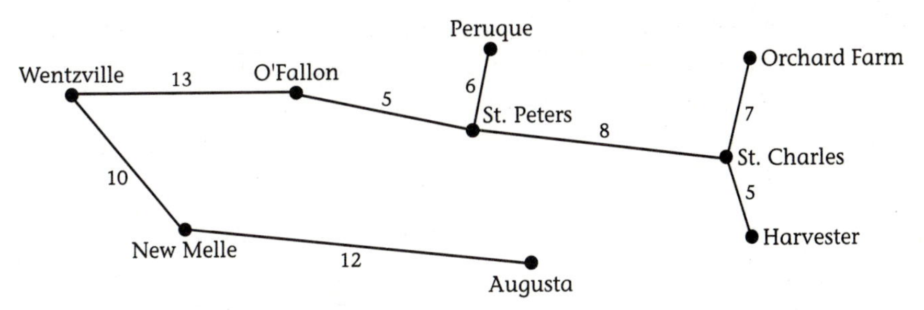

Figure 5.19 Spanning tree of minimum weight for the towns in St. Charles County.

number of miles of road that would need to be repaired in your plan. Compare this with the plans of others in your class. Which plan or plans yield the minimum number of miles?

For this particular problem the minimum possible number of miles of road is 66, and the spanning tree with that total weight is shown in Figure 5.19.

A spanning tree of least or minimal weight is called a **minimum spanning tree.** One algorithm for finding the minimum spanning tree for a graph is known as *Kruskal's algorithm.*

Kruskal's Minimum Spanning Tree Algorithm

1. Examine the graph. If it is not connected, there will be no minimum spanning tree.

2. List the edges in order from shortest to longest. Ties are broken arbitrarily.

3. Darken the first edge on the list.

4. Select the next edge on the list. If it does not form a cycle with the darkened edges, darken it.

5. For a graph with n vertices, continue Step 4 until $n - 1$ edges of the graph have been darkened. The vertices and the darkened edges are a minimum spanning tree for the graph.

Example

Use Kruskal's algorithm to find a minimum spanning tree for the graph below.

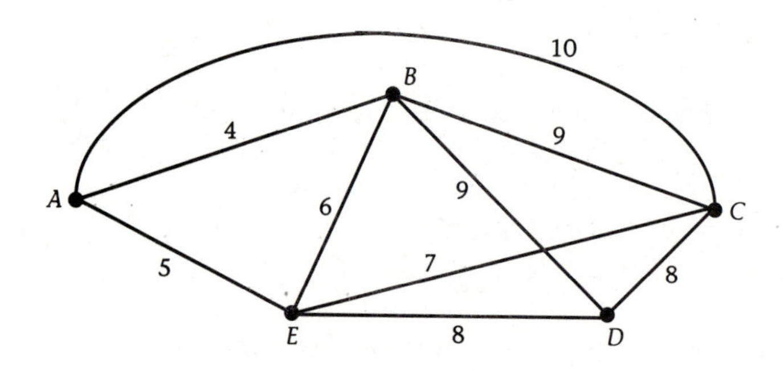

There are five vertices in the graph, so four edges must be chosen. List the edges from shortest to longest. First on the list is $AB(4)$. Darken it. Then darken $AE(5)$. The next shortest edge is BE, but if picked, it will form a cycle. So pick $EC(7)$. For the last edge there are two edges of length 8. Either CD or ED can be darkened. The darkened edges of the graph below are one of the minimum spanning trees of the graph. It has a minimal weight of $4 + 5 + 7 + 8 = 24$.

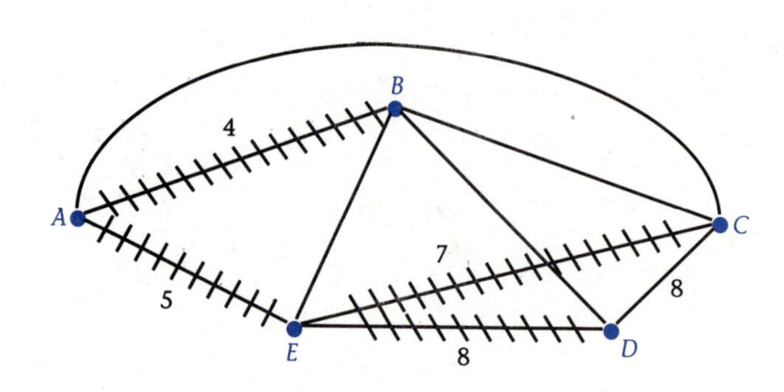

Exercises

In Exercises 1 through 4, find a spanning tree for each graph if one exists.

1.

2.

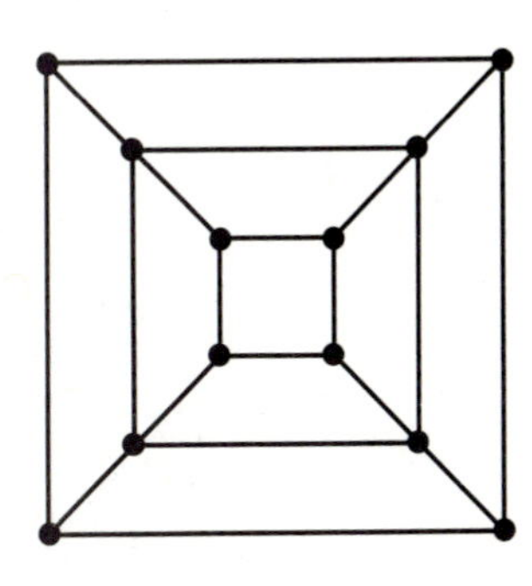

3.

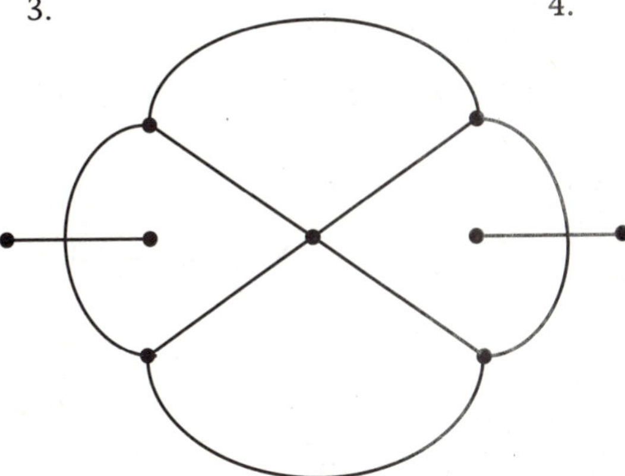

4.

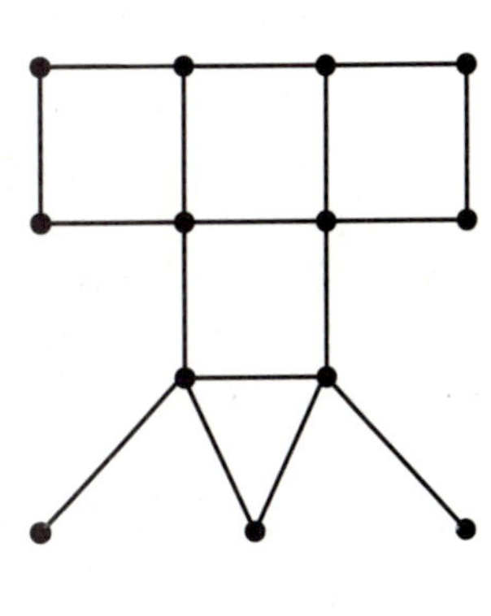

5. Draw a spanning tree for a K_4 graph.

6.

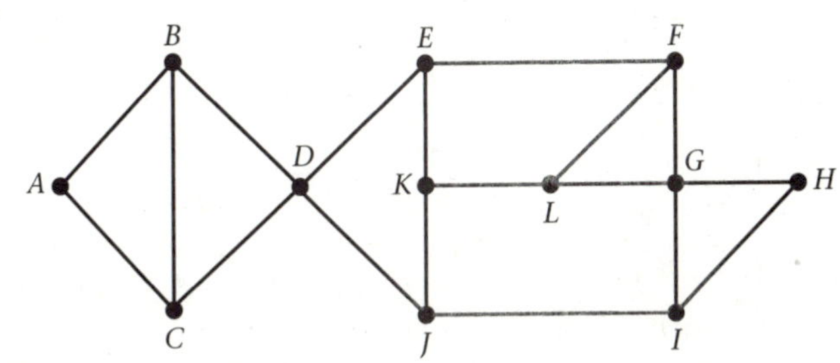

Sid started using the breadth-first search algorithm to try to find a spanning tree for the graph above. He began with vertex *A*, labeled it with a 0, and labeled *B* and *C* with 1s. He then darkened edges *AB* and *AC*, looked for vertices adjacent to the 1s and selected vertex *D*. He labeled it with a 2 and darkened edge *BD*.

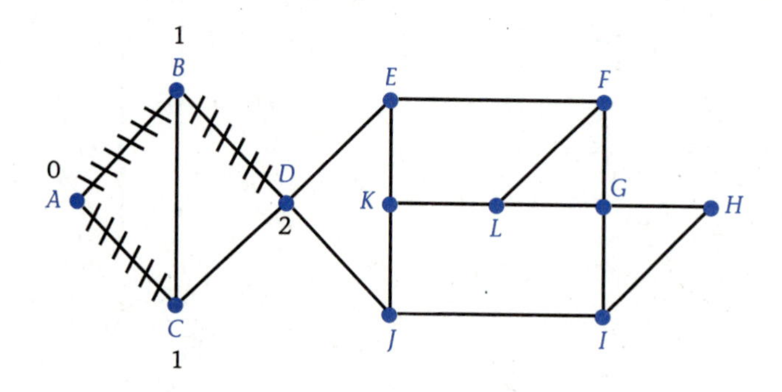

a. Could Sid have darkened *CD* instead of *BD*?

Complete the search for Sid by answering the following questions.

b. Which vertices will receive 3s for labels?
c. Which edges will subsequently be darkened?
d. Three vertices should be labeled 4. Which ones? Darken the appropriate edges. Your graph could look like the following:

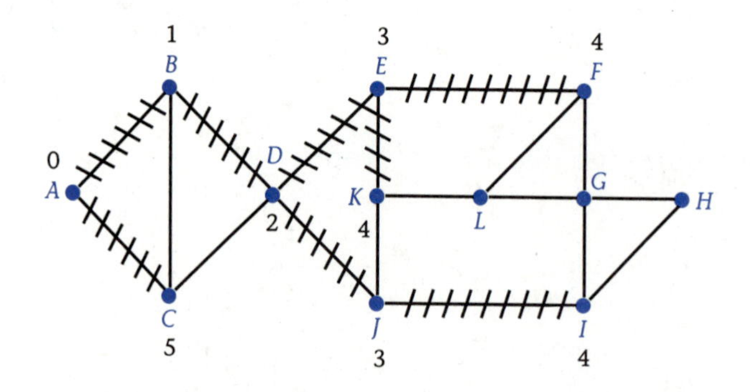

Continue the algorithm until all vertices are labeled. Check your darkened edges to make sure they form a spanning tree.

7.

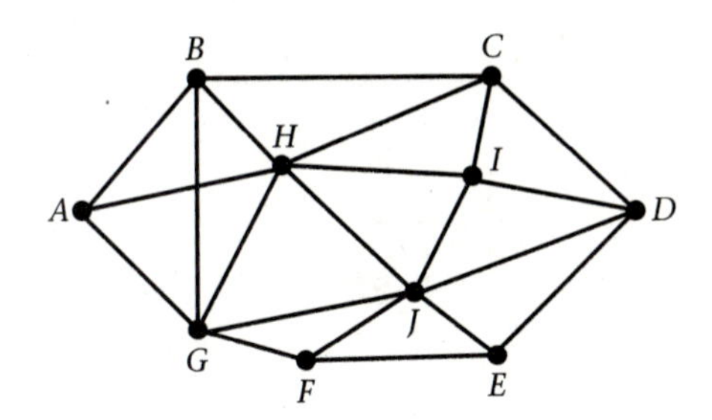

Use the breadth-first search algorithm to find a spanning tree for this graph. Begin at vertex A.

8. The breadth-first search algorithm can be applied to digraphs. How should the algorithm be modified so that it can be used with digraphs? Apply your modified breadth-first search algorithm to the digraph below.

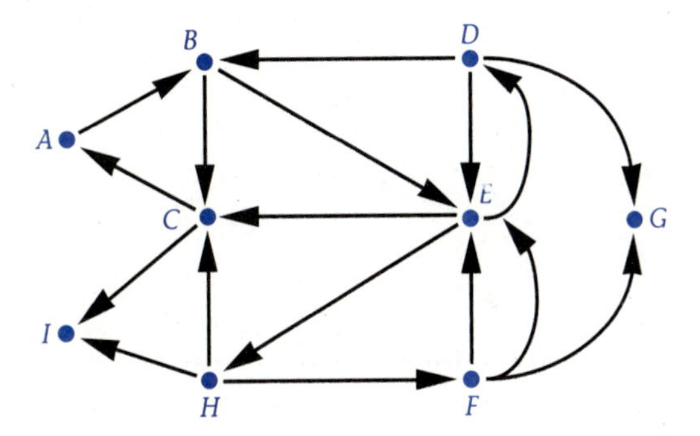

Use Kruskal's algorithm to find a minimum spanning tree for the graphs in Exercises 9 through 11. What is the minimal weight in each case?

9.

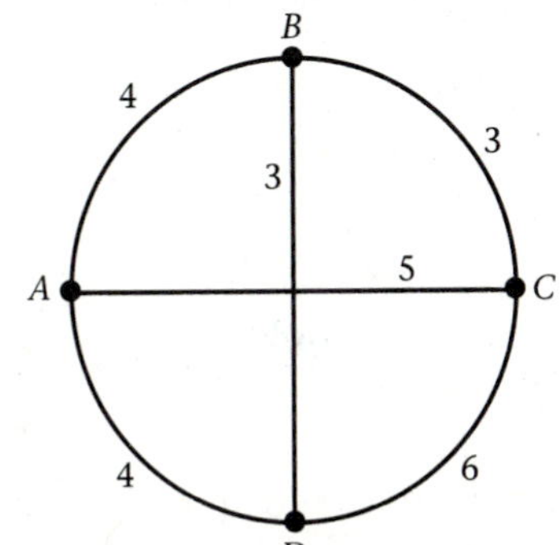

10.

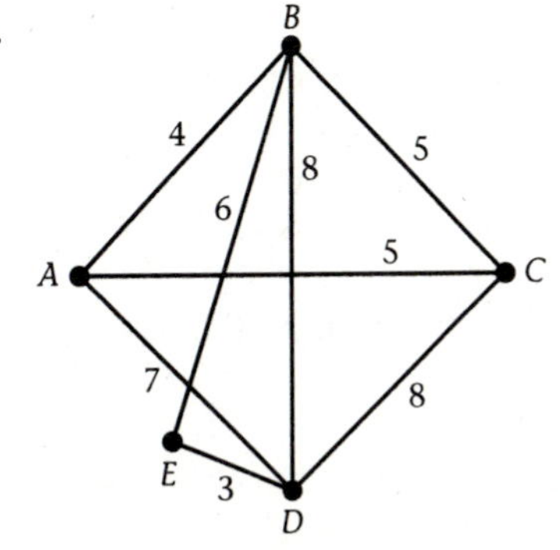

11.

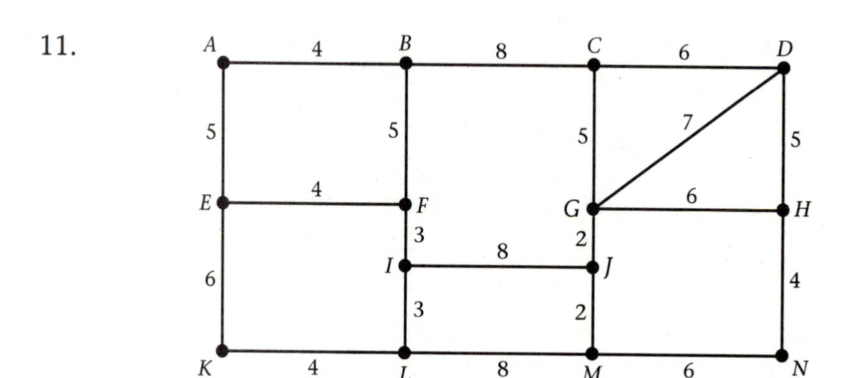

12. The computers in each of the offices at Pattonville High School need to be linked by cable. The map below shows the cost of each link in hundreds of dollars. What is the minimum cost of linking the five offices?

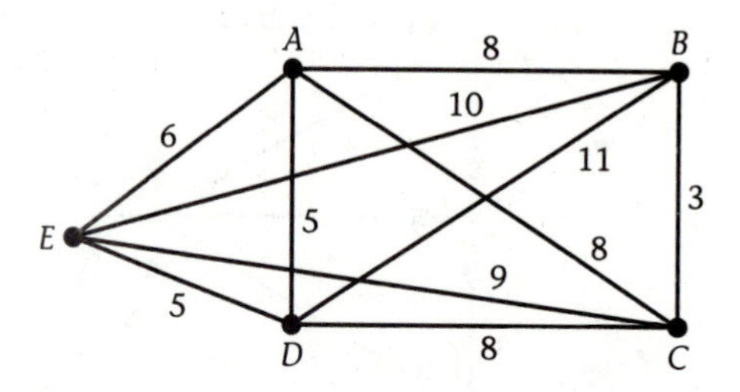

13. Suppose that the cable in Exercise 13 was installed by a disreputable firm that used only the most expensive links. What would be the maximum cost for the four links?

14. How might Kruskal's minimum spanning tree algorithm be modified to make it a maximum spanning tree algorithm?

Another algorithm for a minimum spanning tree is known as *Prim's algorithm.*

Prim's Minimum Spanning Tree Algorithm

1. Find the shortest edge of the graph. Darken it and circle its two vertices. Ties are broken arbitrarily.

2. Find the shortest remaining undarkened edge having one circled vertex and one uncircled vertex. Darken this edge and circle its uncircled vertex.

3. Repeat Step 2 until all vertices are circled.

15. Use Prim's algorithm to find the minimum spanning tree for Exercise 10 and then for Exercise 11.

16. When the shortest path algorithm is applied until all vertices of a graph are used, it yields a spanning tree of the graph. Is it always a minimum spanning tree?

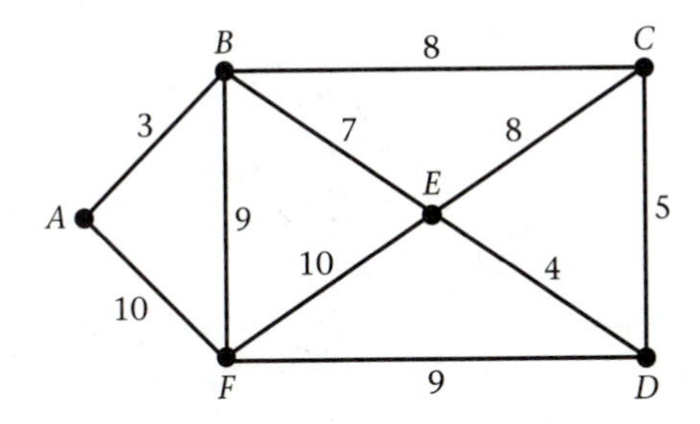

Check your answer to this question by doing the following:

a. Find a minimum spanning tree of the graph above using either Kruskal's or Prim's algorithm. What is the total weight of the minimum spanning tree?

b. Find the shortest route from *A* to each of the other vertices using the shortest path algorithm (Lesson 5.3 on page 223). Give the lengths of each of these routes.

c. Is the shortest route tree from *A* to each of the other vertices a minimum spanning tree of the graph?

17. Traveling salesperson problems, shortest route problems, and minimum spanning tree problems are often confused because each type of problem can be modeled with a weighted graph and each problem can be solved by finding a subgraph that includes all the vertices of the graph. Compare and contrast what each type of problem asks and when each type of problem is used.

Binary Trees, Expression Trees, and Traversals

Decision trees and family trees are two examples of a special kind of tree known as a *rooted tree*. Rooted trees are used to model situations that are multistaged or hierarchical in structure. Figure 5.20 shows an example of a rooted tree. Vertex R is called the *root* of the tree.

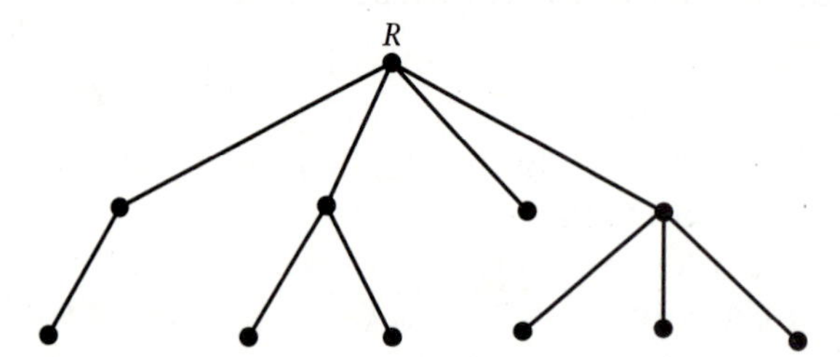

Figure 5.20 Rooted tree.

Example

A couple decides to have three children. What are the possible outcomes? See figure at the top of the following page. The eight possible outcomes are shown on the leaves of the tree.

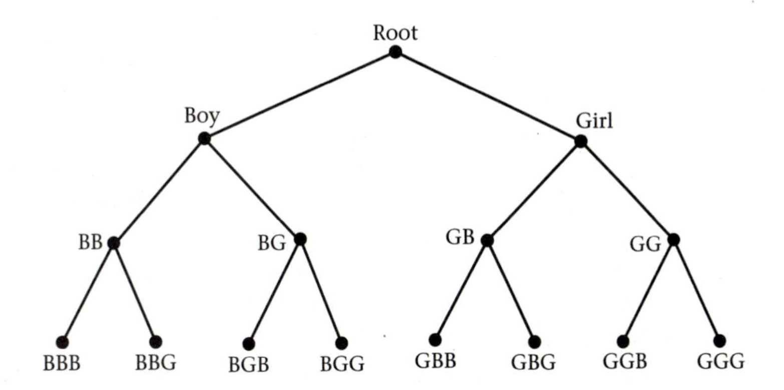

In a rooted tree, a vertex V is said to be at level K if there are K edges on the path from the root to V. The root is at level 0. If a vertex V has level 4, then any vertex adjacent to V at level 3 is called the **parent** of V, and an adjacent vertex of level 5 is called a **child** of V. A rooted tree in which each vertex has at most two children is called a **binary tree.**

Example

Which trees are binary trees? For those that are binary trees, name the parent of V and the children of V.

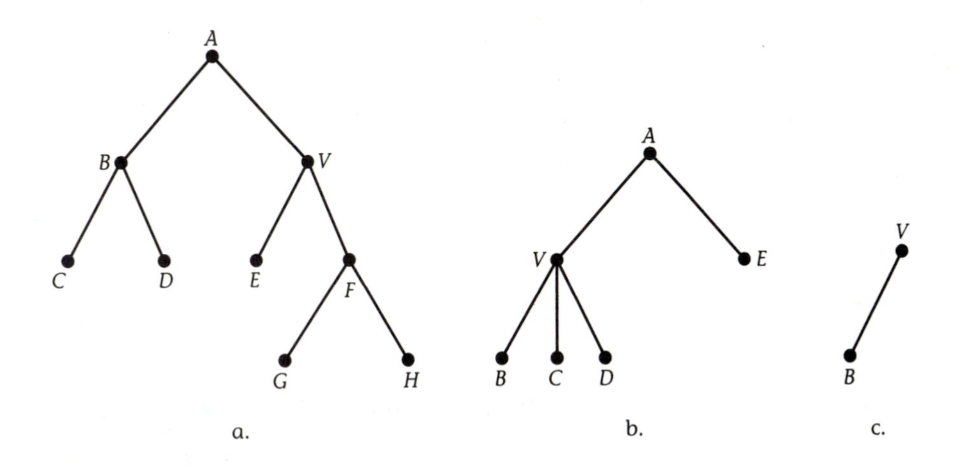

To solve this, note that a and c are binary trees but that b is not. In a, the parent of V is A and the children are E and F. In c, V has no parent and B is the only child.

Binary trees have applications in computer science, in which, for example, they are used to evaluate arithmetic expressions. When we write the expression (4 + 6) * 8 − 4/2, we understand how to evaluate it because of our familiarity with the order of operations for expressions. Unfortunately a computer cannot efficiently imitate our methods. When the expression is represented as a binary tree, the computer can quickly and efficiently evaluate it.

To represent the expression (4 + 6) * 8 − 4/2 as a binary tree, find the operation in the expression to be performed last. Make that operation the root of the tree. The right and left sides of this operation are the

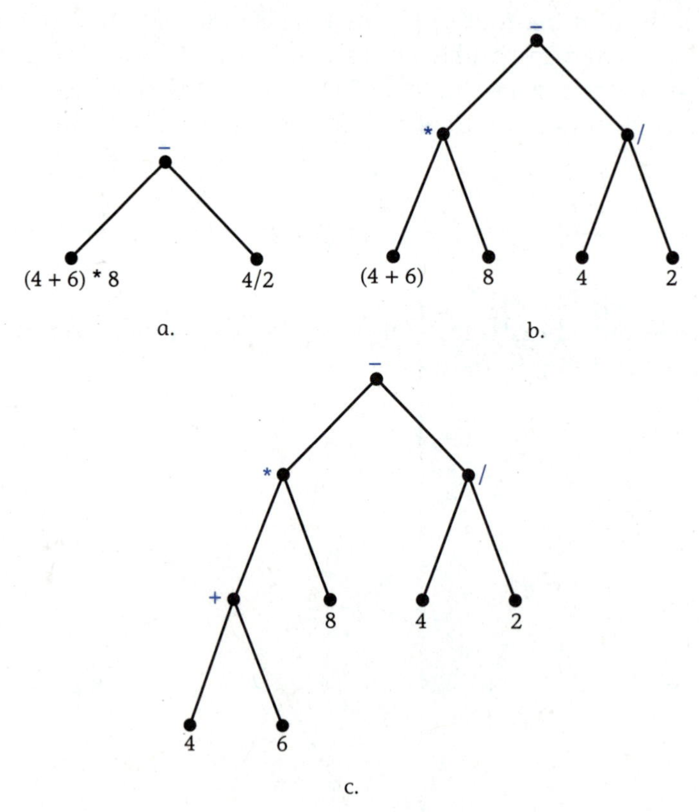

Figure 5.21 The process of placing the expression (4 + 6) * 8 − 4/2 on a tree.

children of the root. Continue this process of placing operations at each internal vertex and putting operands on the leaves until no expression that contains operations appears on the leaves.

The resulting binary tree in Figure 5.21 is called an **expression tree.**

Example

Represent $A/B + C * (D - E)$ as an expression tree.

The solution is as follows:

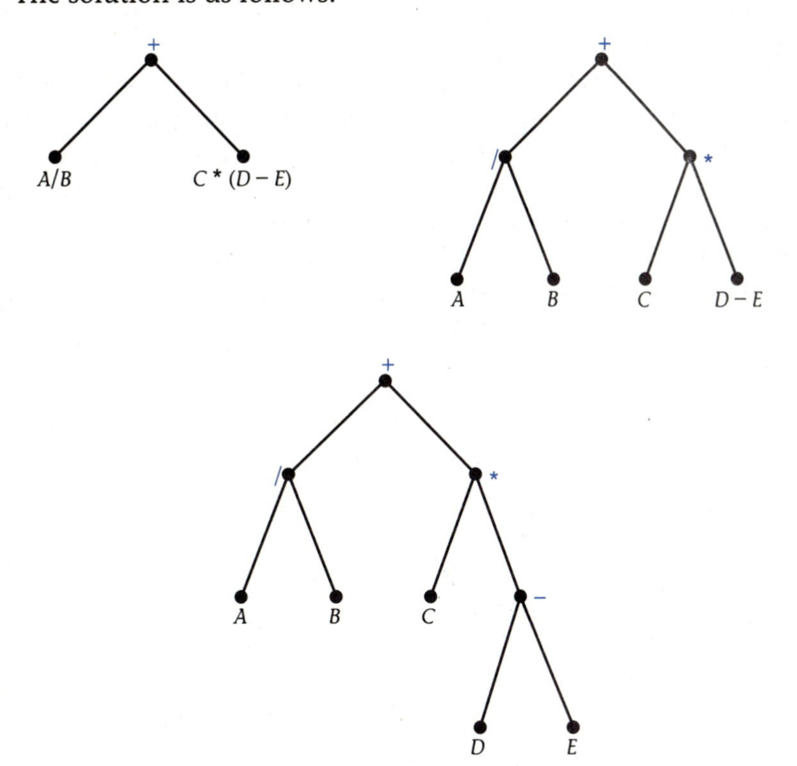

Once an expression is represented by a binary tree, the computer must have some systematic way of "looking at" the tree in order to find the value of the original expression. This organized procedure for obtain-

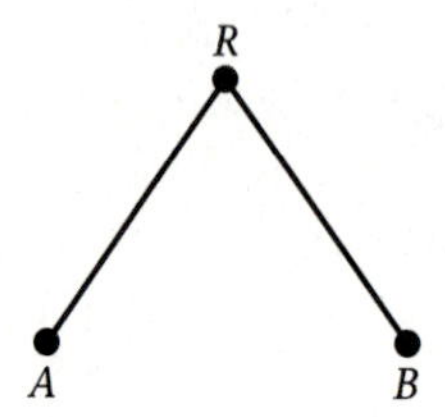

Figure 5.22 Postorder traversal *A, B, R.*

ing information by visiting each vertex of the tree exactly once is called a **traversal** of the graph.

There are several different types of traversals, including one called a **postorder traversal.** This traversal differs from other traversals in that it visits the left child of the tree first, then the right child, and finally the parent or root. (See Figure 5.22 above.)

To find the postorder listing of the vertices of the tree in Figure 5.23, begin by moving to the left subtree of *A* and doing a postorder traversal on the subtree. This requires that we branch to the left subtree of *B* and do a postorder traversal. Since the left subtree of *B* consists of only the vertex *D*, visit that vertex by numbering it with a 1. Now go to the right subtree of *B* and do a postorder traversal. Again this subtree consists of only one vertex. Visit *E* and number it with a 2. Since the left and right children of *B* have been traversed, visit *B*. Number it 3.

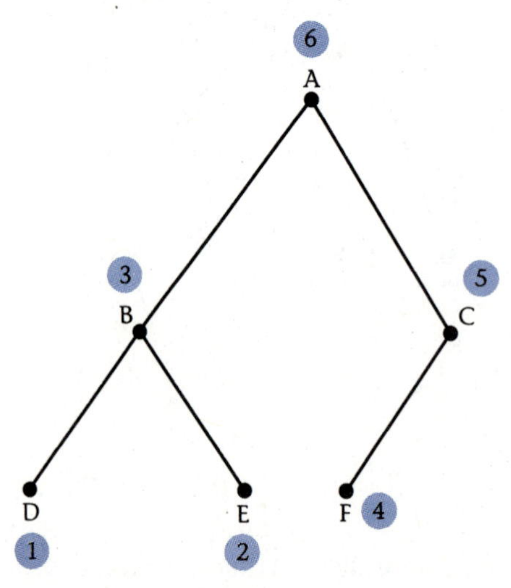

Figure 5.23 Postorder traversal of a binary tree.

The left subtree of A has been traversed. Now traverse the right subtree and do a postorder traversal of C by going to the left subtree of C. Visit F and number it 4. Since there is no right subtree of C, visit the root of the subtree, C, and number it 5. Since both subtrees of the root A have been visited, A can now be visited and numbered with a 6. The postorder traversal is complete and the postorder listing is *DEBFCA*.

Example

Give a postorder listing for the following expression tree:

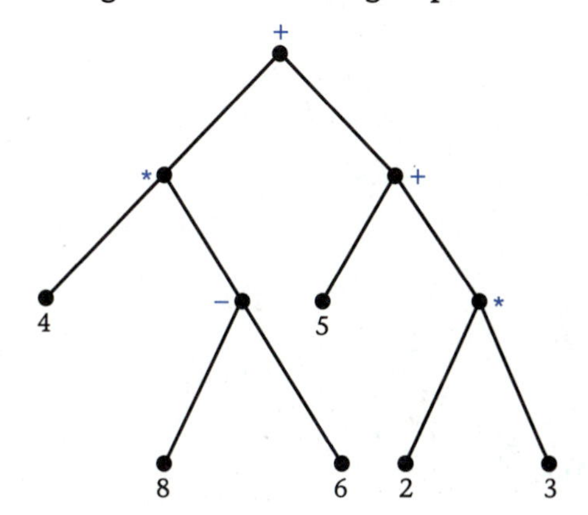

The following is the solution:

Left subtree of A

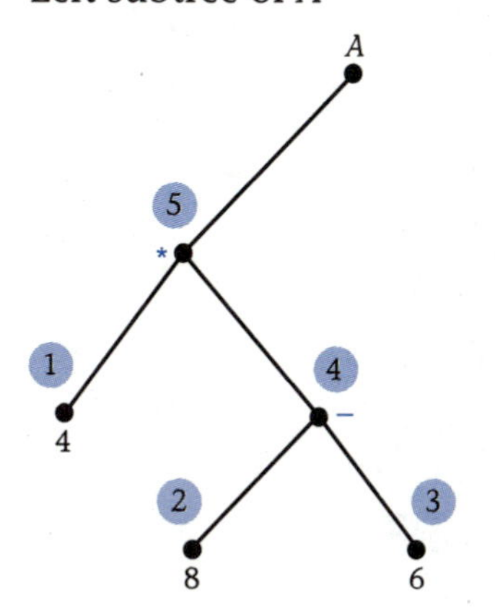

Right subtree of A

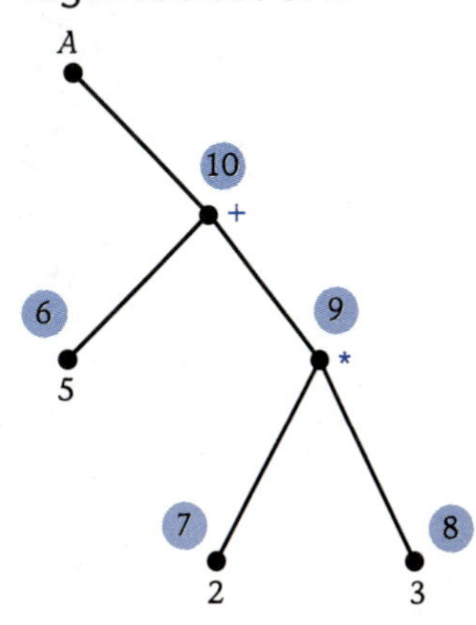

The root is visited last and 11 is assigned to *A*. The postorder listing is 4 8 6 − * 5 2 3 * + + .

The notation obtained by doing a postorder traversal is known as **reverse Polish notation.** The notation was so named because it was introduced by the Polish mathematician Lukasiewiez. It may look strange, but to owners of certain calculators this notation is familiar. Reverse Polish notation works well with calculators and computers because no parentheses are ever needed to indicate the order of operations.

So what is the value of the expression 4 8 6 − * 5 2 3 * + + ? To evaluate reverse Polish notation, scan the expression from the left until

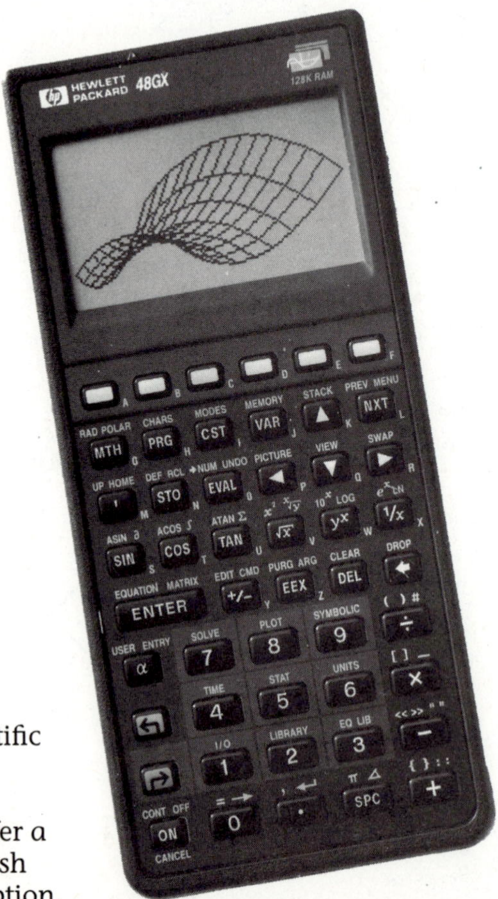

Some scientific calculators by Hewlett Packard offer a reverse Polish notation option.

you find two numbers followed by an operation sign. In this case, 8 6 −. Take the 8 and 6 and subtract. Substitute the result, 2, back in the expression. Repeat this until you have evaluated the expression.

4 (8 6 −) * 5 2 3 * + +

(4 2 *) 5 2 3 * + +

8 5 (2 3 *) + +

8 (5 6 +) +

(8 11 +)

19

People who become accustomed to using this type of notation find it very quick and convenient to use because there is never a question about the order in which to perform the operations.

Exercises

1. Tony wants to buy a car. He has the options of two different brands of radios and four different exterior colors. Draw a tree diagram to show the possible outcomes of choosing a radio and a color for the car.

2. A coin is tossed three times. Draw a tree diagram to show the possible outcomes.

Examine the trees in Exercises 3 through 6. If the tree is a binary tree, (a) give the level of vertex *V*, (b) name the parent of *V*, and (c) name the children of *V*.

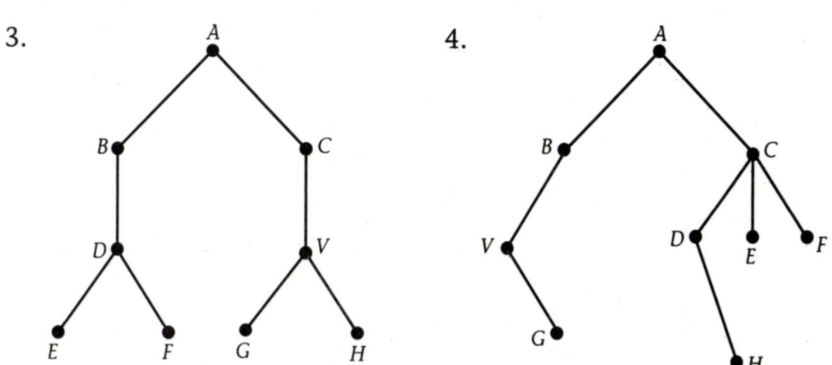

5. 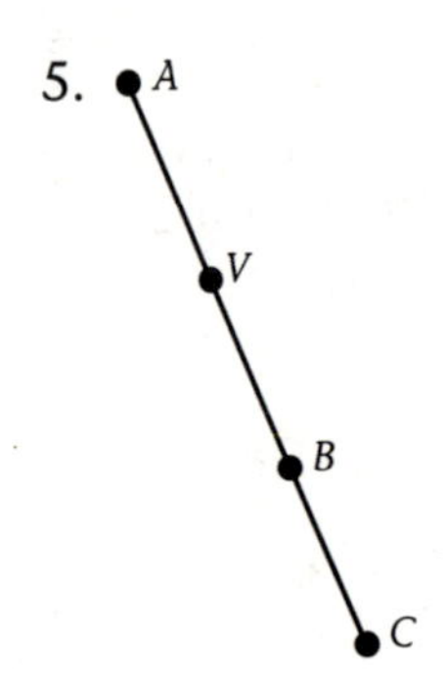

6. 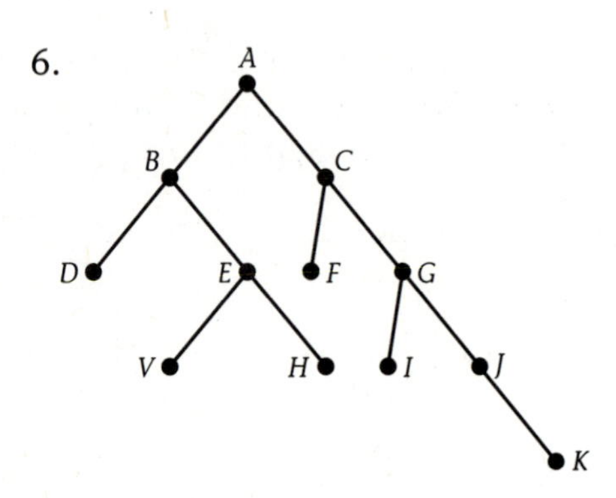

In Exercises 7 through 10, represent the expression as a binary expression tree.

7. (2 − 5) * (4 + 7)

8. (2 + 3) * 4

9. 2 + 3 * 4 − 6/2

10. $A * B + (C - D/E)$

Find the postorder listings for the binary trees in Exercises 11 through 13.

11. 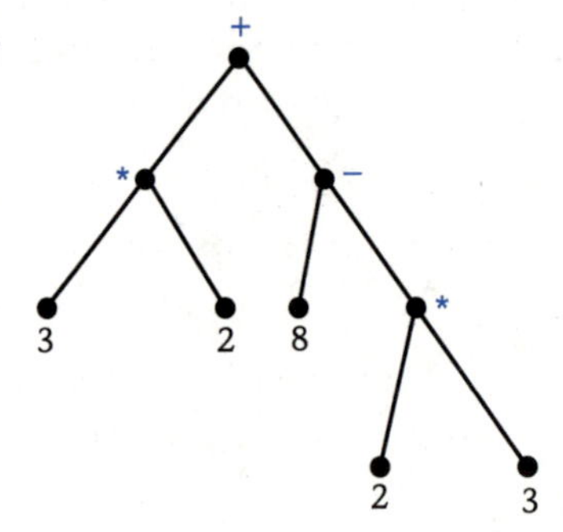

12. 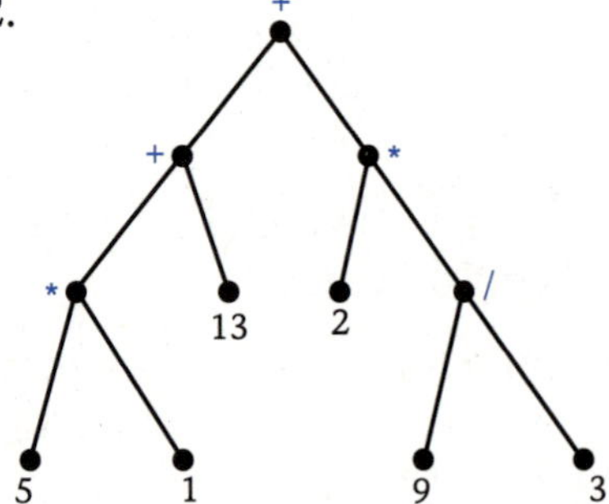

13. 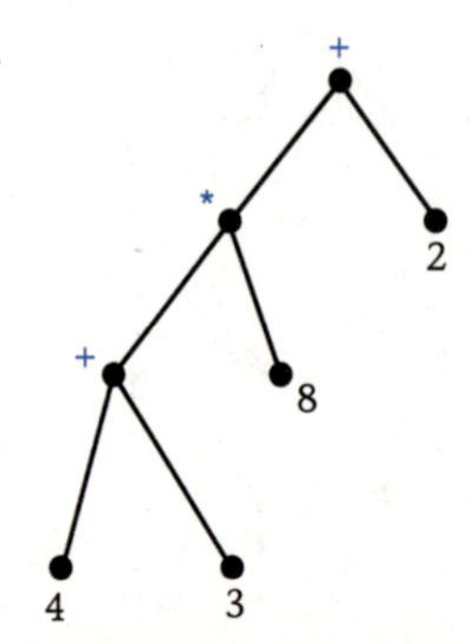

14. Evaluate the following reverse Polish notations:
    a. 6 2 − 7 * 3 2 ++
    b. 6 5 4 3 2 − +/+
    c. 1 2 + 4 3 − + 6 2/2 ++
    d. 4 3 + 8 2 − + 4 + 3 −

15. Give the reverse Polish notation for the following expressions:
    a. 2 + 3 * 6 − (4 + 1)
    b. (5 − 3) * 2 + (7 − 6/2)

16. Construct an expression tree for the reverse Polish notation *A B* * *C D* + *E* − +.

17. Construct a binary tree that has *ABC* as its postorder listing. Is your answer unique? If not, construct additional tree(s).

18. A traversal that visits first the parent or root of the tree, then the left child, and finally the right child is called a **preorder traversal.**

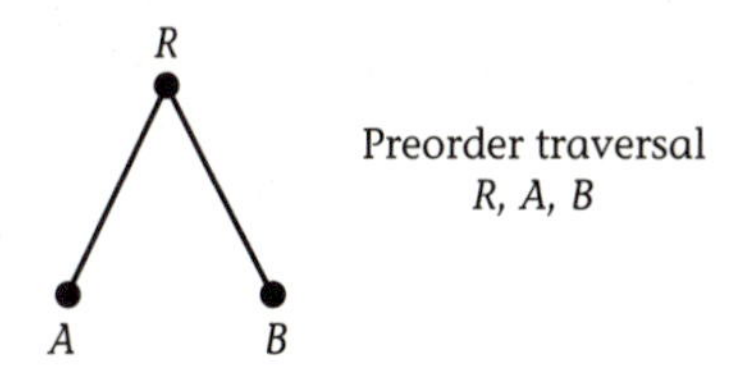

The preorder listing for the following binary tree is *ABDECFG:*

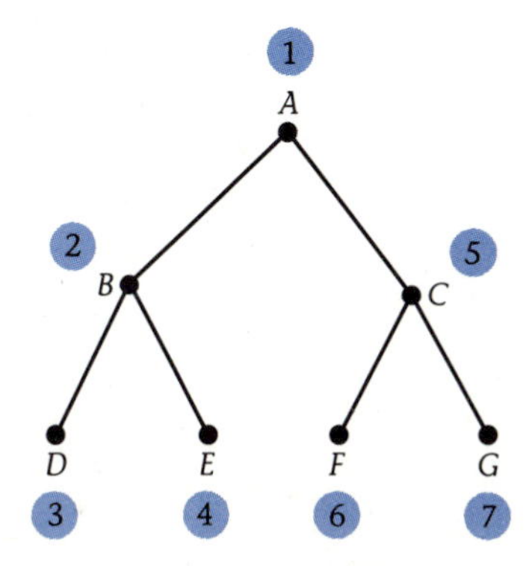

Find the preorder listing for the following binary tree:

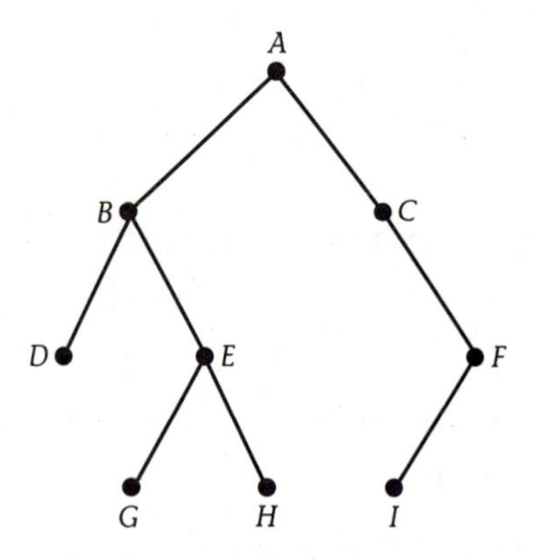

19. Find the preorder listings for the binary trees in Exercises 11 through 13.

20. The notation obtained from a preorder traversal is called **Polish notation.** To evaluate the expression, scan it from the left until you come to an operation followed by two numbers. Perform that operation, place the result back in the expression, and continue.
    Complete the following evaluation:

+ + 4 * 2 3 + 5 / 6 3

+ + 4 6 + 5 / 6 3

+ 10 + 5 / 6 3

21. Evaluate the following Polish notations:
    a. + * 32 − 8 * 2 3
    b. + / 6 3 + 4 3

1. Make a planar drawing of the graph described by the adjacency matrix below.

$$\begin{array}{c} \\ A \\ B \\ C \\ D \\ E \end{array} \begin{array}{c} \begin{array}{ccccc} A & B & C & D & E \end{array} \\ \begin{bmatrix} 0 & 1 & 1 & 1 & 0 \\ 1 & 0 & 0 & 1 & 1 \\ 1 & 0 & 0 & 1 & 1 \\ 1 & 1 & 1 & 0 & 1 \\ 0 & 1 & 1 & 1 & 0 \end{bmatrix} \end{array}$$

2. What is the chromatic number for a tree with five vertices? Any odd number of vertices? Any even number of vertices? Any tree with two or more vertices?

3.

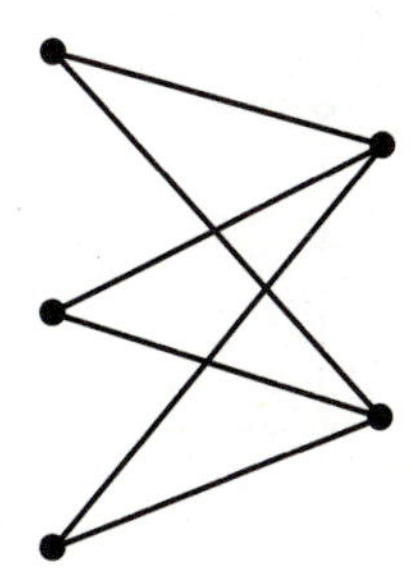

a. Is this graph bipartite? Why or why not?
b. Is this graph a complete bipartite graph? Why or why not?
c. Is this graph planar? If so, find a planar drawing of the graph.
d. What is the chromatic number of this graph?

4. Mr. Gonzalez, the principal at Central High School, leaves his office once an hour to visit the math, science, and civics classrooms, and then he returns to his office. The distances between rooms are shown on the graph below.

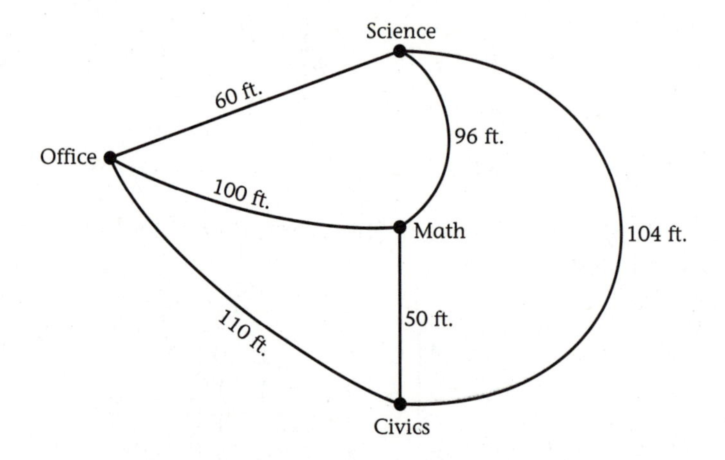

   a. What is the shortest path for Mr. Gonzalez to travel?
   b. What is the total distance that he travels on that route?
   c. What kind of a circuit does Mr. Gonzalez make?

5. Find a spanning tree for the graph below if one exists.

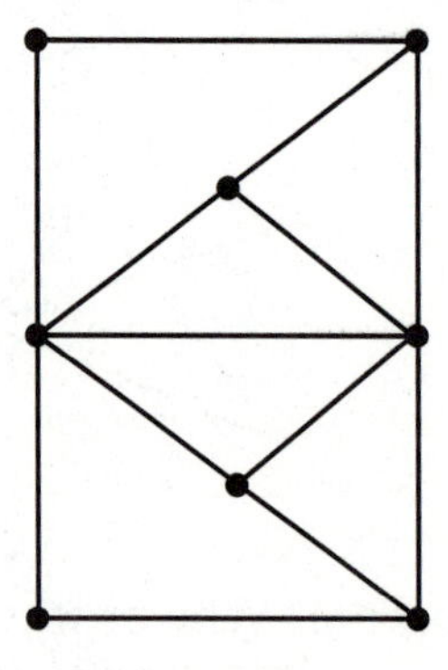

6. When given the position where it is currently located and its destination, a certain robot car is programmed to figure the shortest

path for the trip. The routes that the car can travel are shown on the graph below.

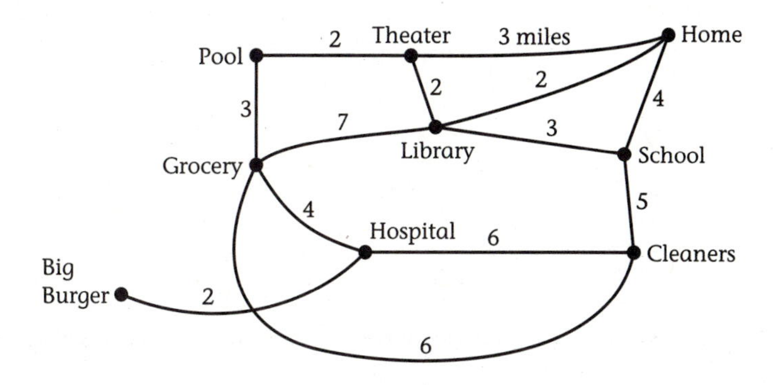

a. Use inspection to find the shortest path from Home to Big Burger.
b. What is the minimum distance from Home to Big Burger?
c. Use the shortest path algorithm from Lesson 5.3 on page 223 to find the shortest paths from Home to each of the other locations on the graph.

7. All the locations represented by the graph in Exercise 6 need to be connected by cable. Find the minimum amount of cable needed to link the nine locations.

8. Are the following graphs trees?

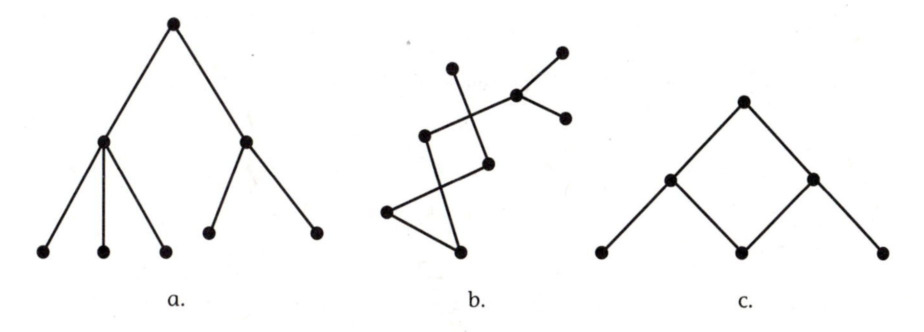

9. Use the breadth-first search algorithm on page 240 to find a spanning tree for the graph at the top of the following page. Begin the algorithm at the vertex labeled S.

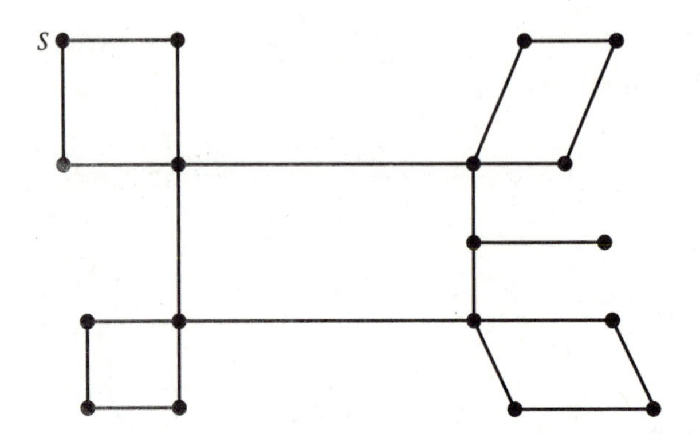

10. Draw a tree with eight vertices that has exactly four vertices of degree 1.

11. a. Using one of the spanning tree algorithms, find a minimum spanning tree for the graph below.
    b. What is the total weight of the minimum spanning tree?

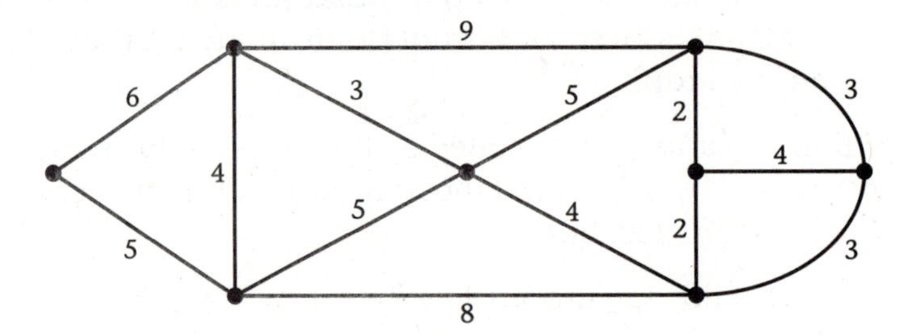

12. Represent (4 − 3) * 8 + 5 as a binary expression tree.

13. Evaluate the following reverse Polish notation:

7 1 + 3 * 2 4 + −

Bibliography

Chartrand, Gary. 1977. *Introductory Graph Theory*. New York: Dover Publications.

Chavey, Darrah. 1992. *Drawing Pictures with One Line*. (HistoMAP Module #21). Lexington, MA: COMAP.

COMAP. 1991. *For All Practical Purposes: Introduction to Contemporary Mathematics.* 2nd ed. New York: Freeman.

Copes, W., C. Sloyer, R. Stark, and W. Sacco. 1987. *Graph Theory: Euler's Rich Legacy.* Providence, RI: Janson Publications.

Cozzens, Margaret B., and R. Porter. 1987. *Mathematics and Its Applications.* Lexington, MA: COMAP.

Cozzens, Margaret B., and R. Porter. 1987. *Problem Solving Using Graphs.* Lexington, MA: COMAP.

Crisler, Nancy, and Walter Meyer. 1993. *Shortest Paths.* Lexington, MA: COMAP.

Dossey, John, A. Otto, L. Spence, and C. Vanden Eynden. 1992. *Discrete Mathematics.* 2nd ed. Glenview, IL: Scott, Foresman.

Francis, Richard L. 1989. *The Mathematician's Coloring Book.* (HiMAP Module #10). Lexington, MA: COMAP.

Ore, O. 1990. *Graphs and Their Uses.* Washington, DC: Mathematical Association of America.

NEW YORK
PICK 10

Counting and Probability

Lotteries are quite popular in the United States. Each week millions of lottery tickets are sold to people who pick several numbers in hopes that the ones they choose will match those generated by a random process. The lucky few who match several or all of the numbers win anywhere from a few dollars to several million dollars.

In how many ways can a lottery participant choose several numbers from those on a lottery ticket? What is the probability of winning the jackpot in a lottery? How can the methods that mathematicians use to determine the probability of winning a gambling game also be used to determine the probability that a medical test will give correct results? How can proper understanding of probability improve the reliability of a space shuttle launch?

A Counting Activity

The members of the Central High School student council are debating fund-raising activities for the organization. Three proposals are under discussion.

Pierre suggested that the group operate a game involving the school's team name, the Lions, at the annual school fair. His idea is to write each of the letters of the word on a Ping-Pong ball and to allow students to draw two of them from an opaque container. If the two letters drawn spell (in order) a legal word, the participant will win a prize. His plan has been criticized by other members who feel it would be too easy to win.

Hilary's idea is to print cards with the numbers 1 through 9 displayed in a square matrix and have students select any two numbers. A winning pair would be generated at random and a prize given to any student who matches both the winning numbers. Her scheme has left members uncertain as to how many winners might be expected in a school of 1,000 students.

Chuck also wants to operate a game at the school fair. His game would involve a board with the numbers 1 through 6 displayed. A participant would bet $1 on any one of the numbers, roll two dice, and win a dollar for each time the chosen number appeared. Several members of the council feel that the organization would lose money on such an endeavor.

Try This

At the direction of your instructor, divide your class into groups of three people. Write the numbers 1, 2, or 3 on each of several slips of paper. Have each group draw one of the slips from a bag or box. Each group should consider the proposal listed below that corresponds to the number drawn by the group. After all groups have finished, a spokesperson for each group should present the results of the group's discussion to the class. Have each group that discussed Question 1 report first, and so on. Each member of the class should take notes on each group's analysis.

1. Analyze Pierre's proposal. Try to determine the number of possibilities. How many different two-letter "words" are there? How many are real words? Suppose that each of the school's 1,000 students enters exactly once and pays a $1 entry fee. How many winners might there be? How much should each winner receive if the council needs to raise $500?

2. Analyze Hilary's proposal. In how many ways could a student fill in the entry form? If each of the school's 1,000 students enters exactly once and pays a $1 entry fee, how many winners might there be? How much should each winner receive if the council needs to raise $500?

3. Analyze Chuck's proposal. In how many ways could the two dice fall? How often would the council pay the player $1? $2? How often would the council make $1? Do you think that the council could raise $500 if that is its goal? If you want to try the game, check with your teacher to see if there are dice available in your room.

Exercises

1. Make a list of all possible "words" of two letters that can be made from the word *Lions*. How many different "words" are there?

2. Suppose the Ping-Pong balls are drawn one at a time. How many different Ping-Pong balls can be drawn first? Second? What is the connection between these numbers and the number of "words" in the list you made in Exercise 1? If the school's team name were Tigers, how many "words" of two letters would there be?

3. Make a list of all possible ways of choosing two numbers from the nine available on one of Hilary's cards. How many are there?

4. If you were filling in one of Hilary's cards, in how many ways could you select your first number? After you've picked your first number, in how many ways could you pick your second? How are these two numbers connected to the number of things in the list you made in Exercise 3?

5. Suppose that the two dice rolled by the player in Chuck's game are different colors, say red and green. One possible way the dice could fall is the red die a 3 and the green die a 4. This can be written in shorthand: (3, 4). Another way is the red die a 4 and the green die a 3, which can be written as (4, 3). Make a list of all possible ways the red die and the green die could fall. How many are there?

6. The red die could land in six different ways, as could the green. How are these two sixes connected to the number of things in the list you made in Exercise 5?

In mathematics, a fundamental multiplication principle is that if events A and B can occur in a and b ways separately, then there are $a \times b$ ways that the events can occur together. To use the principle, mark a blank for each of the events and write the number of possibilities for each event in each blank. Then multiply these numbers. For example, to determine the number of ways that a die and a coin can fall together, make two blanks: _____ _____, then write the number of possibilities for the die and the coin in the blanks: 6 2 , and multiply to get 12.

If a full list of the 12 is needed, a systematic way of ensuring that all items are listed is to make a tree diagram like that shown below.

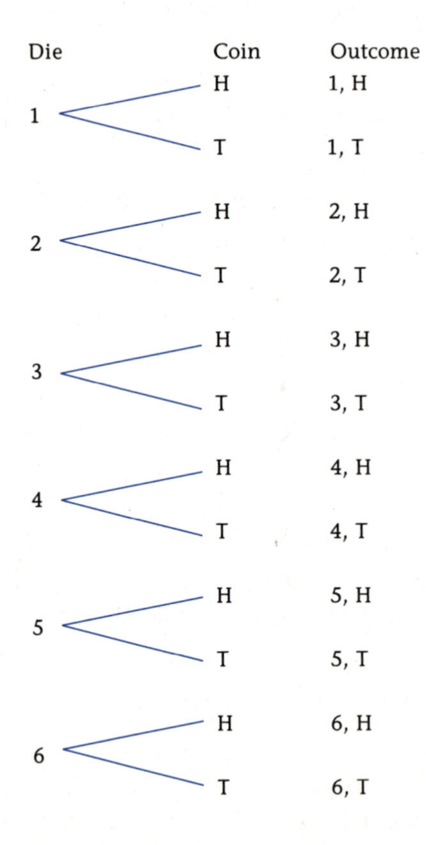

7. Explain how the multiplication principle can be applied to the problem of determining the number of different "words" of two letters that can be made from the letters of *Lions*.

8. A utility company in North Dakota once sponsored a contest to promote energy conservation. The contest was to find all the words

that could be made from the letters of the word *insulate.* Use the multiplication principle to determine the number of "words" of two letters that are possible. How can the principle be extended to determine the number of "words" of three letters that are possible?

9. It is possible to modify the multiplication principle to determine the number of ways of selecting two numbers on one of Hilary's cards. Explain how this can be done.

10. Why was it necessary to modify the multiplication principle the way you did in the previous exercise?

11. State lotteries often select several numbers from a collection of numbers printed on a card. Suppose that a state lottery card has the numbers 1 through 25 in a square matrix. In how many ways can you select two of them?

12. Explain how the multiplication principle can be used to determine the number of ways in which two dice can fall.

13. Make a tree diagram to show all the outcomes when a red die and a green die are rolled together.

14. Make a tree diagram to show all the possibilities when filling out one of Hilary's cards.

15. You are playing Chuck's game and decide to bet on the number 5. Use the tree diagram you made in Exercise 13 to count the number of ways in which you could win or lose. In how many ways could you win \$1? \$2? In how many ways could you lose \$1? In the long run, if you played this game many times, do you think you would win or lose money?

16. In a common carnival dice game three dice are rolled. Use the multiplication principle to determine the number of ways in which three dice can fall.

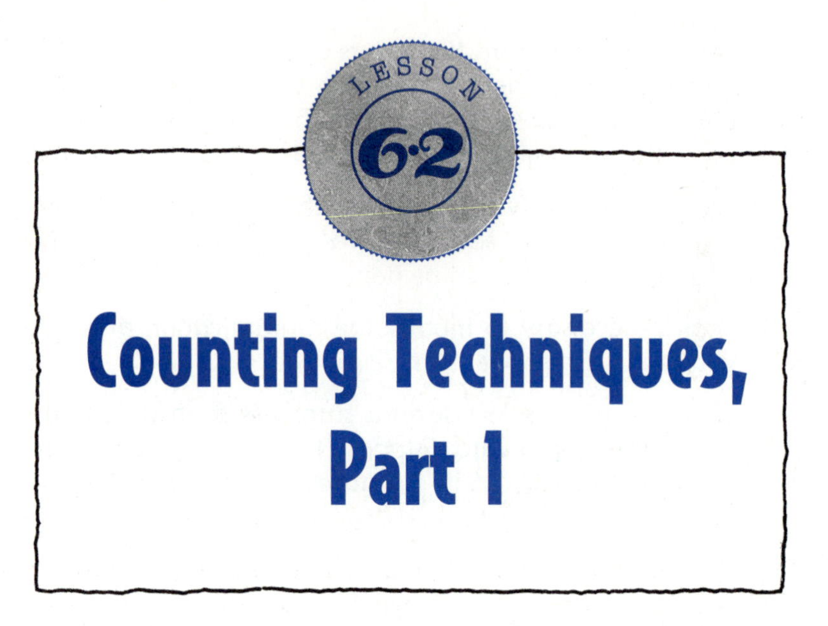

Counting Techniques, Part 1

The previous lesson examined several situations in which the answer to the question, In how many ways can this be done? is important. There are many instances in discrete mathematics when the answer to this question is crucial.

The game suggested by Pierre is one in which the order of selection makes a difference. The words *is* and *si* are not the same. A situation in which the order matters is called a **permutation,** and one in which the order does not matter is called a **combination.** This lesson is devoted to situations in which order matters.

Two important counting techniques in the previous lesson are the multiplication principle and tree diagrams.

> The multiplication principle states that if events A and B can occur in a and b ways, respectively, then events A and B can occur together in $a \times b$ ways. Another important counting principle is the **addition principle,** which states that if events A and B can occur in a and b ways, respectively, then either event A or event B can occur in $a + b$ ways.

For example, the student council at Central High consists of 17 members, 9 girls and 8 boys. If one girl *and* one boy are to be selected to hold two different offices on the council, then there will be $9 \times 8 = 72$ ways of making the selection. If a single student, who may be either a boy *or* a girl, is to be selected to hold a single office, then there will be $9 + 8 = 17$ ways of making the selection.

The word *and* in the description of an event often indicates that the multiplication principle should be used, and the word *or* often indicates that the addition principle should be used.

The events "selecting a boy" and "selecting a girl" are called **mutually exclusive** or **disjoint** because there are no people who are both boy and girl. On the other hand, events such as "selecting a member of your school's football team" and "selecting a member of your school's basketball team" probably have at least one member in common and are therefore not mutually exclusive. In such cases, the addition principle requires a modification, which will be considered in this lesson's exercises.

The multiplication and addition principles are often used together. For example, the students at Central High are considering a contest in which words of any length made from the team name *Lions* may be used.

A word of one letter, of course, may be composed in only five ways: l, i, o, n, s. A word of two letters requires a first letter *and* a second, so there are $5 \times 4 = 20$ ways of composing a word of two letters. A word of three letters requires a first letter *and* a second letter *and* a third, so there are $5 \times 4 \times 3 = 60$ ways of composing a word of three letters. Similarly, there are $5 \times 4 \times 3 \times 2 = 120$ ways of composing a word of four letters and $5 \times 4 \times 3 \times 2 \times 1 = 120$ ways of composing a word of five letters.

A word may be composed by using a single letter *or* by using two letters *or* by using three letters *or* by using four letters *or* by using five letters. Thus, the total number of words is $5 + 20 + 60 + 120 + 120 = 325$.

The calculation of the number of words of five letters that can be made from the letters of *Lions* requires multiplying all integers from 1 through 5, inclusive. This product is known as the **factorial** of 5 or, more simply, 5 factorial. It is symbolized with an exclamation mark: $5!$. Your scientific calculator has a factorial key. If you have never used it, press the digit 5, followed by the factorial key. The result should be 120.

Permutations can also be computed by using the calculator's factorial key. For example, to find the number of words of three letters that are possible, divide $5!$ by $2!$. Note that this works because $5 \times 4 \times 3 = (5 \times 4 \times 3 \times 2 \times 1)/(2 \times 1)$. A permutation, or arrangement, of three things selected from a group of five is expressed symbolically as $P(5, 3)$ and is calculated by evaluating the expression $5!/(5 - 3)!$.

In general, $P(n, m)$ is calculated by evaluating the expression $\frac{n!}{(n-m)!}$.

Counting techniques are often used to determine the likelihood or **probability** of an event. Suppose, for example, that Pierre's word game allows participants to draw any number of Ping-Pong balls they choose. What is the probability that a participant will form a word of two letters from the 325 possible words of any length?

The probability is simply the ratio of the number of two-letter words to the total number of possible words: 20/325. Probabilities may be expressed as fractions, decimals, or percentages. This probability expressed as a decimal is .0615, and so we would expect a two-letter word to be formed about 6 times out of 100.

Because the numerator of a probability is never smaller than 0 and never larger than the denominator, probabilities always range between 0 and 1, inclusive. An event that can't happen has a probability of 0 and an event that is certain to happen has a probability of 1.

You now have several different counting techniques at your disposal:

1. The multiplication principle and the related permutation formula.

2. The addition principle.

3. Tree diagrams as an aid to making a list of all possibilities.

Skill at applying these principles does not come without a good deal of practice. The following exercises help develop that skill and also show some refinements of these three basic techniques.

Exercises

1. Which is equivalent to $P(10, 4)$, $10!/4!$ or $10!/6!$? Find the value of $P(10, 4)$.

2. Shown below are the final *USA Today* 1991–1992 season rankings of high school wrestling teams. Suppose that you are a sportswriter voting for the top teams. If you can rank only your top 5 and must choose them from the 25 shown here, in how many ways can you make your selection?

3. A multispeed bicycle has three sprockets in front and five sprockets on the rear wheel. The rider uses the bicycle's front and rear shift mechanism to move the chain from one front or rear sprocket to another.

NEWS CLIP

High School Wrestling Rankings

School	Dual Meet Record	1992 Finish
1. Bismarck (N.D.)	12–0	Class A state champ
2. Lakewood (Ohio) St. Edward	14–0	Div. I state champ
3. Clovis (Calif.)	16–0	State team champion
4. Poway (Calif.)	18–0	State runner-up
5. Grundy (Va.)	19–0	Class AA champion
6. Del City (Okla.)	13–0	Class 5A dual champ
7. South Chesapeake (Va.) Great Bridge	22–0	Class AAA champs
8. Oak Ridge (N.J.) Jefferson Township	19–0	No team champion
9. Brandon (Fla.)	14–0	Class 4A champ
10. Lawton (Okla.)	14–2	5A individual champ
11. Middle Island (N.Y.) Longwood	13–0	3 individual champions
12. Lake Ronkonkoma (N.Y.) Sachem	12–1	2 individual champions
13. Vernal (Utah) Uintah	10–2	Class 3A champion
14. Canonsburg (Pa.) Canon-McMillan	20–1	Class 3A champion
15. Midwest City (Okla.)	14–3	5A runner-up
16. Stow (Ohio) Walsh Jesuit	9–1	Div. I runner-up
17. Temperance (Mich.) Bedford	20–2	Class A champion
18. Chicago Mount Carmel	18–2	Class AA dual champ
19. Butte (Mont.)	19–0	Class AA champ
20. Millersville (Md.) Old Mill	15–0	Class 4A-3A champ
21. Wilmington (Del.) St. Mark's	11–2	State champ
22. Bennington (Vt.) Mt. Anthony Union	14–0	State champ
23. Aztec (N.M.)	13–0	Class 3A champs
24. Indianapolis Lawrence North	16–1	State champ
25. Greenville (Pa.) Reynolds	14–0	State runner-up

Ranked for USA TODAY by Rob Sherrill, publisher of *Center Mat* magazine.

a. How many speeds does the bicycle have?
b. Is it correct to say that a particular speed requires a particular front sprocket and a particular rear sprocket, or is it correct to say that a particular speed requires a particular front sprocket or a particular rear sprocket?

4. (See Exercise 8 of Lesson 6.1, p. 270.)
   a. How many different words of any length can be made from the letters of the word *insulate?* (Hint: You can make a word of one letter or a word of two letters or a word of three letters . . . or a word of eight letters.)
   b. A group of students is considering entering the contest involving the words made from *insulate* by programming a computer to print all possible words and then checking the list against an unabridged dictionary. Suppose the computer prints the words in four columns of 50 words on each page of paper. How many pages would there be?

5. Many states have vehicle license numbers that consist of three letters followed by three digits. Often the letters I, O, and Z are not used because they can be confused with the digits 1, 0, and 2.

a. If these are the only characters that may not be used and if letters and digits may be repeated, how many different license plates are possible?
b. What is the probability that a vehicle selected at random would have a license number that begins with CAT?

6. a. In how many ways can the coach of a baseball team arrange the batting order of the nine starters?
b. A sportscaster once suggested that a baseball team try every possible batting order for its nine starters in order to determine which one worked best. Suppose that the team decides to do just that and plays one game a day every day of the week. How long would it take to try every possible order if a different order were used in each game?

7. Three math students and three science students are taking final exams. They must be seated in six desks so that no two math students are next to each other and no two science students are next to each others.
a. In how many ways can this be done if the desks are in a single row? (Hint: Draw six blanks and use the multiplication principle.)
b. What is the probability that a math student will occupy the first seat?
c. What is the probability that math students will occupy the first seat and the last seat?
d. What is the probability that a math student will occupy either the first seat or the last seat?

8. The multiplication principle states that the number of permutations of the letters of the word *math* is 4!. The permutation formula says that $P(4, 4)$ is $4!/(4 - 4)!$. The denominator of this expression is 0!, which is meaningless. However, in order for these two methods to yield the same result, what value must 0! have?

9. The U.S. Postal Service originated five-digit zip codes in 1963. Every post office was given its own code, which ranged from 00601 in Adjuntas, Puerto Rico, to 99950 in Ketchikan, Alaska.
a. If the only five-digit zip code that may not be used is 00000, how many are possible?

b. Some five-digit zip codes are prone to errors because they still are legal five-digit zip codes when read upside down. When this happens, a letter goes to the wrong post office and must be sent back. How many zip codes are legal when read upside down? (Hint: Draw five blanks, think carefully about which digits could go in each blank, and apply the multiplication principle.)

c. How many of these zip codes remain the same when read upside down?

10. In how many ways can a person draw 2 cards from a standard 52-card deck if the first card is returned to the deck? (Assume that the order in which the cards are drawn makes a difference.) In how many ways can this be done if the first card is not put back?

11. The addition principle of this lesson cannot be used if the two events have at least one member in common. For example, there are 20 people on your school's basketball team and 45 people on the football team. Are there 65 different ways of choosing someone from either group? Only if there are no people who are on both teams, which isn't very likely.

a. There are 10 people who play both football and basketball. In how many ways can a person be selected from either team?

b. Write the appropriate number of people in each of the three regions of the diagram shown below.

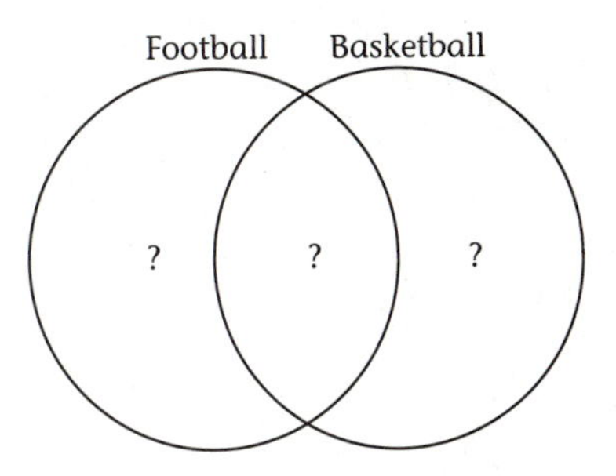

c. Describe how the addition principle is applied when two events are mutually exclusive and how it is applied when the two events are not mutually exclusive.

d. Event A and event B can occur in a and b ways, respectively, and events A and B have c items in common. Write an algebraic expression for the number of ways in which event A or B can occur.

e. Central High's football team has 38 members, and its basketball team has 13 members. If there are a total of 42 students involved, how many are on both teams?

12. Before the beginning of the 1992 major league baseball season, St. Louis Cardinal manager Joe Torre said that he'd picked his starting lineup. He also said that he'd determined his first three batters, but not in which order they would bat. In how many ways can Joe arrange his batting order if the pitcher must bat last? (Hint: Draw nine blanks and apply the multiplication principle.)

13. a. In how many different ways can a teacher arrange 30 students in a classroom with 30 desks?

    b. The radius of the earth is approximately 6,370 kilometers, and a standard medical drop is one-tenth of a cubic centimeter. Use the formula for the volume of a sphere, $V = \frac{4}{3}\pi r^3$, to find the volume of the earth in drops of water. Compare this with the number of seating arrangements in part a.

14. There are three highways from Claremont to Upland and two highways from Upland to Pasadena.

    a. In how many ways can a driver select a route from Claremont to Pasadena?

    b. Is it correct to say that a trip from Claremont to Pasadena requires a road from Claremont to Upland and a road from Upland to Pasadena, or is it correct to say that it requires a road from Claremont to Upland or a road from Upland to Pasadena?

    c. In how many ways can a driver plan a round-trip from Claremont to Upland and back?

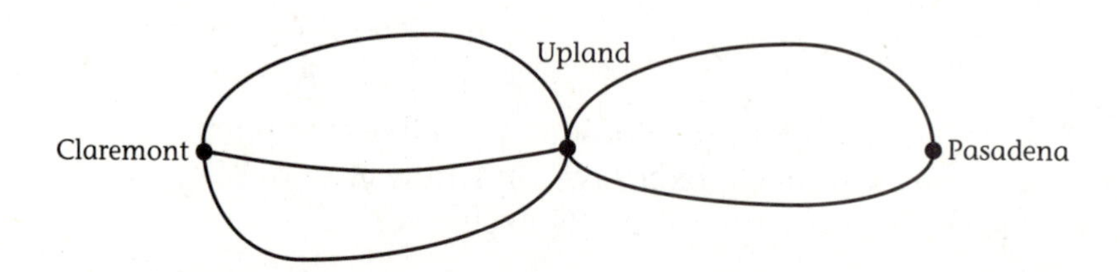

15. Radio station call letters in the United States consist of three or four letters, of which the first must be either a K or a W. Assuming that letters may be repeated, determine the number of call letters possible.

16. Six different prizes are given by drawing names from the 72 Central High orchestra members attending the orchestra's annual picnic. In how many ways can the prizes be given if no one can receive more than one prize?

17. Telephone area codes consist of three digits, of which the first may not be a 0 or a 1, the second may be only a 0 or a 1, and the third may be any digit except 0.
    a. How many area codes are possible?
    b. Because of a shortage of area codes, beginning in 1995, any digit will be a legal area code second digit. How many area codes will be possible in 1995?

18. UPC bar codes consist of two groups of five digits each. One group represents the manufacturer, as assigned by the Uniform Code Council in Dayton, Ohio, and the other group represents the products of that manufacturer. How many different manufacturers can be encoded? How many products can each manufacturer encode?

19. Many newspapers publish weekly word puzzles that offer an accumulating cash prize. These puzzles are modified crosswords in which clues are given and usually only two choices for the correct answer to any question are offered.
    a. Suppose a prize word contains 20 questions, each having two possible answers. How many different entries are possible? (Hint: Imagine 20 blanks and use the multiplication principle.)
    b. Suppose that you embark on the ambitious project of submitting every possible entry. Further suppose that it takes you five minutes to do an entry and that the piece of paper on which you submit an entry is 0.003 inch thick. How long would it take you to prepare the entries, and how thick a stack of paper would you deliver to the newspaper?

20. The carnival game Chuk-a-Luk is similar to the one proposed by Chuck in the previous lesson, except that three dice are used.
    a. In how many ways can three dice fall?
    b. Determine the number of ways you could win $1, win $2, win $3, or lose $1 in the game of Chuk-a-Luk if you win a dollar for each time your number shows. (Hint: You win $2 if the first and second dice show your number and the third die doesn't, or if the first and third dice show your number and the second die doesn't, or if the second and third dice show your number and the first die doesn't. Draw several sets of three blanks and then use the multiplication and addition principles.)
    c. In the long run, do you think a player would win or lose money at the game of Chuk-a-Luk?

Projects

21. Research the history of probability. How did it begin? What roles did Jerome Cardan, Blaise Pascal, and Pierre Fermat play? What problems interested them?

22. Investigate the number of permutations of several objects, of which some look alike. For example, if the letters of *math* can be arranged in 24 ways, how many different permutations can there be of the letters of *look?* What if there are several sets of identical

letters, such as in "Mississippi"? Write a summary that includes a general algebraic principle for handling such situations and several examples.

23. Investigate the number of permutations of several objects arranged in a circular, rather than a linear, fashion. For example, Ann, Sean, Juanita, and Herb can be seated along one side of a rectangular table in 24 ways. In how many different ways can they be seated around a circular table? Write a summary that includes a general algebraic principle for handling such situations and several examples. Also explain your interpretation of the meaning of the word *different*.

24. Investigate the use of the addition principle with three events that are not mutually exclusive. Suppose, for example, that the football, basketball, and track teams of Central High have 41, 15, and 34 members, respectively. If 6 people play both football and basketball, 7 are on the basketball and track teams, 15 are on both the football and track teams, and 4 play all three sports, how many people are involved in one sport or another? Write a general algebraic principle for handling situations of this type, and draw a diagram to represent it. Can the principle be extended to four or more events? How?

LESSON 6.3

Counting Techniques, Part 2

The game proposed by Hilary in Lesson 6.1 is an example of a lottery. Hilary's game requires the selection of two numbers from nine printed on a card. The order in which these numbers are selected is not important. That is, if the winning numbers are 2 and 6, then it does not matter whether the owner of a winning ticket selected 2 or 6 first.

A situation in which the order does not matter is called a **combination.** This lesson is primarily concerned with counting situations in which the order does not matter. But if the order of selection of numbers on one of Hilary's tickets did make a difference, then the number of ways of filling out a ticket would be counted as a permutation: $P(9, 2) = 9!/7! = 72$. Because this counts the selection of a pair such as 2 and 6 as different from the selection of 6 and 2, every possible pair is counted twice. Thus, the number of combinations of two things selected from a group of 9 is $72/2 = 36$, which is expressed symbolically as $C(9, 2)$.

What if Hilary's game required the selection of three numbers? Although $P(9, 3) = 504$, this is too large because it considers the order of selection important. Suppose that the winning numbers are 2, 5, and 8. The number of different ways of arranging these three numbers is $3 \times 2 \times 1 = 3! = 6$. Because $P(9, 3)$ counts every one of these as different, it is six times as large as it should be. Therefore, $C(9, 3) = P(9, 3)/3! = 504/6 = 84$.

In general, $C(n, m)$ is calculated by evaluating the expression $P(n, m)/m!$. But, $P(n, m) = \frac{n!}{(n-m)!}$, so

$$C(n, m) = \frac{n!}{(n-m)!m!}.$$

One way to evaluate a combination such as $C(9, 3)$ on a calculator is to divide 9! by 6!, and then to divide again by 3!.

The probability that a single ticket will win Hilary's lottery is 1/36, or about .028. If 1,000 tickets are sold, one can expect about 1,000 × .028 = 28 winners. If Hilary's game requires the selection of three numbers, the probability that a single ticket will win is 1/84, or about .012. If 1,000 tickets are sold, about 1,000 × .012 = 12 winners can be expected.

Combinations are sometimes used along with other counting techniques. For example, suppose that the 17-member student council at Central High consists of 9 girls and 8 boys. A committee of 4 council members is being selected. If the positions on the committee are not different, then the number of ways the committee can be selected is $C(17, 4) = 17!/(13!4!) = 2,380$.

Suppose, however, that the committee must consist of 2 girls and 2 boys. There are $C(9, 2) = 9!/(7!2!) = 36$ ways of selecting the 2 girls and $C(8, 2) = 8!/(6!2!) = 28$ ways of selecting the 2 boys. Because the committee must consist of 2 boys *and* 2 girls, there are 36 × 28 = 1,008 ways of forming the committee. If the 4 committee members are selected at random, the probability that the committee will consist of 2 boys and 2 girls is 1,008/2,380, or about .424.

Suppose that the committee must consist of either all boys or all girls. There are $C(9, 4) = 9!/(5!4!) = 126$ ways of selecting 4 girls and $C(8, 4) = 8!/(4!4!) = 70$ ways of selecting the boys. Because the committee must consist of either 4 girls *or* 4 boys, there are 126 + 70 = 196 ways of forming the committee. Again, if the 4 committee members are selected at random, the probability that the committee will consist of either all boys or all girls is 196/2,380, or about .082.

Australian Group Wins the Virginia Lottery

In February 1992, the Melbourne, Australia–based International Lotto Fund sent representatives to Virginia to purchase tickets in that state's lottery. The representatives spread their purchases among eight retail chains that had a total of 125 outlets, but one representative bought a total of 2.4 million tickets at a single retail chain headquarters. When time ran out, the group had purchased 5 million tickets, or about 70% of all possible combinations in the lottery.

NEWS CLIP

Cornering a Lottery: Foolproof, or Foolish?

By Albert B. Crenshaw, WASHINGTON POST, February 28, 1992

You might think that betting on every possible number combination in a state lottery would guarantee a big payoff.

But it doesn't.

A look at what happened in Virginia last week, when an Australian syndicate apparently attempted to corner the lottery, indicates that to be successful, players must not only bet every number, but they must also be lucky enough to be the only winners.

The Australians' system is far from a sure-fire moneymaker, and as soon as more than one group tries it, the system is almost certain to be a money-loser. . .

The difference between the size of the investment—around $7 million to cover all the numbers in Virginia—and the $27 million jackpot is not as great as it seems. The grand-prize winnings in most lotteries are paid out over a long time, making them far less valuable than if they were paid all at once.

In fact, after adjusting for inflation and potential investment earnings, $27 million paid out over 20 years is the equivalent of only about $9 million to $12 million if it were paid out today.

One of the tickets purchased by the group matched the winning combination of 8, 11, 13, 15, 19, and 20 generated by the lottery's computer. After some investigation into the legality of the purchase, the lottery decided to award the $27 million jackpot to the Australian group. The fund represented about 2,500 investors who paid an average of $3,000 each. Each stood to receive an average of $10,800, at the rate of $540 a year over the 20-year payment period.

Exercises

1. Which is larger: $C(10, 2)$ or $C(10, 8)$?

2. Find the sum of all possible combinations of four things. That is, find $C(4, 0) + C(4, 1) + C(4, 2) + C(4, 3) + C(4, 4)$. Do the same for all possible combinations of three things and all possible combinations of five things. Can you guess the sum for all possible combinations o six things?

3. In this lesson the number of all-boy four-person committees on the Central High student council was calculated as $C(8, 4) = 70$; the number of all-girl committees was calculated as $C(9,4) = 126$; and the number of committees that were half boys and half girls was calculated as $C(8, 2) \times C(9, 2) = 1,008$.
   a. How many committees consist of three girls and one boy?
   b. How many committees consist of one girl and three boys?
   c. Find the sum of the numbers of committees that consist of four boys, no boys, two boys, three boys, and one boy. Compare this with the total of committees calculated by $C(17, 4)$ in the lesson (see page 285).

4. Darrell Dewey has just left his Central High social studies class and bumped into his friend, Carla Cheetham, who is on her way to the same class. Darrell informs Carla that Ms. Howe is giving a 10-question true/false drop quiz today. When Carla asks about the quiz, Darrell says that he thought it was easy and that he thought four of the answers were false.

a. When Carla takes the quiz, in how many ways can she select four questions to mark false?
b. In how many ways can Carla select six questions to mark true?
c. In how many ways can Carla fill in the quiz if she ignores Darrell's hint?

5. a. In how many ways can 2 cards be dealt from a standard 52-card deck?
   b. In how many ways can 2 red cards be dealt from a standard 52-card deck?
   c. What is the probability that 2 cards dealt from a standard 52-card deck both are red?

6. Maria has a part-time summer job selling ice cream from a small vehicle that she drives through residential areas of her community. She carries six different flavors of ice cream and sells a double-dip cone for $1.40.
   a. How many double-dip cones are possible if both scoops are the same flavor?
   b. How many double-dip cones are possible if each scoop is a different flavor?
   c. All together, how many double-dip cones are possible?

7. Hedy Foans, who writes a music column in the Central High Scribbler, has decided to poll students on their favorite songs. She has prepared a list of 10 current favorites, from which students will be asked to rank their top 3. In how many ways can a student pick a first, second, and third choice from Hedy's 10?

42 *The Billboard's Music Popularity Charts . . . POP RECORDS* APRIL 25, 1960

FOR WEEK ENDING MAY 1

The Billboard HOT 100

★ STAR PERFORMERS showed the greatest upward progress on Hot 100 this week.
[S] Indicates that 45 r.p.m. stereo single version is available.
△S Indicates that 33⅓ r.p.m. stereo single version is available.

This Week	One Week Ago	Two Weeks Ago	Three Weeks Ago	Title — Artist, Company Record No.	Stereo	Weeks on Chart
★1	6	17	84	STUCK ON YOU — Elvis Presley, RCA Victor 7740	S	4
2	2	5	12	GREENFIELDS — Brothers Four, Columbia 41571	△S	10
3	5	7	6	SINK THE BISMARCK — Johnny Horton, Columbia 41568		8
4	1	1	1	THEME FROM A SUMMER PLACE — Percy Faith, Columbia 41490	△S	16
5	8	3	4	HE'LL HAVE TO GO — Jim Reeves, RCA Victor 7643	S	18
6	9	11	16	SIXTEEN REASONS — Connie Stevens, Warner Bros. 5137		13
7	3	2	2	PUPPY LOVE — Paul Anka, ABC-Paramount 10082	S	10
34	20	15	8	HARBOR LIGHTS — The Platters, Mercury 71563	S	14
★35	51	55	92	CHERRY PIE — Skip and Flip, Brent 7010		4
★36	47	63	86	WHAT AM I LIVING FOR — Conway Twitty, M-G-M 12886		5
37	32	33	29	LITTLE BITTY GIRL — Bobby Rydell, Cameo 171		13
38	17	13	10	BABY — Brook Benton and Dinah Washington, Mercury 71565	S	14
39	41	52	85	LOVE YOU SO — Rod Holden, Donna 1315		4
40	39	44	61	MOUNTAIN OF LOVE — Harold Dorman, Rita 1003		9
68	38	29	21	THIS MAGIC MOMENT — Drifters, Atlantic 2050		10
69	73	84	90	THINK ME A KISS — Clyde McPhatter, M-G-M 12877		4
70	76	—	—	WAY OF A CLOWN — Teddy Randazzo, ABC-Paramount 10088		2
71	82	—	—	NOBODY LOVES ME LIKE YOU — Flamingos, End 1068		2
72	69	32	20	LADY LUCK — Lloyd Price, ABC-Paramount 10075	S	13
73	72	56	59	CHINA DOLL — Ames Brothers, RCA Victor 7655	S	13
★74	94	—	—	CATHY'S CLOWN — Everly Brothers, Warner Bros. 5151	S	2

8. Ms. Howe has a planter in one of her classroom windows that is subdivided into five sections. She has purchased two geraniums and three marigolds to plant in the five spaces.
   a. In how many ways can Ms. Howe select the two sections in which to plant the geraniums?
   b. In how many ways can Ms. Howe select the three sections in which to plant the marigolds?

9. a. A Virginia lottery ticket contains the numbers 1 through 44, from which a participant selects six. In how many way can this be done?
   b. If it takes 5 seconds to fill out a Virginia lottery ticket, how long would it take one person working 40 hours a week to fill out all possible tickets?
   c. If each Virginia lottery form has space for five entries and if each form has a thickness of 0.003 inch, how thick would a stack of forms of all possible entries be?
   d. A Florida lottery ticket contains the numbers 1 through 49, from which a participant selects six. In how many ways can this be done?
   e. In 1990, a man bought 80,000 tickets in the Florida lottery. What was his probability of winning a share of the $94 million jackpot?
   f. How does the probability of winning the jackpot in the Virginia lottery compare with the probability of winning the jackpot in the Florida lottery?

10. Most lotteries include several prizes besides the jackpot. Suppose that the Virginia lottery gives second prize to tickets that match 5 of the 6 winning numbers, third to those that match 4 of the 6, and fourth to those that match 3 of the 6.
    a. How many different tickets can share the second prize? (Hint: The ticket must match 5 of the 6 winning numbers and 1 of the 38 nonwinning numbers.)
    b. How many different tickets can share third prize?
    c. How many different tickets can share fourth prize?

11. Chapter 1 discussed various ways of voting and of determining the winner in elections. Suppose there are seven choices on a ballot.

a. In how many ways can you rank the seven choices?
b. Recall that when approval voting is used, the choices are not ranked. In how many ways can you select three choices of which to approve?

12. Dee Knote, the director of Central High's music department, is holding tryouts for the school's jazz band. Seven students are competing for the three saxophone positions, 8 for the two piano spots, 5 for the two percussion positions, and 12 for three places as guitarists. In how many ways can Dee select her band?

13. The figure below was drawn by marking nine points at equal distances on the circumference of a circle and connecting every pair of points.
    a. How many chords are there?
    b. Number the points from 1 through 9, and explain why drawing the chords is analogous to filling out every possible ticket in Hillary's lottery.

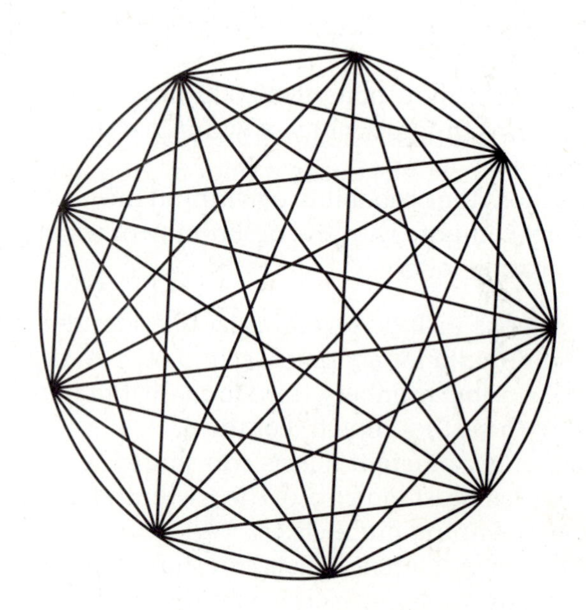

14. Emily's Pizza Emporium canprepare a pizza with any one or more of eight available ingredients. In how many different ways can a

pizza be ordered at Emily's? (Hint: A pizza can be ordered with one ingredient, or two ingredients, or three ingredients. . . .)

15. College Inn Pizza claims that it offers 105 different two-topping pizzas. How many different toppings do you think it uses? Explain.

16. Carl Burns, coach of the Central High Lions basketball team, has 12 players on his squad. Of these, 3 are centers, 4 are forwards, and 5 are guards.
    a. Is it correct to say that a team requires a center and 2 forwards and 2 guards, or is it correct to say that a team requires a center or 2 forwards or 2 guards?
    b. In how many ways can Coach Burns select his starting team?

17. A telephone exchange consists of all phone numbers with the same three-digit prefix.
    a. How many different phone numbers are possible in a given exchange?
    b. If a community has 95,000 telephone subscribers, what is the minimum number of exchanges needed?

18. Allison Gerber, a math teacher at Central High, gives prizes to the students in her class who have improved their average. At the end of each term she places the names of all qualifying students in a container and draws 3.
    a. Suppose that this term there are 19 such students in Ms. Gerber's classes and the prizes are three Central High Lions T-shirts. In how many ways can the prizes be given?
    b. Again suppose that there are 19 such students but that the prizes are a new calculator, a Lions T-shirt, and a book on discrete mathematics. In how many ways can the prizes be given?

19. Dominoes come in different-sized sets, but a double-six set is the most common. In a double-six set, each half of a domino may have any number of spots from 0 to 6. The two halves of a given domino in the set pair a number of spots with itself or with another number of spots.
    a. How many dominoes with the same number of spots on each half are there in a double-six set?

b. If every possible pairing is included in the set, how many dominoes with a different number of spots on each half are there in a double-six set?
c. What is the total number of dominoes in a double-six set?
d. If you select a domino at random from a double-six set, what is the probability that it will have the same number of spots on each half?
e. How many dominoes are there in a double-twelve set?

20. How many different sums of money can be made from a $1 bill, a $5 bill, a $10 bill, and a $20 bill? (Hint: Use one bill at a time or two bills at a time or three bills at a time or four bills at a time.)

21. Many card games involve hands of 5 cards.
   a. How many different 5-card hands can be dealt from a standard 52-card deck?
   b. In how many ways can 3 aces be selected from the 4 that are found in a standard deck?
   c. In how many ways can 3 cards of the same kind (aces, twos, threes, etc.) be dealt from a standard deck? (Hint: You can deal 3 aces or 3 twos or 3 threes. . . .)
   d. Repeat the last question for 2 cards of the same kind.
   e. In how many ways can a hand consisting of 3 of one kind and 2 of another be dealt from a standard deck?

Projects

22. Research one or more of the lotteries in your area. How large are the jackpots? How many tickets are usually sold? What portion of the proceeds goes to the players? What happens to the rest of the money? Are there any rules to prevent the kind of purchase made by the Australian group in the Virginia lottery? What kinds of strategies are known to be used by players?

23. Investigate probabilities of common card hands. Show how to calculate as many as possible.

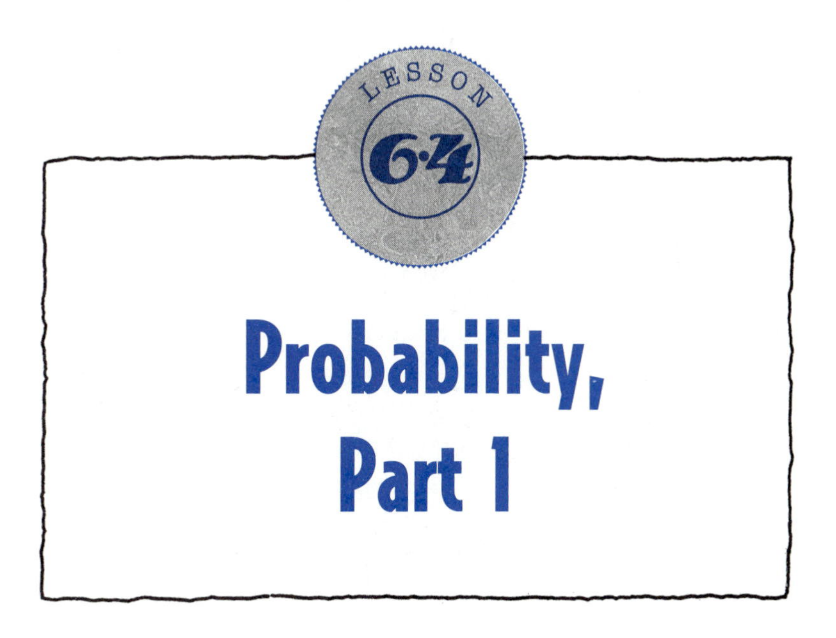

Probability, Part 1

In this and the next lesson we consider some of the basic laws that govern calculations with probabilities. First, recall that the probability of an event is the ratio of the number of ways in which the event can occur to the total number of possibilities. For example, the probability that a die will fall with an even number showing is $\frac{3}{6}$ because three of the six possibilities are even. For convenience, the statement "the probability that a die will fall even" is sometimes abbreviated *p*(a die will fall even).

In the previous lessons of this chapter, we used the addition and multiplication principles to determine the number of ways in which events can occur. Similar principles govern calculations with probabilities.

> The addition principle for probabilities states that for two events A and B, $p(\text{A or B}) = p(\text{A}) + p(\text{B}) - p(\text{A and B})$. If the events are disjoint, then $p(\text{A or B}) = p(\text{A}) + p(\text{B})$.

Consider, for example, Exercise 11 of Lesson 6.2 on page 279. The number of people on a school football team was 45, the number on the

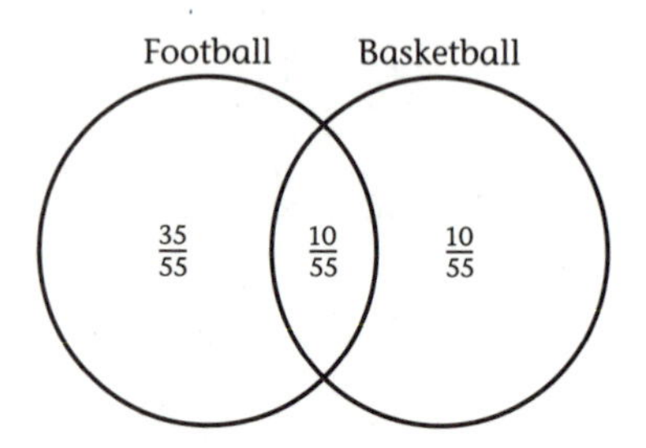

Figure 6.1 The addition principle for probabilities.

basketball team was 20, and the number of people on both was 10. To determine the total number of people involved, you added 20 and 45 and then subtracted 10 to get 55.

Now suppose that a person is chosen at random from this group. The probability of choosing a football player is 45/55; the probability of choosing a basketball player is 20/55; and the probability of choosing someone who is both a football and a basketball player is 10/55. To determine the probability of selecting either a football or basketball player, perform a calculation similar to the one in the previous paragraph: $45/55 + 20/55 - 10/55 = 55/55$ or 1 (see Figure 6.1). Without the subtraction of 10/55, the probability would exceed 1, which is impossible.

Recall that when events have nothing in common, they are called disjoint. When events are disjoint, it is not necessary to include the subtraction in this calculation because the probability of both events occurring is 0.

In order to apply the addition principle properly, it is necessary to know whether events are disjoint. Next we consider the multiplication principle and discover that in order to apply it properly, it is necessary to know whether events are independent.

As an example, examine the following data on the student population at Central High, which contains exactly 1,000 students:

	Male	Female	Total
Seniors	156	144	300 (30%)
Juniors	168	172	340 (34%)
Sophomores	196	164	360 (36%)
Total	520 (52%)	480 (48%)	

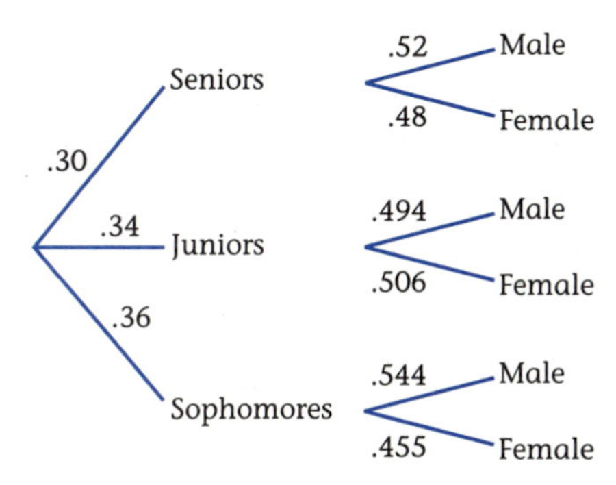

Figure 6.2 Organizing probabilities with a tree diagram.

Many probabilities can be calculated from this set of data. For example, the probability of selecting a junior is .34, and the probability of selecting a girl from the junior class is 172/340 = .5059. A tree diagram with the appropriate probabilities written along the branches is a convenient way to organize these probabilities (see Figure 6.2).

The probability of selecting a girl from the junior class is called a **conditional probability.** It does not state the probability of selecting a girl from the student body as a whole but, rather, the probability of selecting a girl with the condition that the person selected be a junior. Conditional probabilities are often stated with the word *from.* (Other commonly used words are *if, when,* and *given that.*)

Unlike probability statements that use *and* and *or,* conditional probabilities change when the order of the events is reversed. The probability of selecting a girl from the juniors (172/340 = .5059), for example, is not the same as the probability of selecting a junior from the girls (172/480 = .3583).

Refer back to the table and note that 172 of the 1,000 students are both girls and juniors. Thus, the probability of selecting a student who is both a junior and a girl is .172. Now refer to the tree diagram and note the probabilities written along the junior branch and the female branch that follows it. These are .34 and .506, respectively. The product of .34 and .506 is .172. Therefore the probability of selecting a student who is a junior and a girl is the same as the product of the probability of selecting a junior and the probability of selecting a girl from the juniors.

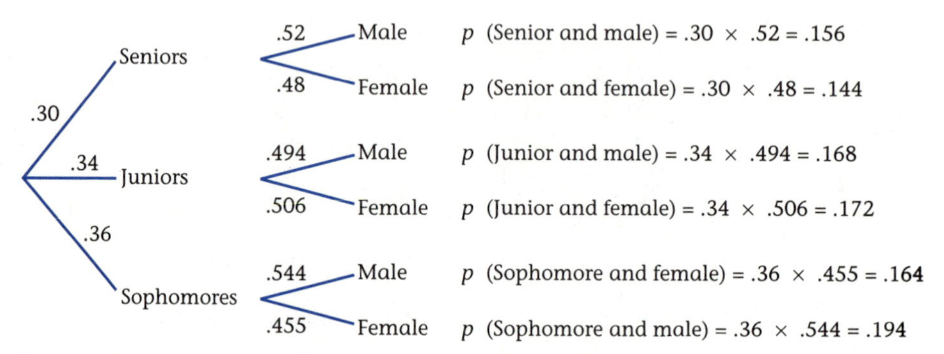

Figure 6.3 The multiplication principle for probabilities.

> The multiplication principle for probabilities states that for two events A and B, $p(\text{A and B}) = p(\text{A}) \times p(\text{B from A})$.

The products that reslt from applying the multiplication principle are often written along with the associated events to the right of the tree diagram, as shown in Figure 6.3.

Refer back to the table and note that 48% of the students are girls. Look at the tree diagram and you'll note something a bit unusual about the senior class. Exactly the same percentage of seniors are girls as there are in the whole school. For this reason, the events selecting a girl and selecting a senior are examples of independent events.

When events are independent, the probability that one occurs is not affected by the occurrence of the other. If you want to know the probability of selecting a girl, it makes no difference whether she is selected from the entire school or from the senior class—the probability will still be .48. Selecting a girl is independent of selecting a senior. The probability of selecting a girl does, however, depend on whether she is selected from the entire school (480/1,000 = .48) or from the junior class (172/340 = .5059). Selecting a girl, therefore, is not independent of selecting a junior.

> Two events A and B are **independent** if $p(\text{B from A})$ is the same as $p(\text{B})$ or if $p(\text{A from B})$ is the same as $p(\text{A})$.

Consider each of the following pairs of events and determine whether or not they are independent:

1. Selecting a senior and selecting a male.

2. Selecting a junior and selecting a male.

3. Selecting a sophomore and selecting a female.

Did you say that the events in the first pair are independent but that those in the last two pairs are not?

When events are independent, a modified form of the multiplication principle can be used, because for independent events only, p(B) is the same as p(B from A), and p(B) can be used as a replacement for p(B from A).

> The multiplication principle for probabilities states that $p(\text{A and B}) = p(\text{A}) \times p(\text{B})$ only when A and B are independent.

In practice, it is often obvious whether or not two events are independent. Consecutive tosses of a coin, for example, have no effect on each other. The probability that a particular toss is heads is $\frac{1}{2}$ regardless of the outcome of the previous toss. This independence means that it is correct to calculate the probability of obtaining two heads in a row by multiplying $\frac{1}{2}$ by $\frac{1}{2}$ to obtain $\frac{1}{4}$.

A man who has a beard is more likely to have a moustache than are men in general, so having a beard and having a moustache are not independent. It would therefore be incorrect to calculate the probability that a man has both a beard and a moustache by multiplying the probability that a man has a beard by the probability that a man has a moustache.

In other cases, it is not possible to determine whether two events are independent without inspecting data to resolve the matter. For example, are people who own poodles more or less likely to own pink cars than are people in general? If either is the case, then the events owning a poodle and owning a pink car are not independent.

There are two ways to examine events A and B for independence:

1. Check the probability of A occurring alone and the probability of A occurring from B only. If they are the same, A and B are independent. (You can also compare p(B) with p(B from A), but it is not necessary to make both comparisons.)

2. Multiply the probability of A by the probability of B. If the result is the same as the probability of both A and B occurring, A and B are independent.

Example

The following table represents ownership of pink cars and poodles in a community. Are owning a pink car and owning a poodle independent in this community?

	Own Poodles	Don't Own Poodles
Own pink cars	250	450
Don't own pink cars	1,250	18,350
Totals	1,500	18,800

The probability of selecting someone who owns a pink car is 700/20,300 = .0345, and the probability of selecting someone who owns a pink car from the poodle owners is 250/1,500 = .167. Because these are not equal, owning a pink car and owning a poodle are not independent.

To use the second method of checking for independence, note the following probabilities: p(owning a poodle) = 1,500/20,300 = .0739, p(owning a pink car) = 700/20,300 = .0345, and p(owning a poodle and owning a pink car) = 250/20,300 = .0123. Calculate the product of the first two probabilities and compare it with the third: $.0739 \times .0345 = .00255 \neq .0123$.

The notion of independence is a subtle one that sometimes requires careful thinking. Failing to pay attention to independence can have serious consequences, as you'll see in the exercises that follow.

Exercises

1. Use the table of Central High student population data given on page 294.
   a. What is the probability of selecting a male?
   b. What is the probability of selecting a male from the sophomore class only?
   c. Use your answers to parts a and b to determine whether the events selecting a male and selecting a sophomore are independent.

d. What is the probability of selecting a sophomore?
e. What is the probability of selecting someone who is both male and a sophomore? Compare this probability with the product of the probability of selecting a male and the probability of selecting a sophomore.
f. Multiply the probability of selecting a male from the sophomore class by the probability of selecting a sophomore. Compare this with the probability of selecting someone who is both a male and a sophomore.
g. Is the probability of selecting a male from the sophomore class the same as the probability of selecting a sophomore from the males?

2. Use the table of Central High student population data given on page 294.
   a. What is the probability of selecting someone who is either a male or a sophomore?
   b. Add the probability of selecting a sophomore and the probability of selecting a male. Compare this with the probability of selecting someone who is either a sophomore or a male.
   c. Are the events selecting a male and selecting a sophomore disjoint?

3. The tree diagram on page 301 represents the Central High data of this lesson, with the events selecting a male and selecting a female drawn first.
   a. Write the correct probabilities along each branch, and complete the calculations at the right.
   b. Find the sum of the probabilities you calculated and wrote on the right.

4. A card is drawn from a standard 52-card deck.
   a. What is the probability of drawing an ace?
   b. What is the probability of drawing an ace from the diamonds only?
   c. Are the events drawing an ace and drawing a diamond independent?
   d. What is the probability of drawing a diamond?

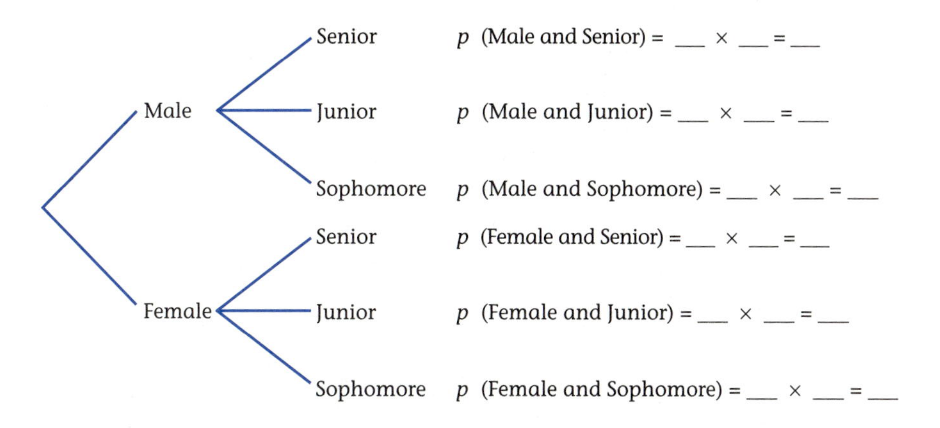

e. What is the probability of drawing a card that is both a diamond and an ace? Compare this probabili with the product of the probability of drawing an ace and the probability of drawing a diamond.
f. What is the probability of drawing a card that is either an ace or a diamond?
g. Are the events drawing an ace and drawing a diamond disjoint?
h. Analyze the events drawing a king and drawing a face card (jack, queen, king). Are they independent? Are they disjoint? Explain your answers.

5. Suppose that two cards are drawn from a standard deck.
   a. What is the probability that the first card will be red?
   b. What is the probability that the second card will be red if the first card is red and is not put back in the deck?
   c. What is the probability that the first card will be red and the second card will be red if the first card is not put back in the deck?
   d. Compare your result in the previous question with your result in Exercise 5 of Lesson 6.3, p. 288.
   e. The tree diagram on page 302 represents the colors of cards when two cards are drawn in succession from a standard deck without the first card's being put back. Write the correct probabilities along each branch, and complete the calculations at the right.

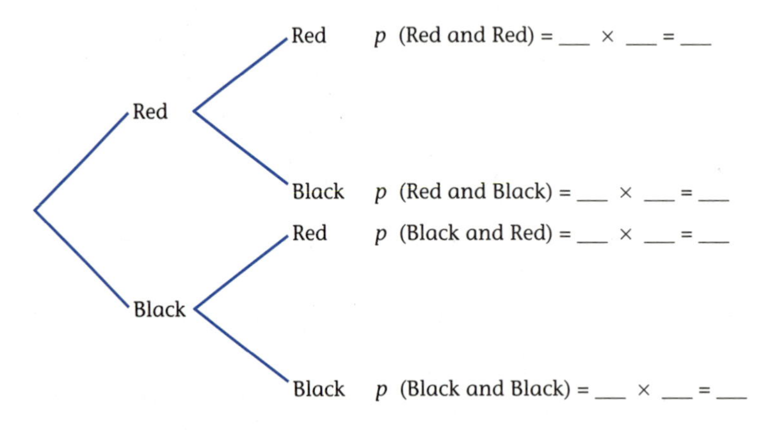

f. What is the probability that one or the other of the two cards will be red? This will occur if the first card is red and the second card is black or if the first card is black and the second card is red. Find this probability by adding two of the probabilities you calculated and wrote to the right of the tree diagram.

6. Again, two cards are drawn from a standard deck.
   a. What is the probability that the first card will be red?
   b. What is the probability that the second card will be red if the first card is red and is put back in the deck?
   c. What is the probability that the first card will be red and the second card will be red if the first card is put back in the deck?
   d. The tree diagram below represents the colors of cards when two cards are drawn in succession from a standard deck, with the first card replaced before the second is drawn. Write the correct

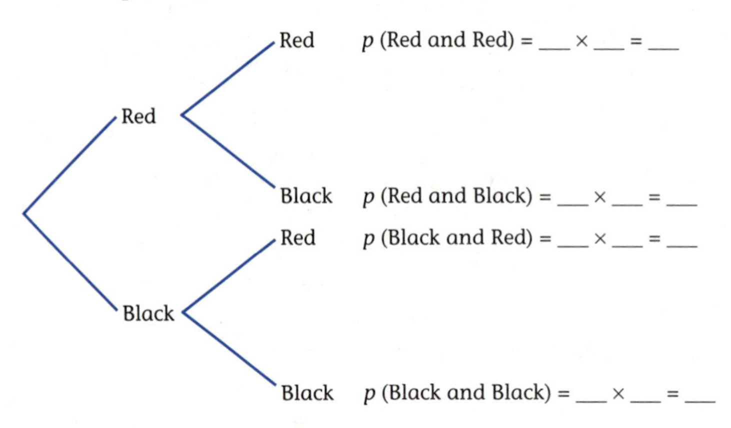

probabilities along each branch, and complete the calculations at the right.

e. What is the probability that one or the other of the two cards will be red?

f. Is the second draw independent of the first in the situation of Exercise 5 or in the situation of this exercise?

7. The probability that Coach Burns's Central High basketball team will win its first game of the season is .9, and the probability the team will win its second game of the season is .6.
   a. What is the probability that the team will win both its first and its second games?
   b. What assumption did you make about the outcomes of the first and second games in answering the previous question? Do you think that this is a safe assumption?

8. The probability of rain today is .3. Also, 40% of all rainy days are followed by rainy days, and 20% of all days without rain are followed by rainy days. The tree diagram below represents the weather for today and tomorrow.
   a. Write the correct probabilities along each branch, and complete the calculations at the right.
   b. What is the probability that it will rain on both days?
   c. What is the probability that it will rain on one of the two days?
   d. What is the probability that it will not rain on either day?
   e. Is tomorrow's weather independent of today's? Explain.

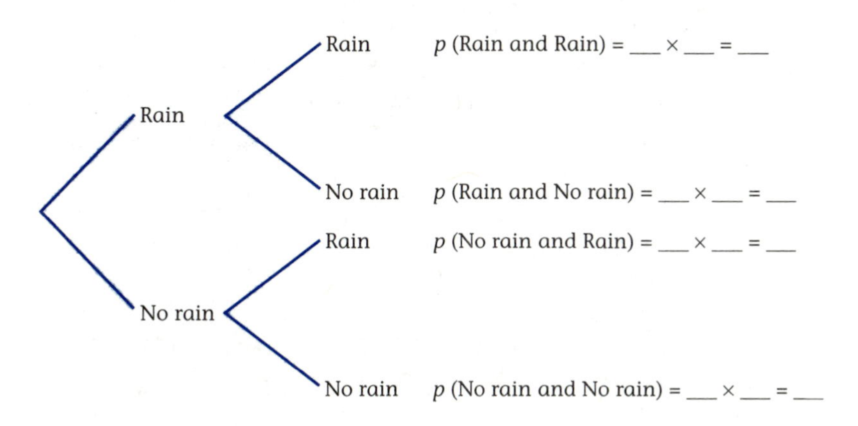

9. Some Americans favor mandatory HIV screening for workers in health care and other professions. If a medical test is 98% accurate—that is, if it correctly reports the existence or absence of a disease in 98% of all cases—then why do so many people who test positive for the disease not have it? Let's consider such a test that fails to report the existence of the disease in only 2% of those who have it, and that incorrectly reports the existence of the disease in 2% of those who don't have it. About 2 people out of 1,000 have the disease.
   a. Write the appropriate probabilities along each branch of the tree diagram, and complete the calculations at the right.

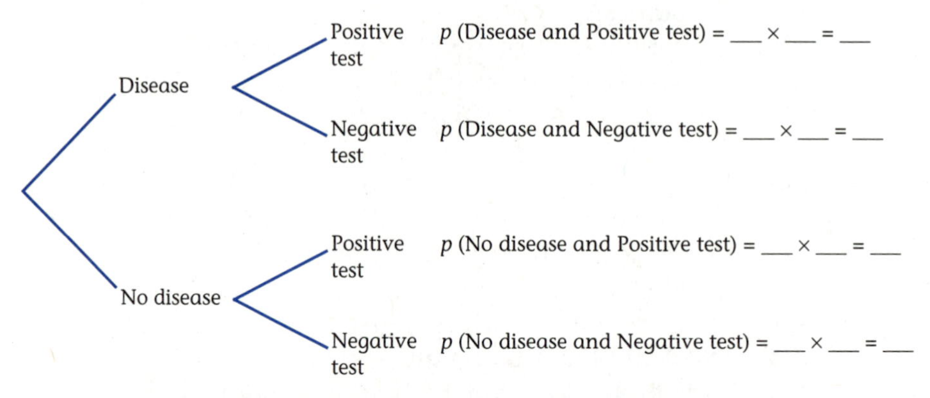

   b. If 100,000 people are screened for the disease, about how many can be expected to test positive for the disease?
   c. Of the 100,000 people, how many people who test positive actually have the disease?
   d. What is the probability that a person who tests positive for the disease actually has it?

10. Consider the dice game proposed by Chuck in the first lesson of this chapter on page 267 and suppose that you have bet on the number five.
   a. Are the outcomes of the two dice independent of each other?
   b. What is the probability of a five's appearing on the first die and on the second?
   c. The tree diagram on page 305 represents the outcomes of the two dice. Write the correct probabilities along each branch, and complete the calculations at the right.

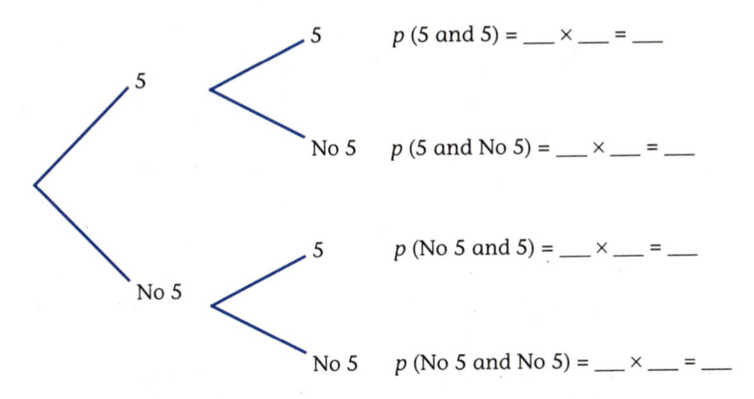

d. A single five can appear if either the first die shows a five or the second shows a five. What is the probability that exactly one five will appear?
e. What is the probability that no fives will appear? How is this probability related to probabilities you found in parts b and d?
f. Do you think you would win or lose money in the long run if you played this game?

11. The multiplication principle for independent events can be used for more than two such events. The carnival game Chuk-a-Luk, for example, is similar to the game proposed by Chuck in Lesson 6.1, except that three dice are rolled. If you bet on the number five, you can calculate the probability that all three dice will show a five by cubing the probability that a single die will show a five. What is the probability that all three dice will show a five?

12. The probability that one of Ms. Howe's plants will bloom is .9. What is the probability that all five will bloom?

13. European roulette wheels contain the numbers 0 through 36. The marble thrown into the spinning wheel has an equal chance of landing in any of them. The British actor Sean Connery once bet on the number 17 three times in a row and the marble landed on 17 all three times. What is the probability that the marble will land on 17 three times in a row?

14. On January 28, 1986, the space shuttle *Challenger* exploded over Florida. Many authorities believe this tragedy could have been

prevented if closer attention had been paid to the laws of probability. The rocket that carried the shuttle aloft was separated into sections that were sealed by large rubber O-rings. Experts believe that at least one of the O-rings leaked burning gases that caused the rocket to explode. Studies after the tragedy found that the probability a single O-ring would work was .977.

a. If the six O-rings were truly independent of one another, what is the probability that all six would function properly?
b. The game of Russian roulette is played by spinning the cylinder of a revolver containing a single cartridge and then pulling the trigger. What is the probability the gun will fire if it has six cylinders?
c. Compare the probability that all six O-rings will function properly with the probability that the gun will not go off in Russian roulette.
d. Each joint was actually sealed by a system of two O-rings that were supposed to be independent of each other but were not. If the probability that one O-ring will fail is $1 - .977 = .023$, what is the probability that both O-rings in a system of two will fail if the two are independent of each other?
e. Subtract your previous answer from 1 to determine the probability that a single joint's seal will work, and find the probability that all six will function properly. Keep in mind that this answer assumes a faulty assumption of independence of the pair of O-rings in a single joint.

15. It has been estimated that about one automobile trip in 100,000 ends in an injury accident.
    a. What is the probability that a given automobile trip will not end in an injury accident?

NEWS CLIP

NASA Faulted on O-Ring Check

Sensors Could Have Monitored Booster Pressure, Experts Say

By Michael Schrage, WASHINGTON POST, March 23, 1986

CAPE CANAVERAL—Despite longstanding concerns about the design of the space shuttle's O-rings, NASA chose not to monitor their performance in flight, even though, according to rocket engineers, it should have done so.

In-flight monitoring could have yielded insights into how well the synthetic rubber O-rings prevented escape of hot gases through joints in the solid-rocket booster.

That information conceivably could have alerted National Aeronautics and Space Administration and Thiokol engineers to potential problems with the O-rings, prime suspects in the Jan. 28 Challenger accident. Investigators say they believe that one or both of the two O-rings in the right solid-rocket booster failed to seal, allowing gas burning at 5,900 degrees Fahrenheit to spew out the side.

"If [the sensors indicated] full pressure between the O-rings, then you know the primary O-ring never worked and the second one could be working," Fitzgerald said. "If there's 100 to 200 pounds-per-square-inch pressure, you'd know the inner O-ring didn't seal immediately" and if pressures changed later, "you'd know there was blowby," or erosion of the O-rings permitting a leak.

b. If you make approximately three automobile trips a day, about how many would you make over a 30-year period?
c. If the outcome of a particular automobile trip is independent of the previous one, what is the probability that all the trips you make over a 30-year period will not end in an injury accident?

16. Forty percent of the inhabitants of Wilderland are Hobbits, and 60% are humans. Furthermore, 20% of all Hobbits wear shoes, and 90% of all humans wear shoes.
    a. Make a tree diagram to show the breakdown of residents into Hobbits who either do or do not wear shoes and humans who either do or do not wear shoes. Write the appropriate probabilities along the branches, and write the appropriate events and calculated probabilities to the right of the diagram.
    b. A group of 10,000 residents are selected at random. About how many would you expect to be Hobbits who wear shoes?
    c. About how many of the 10,000 would you expect to be Hobbits who do not wear shoes?
    d. About how many of the 10,000 would you expect to be humans who wear shoes?
    e. About how many of the 10,000 would you expect to be humans who do not wear shoes?
    f. What percentage of the inhabitants who wear shoes are Hobbits?
    g. If an inhabitant is selected at random, are any of the events selecting a Hobbit, selecting a human, selecting someone who wears shoes, or selecting someone who doesn't wear shoes independent? Are any of them disjoint?

17. Suppose that a witness to a crime sees a man with red hair fleeing the scene and escaping in a blue car. Suppose that 1 man in 10 has red hair and that 1 man in 8 owns a blue car?
    a. What is the probability that a man selected at random has red hair and owns a blue car?
    b. What assumption did you make when you answered the previous question?
    c. The table on the next page represents the men in the community in question. Inspect the data and present an argument either in favor of or against the assumption.

	Blue Car	No Blue Car
Red hair	1,990	14,010
Nonred hair	18,015	125,985

18. Many games involve rolling a pair of dice.
   a. In how many ways can a pair of dice fall?
   b. What is the probability of rolling a pair of sixes?
   c. What is the probability of rolling a pair of sixes twice in succession?
   d. What is the probability of not rolling a pair of sixes?
   e. What is the probability of not rolling a pair of sixes twice in succession?
   f. A New Yorker gambler nicknamed "Fat the Butch" once bet that he could roll at least one pair of sixes in 21 chances. Find the probability of not rolling a pair of sixes in 21 successive rolls of a pair of dice. (By the way, he lost about $50,000 in the course of several bets.)

Projects

19. Gather data on the breakdown of the student population in your school into categories similar to those of the Central High population in this lesson. Determine whether the data exhibit any events that approximate independence.

20. Read the articles on trial by mathematics in the January 8, 1965, and April 26, 1968, issues of *Time* magazine, and report on the results of erroneously multiplying probabilities of events that are not independent.

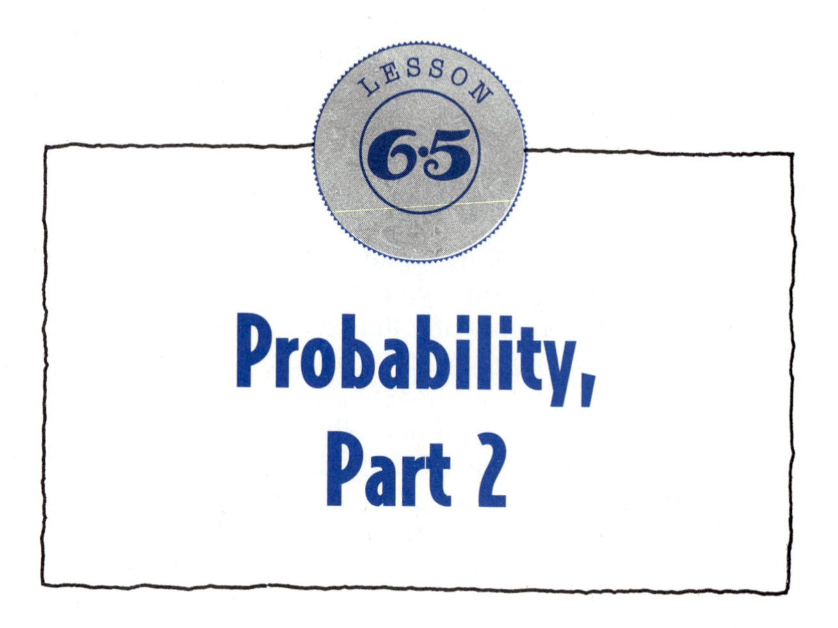

Probability, Part 2

In order to understand games like the ones proposed by Pierre, Hilary, and Chuck in Lesson 6.1, it is necessary to understand probability distributions and the related notion of expectation. These topics are considered in this lesson.

As an example, consider a brief quiz of three questions in which the answers are either true or false. Because there are only two possible outcomes for any one question, the process of answering a single question is called **binomial.** When the process is repeated several times, however, the multiplication principle states that there are more outcomes than the two that are possible in a single trial. With three trials there are $2 \times 2 \times 2 = 8$ outcomes. A tree diagram provides a full list of these (see Figure 6.4).

Because each of the outcomes is as likely as any other, the probability associated with each of the eight is $\frac{1}{8}$. These probabilities can also be calculated by writing a probability of $\frac{1}{2}$ along each branch and multiplying to obtain each individual probability (see Figure 6.5).

The results can be collected into a table called a **probability distribution.** The way in which the table is constructed depends on how

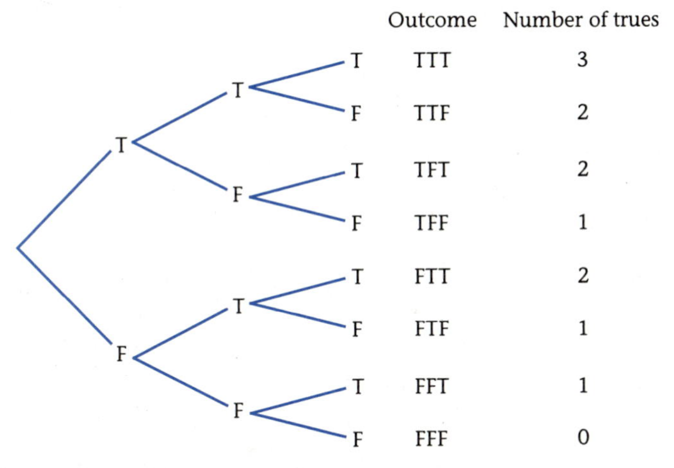

Figure 6.4 A three-question true/false quiz.

the situation is described. If the point of interest is the number of "true" answers, then the table is as follows:

Number of true	0	1	2	3
Probability	$\frac{1}{8}$	$\frac{3}{8}$	$\frac{3}{8}$	$\frac{1}{8}$

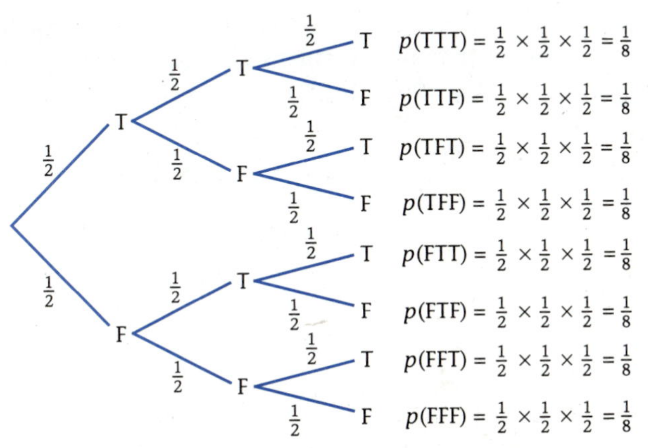

Figure 6.5 Probabilities in a three-question true/false quiz.

The tree diagram approach to obtaining these probabilities is fine if the total number of outcomes is small. In practice, however, this is seldom the case. The probabilities can also be determined by using counting techniques and multiplication of probabilities for independent events.

Consider, for example, the probability that exactly two of the three answers will be true. The blanks below represent the three questions. In how many ways can you select two of them to mark true?

Possibility 1	Possibility 2	Possibility 3
___ ←	___ ←	___
___ ←	___	___ ←
___	___ ←	___ ←

Because the order of selection of the two questions to mark true does not matter, the number of ways of selecting two questions to mark true can be counted as $C(3, 2) = 3!/(2!1!) = 3$. The probability associated with each true answer is $\frac{1}{2}$, and the probability associated with each false answer is also $\frac{1}{2}$, so the probability of two true answers followed by one false is $(\frac{1}{2})^2(\frac{1}{2})$. This probability is multiplied by 3 because there are three different ways that it can occur.

Suppose that the quiz has 10 questions. What is the probability that exactly four are true? If there are four true, there must also be six false, so the probabilities that must be multiplied are four $\frac{1}{2}$s for the true and six $\frac{1}{2}$s for the false, which gives $(\frac{1}{2})^4(\frac{1}{2})^6$. The number of different ways in which the four true questions can be selected is $C(10, 4)$. The correct probability, therefore, is $C(10, 4)(\frac{1}{2})^4(\frac{1}{2})^6 = 210(\frac{1}{16})(1\frac{1}{64})$, or about .205.

$$C(10, 4)\left(\frac{1}{2}\right)^4\left(\frac{1}{2}\right)^6 = \frac{10!}{4!6!}\left(\frac{1}{2}\right)^4\left(\frac{1}{2}\right)^6$$

Note that the denominators of the combination formula match the powers of the probabilities. A sequence of calculator steps that performs this calculation is

$$10! \div 4! \div 6! \times 0.5\ y^x\ 4 \times 0.5\ y^x\ 6 = .$$

When using this technique, the probabilities of a single question's being true or false are multiplied several times. It is essential, therefore, that the individual outcomes be independent of one another. If, for example, there is a reason that the answer to one question depends on the answer to another, then the calculated probabilities will be invalid.

In the previous examples, the probability associated with a given outcome is $\frac{1}{2}$. In many applications, however, this is not the case. Consider, for example, the dice game proposed by Chuck in Lesson 6.1. Because this game involves only two dice, it is not difficult to analyze it with tree diagrams, as we did in previous exercises. The analysis here, however, counts the number of ways of winning and multiplies probabilities.

If you bet on the number 5 in Chuck's game, then you will win $2 if two fives appear, win $1 if a single five appears, or lose $1 if no fives appear. Consider the possibility that one five will appear. There are $C(2, 1)$ ways to select the die that shows a five, and the probability associated with each is $\frac{1}{6} \times \frac{5}{6}$ because one of the dice must be a five and the other must not be. The correct probability is therefore $C(2, 1)(\frac{1}{6})^1(\frac{5}{6})^1$, or about .278.

Similarly, the probabilities of no fives and two fives can be calculated as $C(2, 0)(\frac{1}{6})^0(\frac{5}{6})^2$ and $C(2, 2)(\frac{1}{6})^2(\frac{5}{6})^0$. These are summarized in the distribution table shown below.

Amount won	−1	1	2
Probability	.694	.278	.028

In general, if p is the probability associated with a single binomial outcome, the probability of n successes in m attempts will be $C(m, n)(p)^n(1 - p)^{m-n}$, provided that the individual trials are independent of one anther.

How would a player do in this game in the long run? The player's **expectation** is used to determine this. To calculate expectation, multiply each amount by the probability of winning that amount: $-1(.694) + 1(.278) + 2(.028) = -.36$. The expectation is the average amount per play that a player would win. If the player plays the game 100 times, he can expect to be about 100(.36) = \$36 behind.

If the Central High council decides to run this game at the school's fair, the council's viewpoint will be the opposite of that of the player: It loses \$2 when two fives appear, loses \$1 when one five appears, and makes \$1 when no fives appear. Therefore, the council's expectation is +\$.36. The council can expect to make about \$36 for each 100 times the game is played.

Example

A quality control engineer at the manufacturing plant of an electronics company randomly selects five compact disc players from the assembly line and tests them for defects. If a problem on the assembly line causes the factory to produce 10% defective, what is the probability that it will be detected by the engineer's test? The table below shows the probability distribution for the number of defective players that the engineer found.

Number Defective	Probability Calculation	Probability
0	$C(5, 0)(.1)^0(.9)^5$	.590
1	$C(5, 1)(.1)^1(.9)^4$	.328
2	$C(5, 2)(.1)^2(.9)^3$	.073
3	$C(5, 3)(.1)^3(.9)^2$	.008
4	$C(5, 4)(.1)^4(.9)^1$	.000
5	$C(5, 5)(.1)^5(.9)^0$	.000

The last two are 0 when rounded to three places.

There is about a 59% chance that the engineer will fail to find a defective player if the defective level is 10%. The probability the engineer will find at least one defective player is 41%. This can be found by subtracting 59% from 100% or by adding the last five probabilities in the table.

The engineer's expectation can be calculated as 0(.590) + 1(.328) + 2(.073) + 3(.008) + 4(.000) + 5(.000), or about .498. In this case, the expectation is the average number of defective players that the engineer detects. To put it another way, if the engineer goes through this routine once each day and the defective level is at 10%, the engineer can expect, on the average, to detect about one-half a defective player a day, or one every two days.

Exercises

1. Hale Ault, a student at Central High, is known for occasionally neglecting his studies. When he finds a question on an exam that he cannot answer, he uses one of several random processes as an aid. Examine each of Hale's schemes and discuss, first, whether each outcome has the same probability of occurring as does each of the others and, second, whether several successive applications of the scheme are independent of one another.
   a. On a true/false question, flip a coin and answer true if the result is heads, false if it is tails.
   b. On a three-choice multiple-choice question, flip two coins. Mark the first answer if both are heads, the second if both are tails, and the third if the result is one head and one tail.
   c. On a four-choice multiple-choice question, associate each of your fingers on one hand with one of the choices, slap your fingers against the desk, and select the one that hurts the most.
   d. On a four-choice multiple-choice test that allows the use of scientific calculators, use the calculator's random number generator to display a random number between 0 and 1. Mark the first answer if the number is below .25, the second if it is between .25 and .5, the third if it is between .5 and .75, and the fourth if it is between .75 and 1.

2. Ms. Howe is giving a five-question true/false quiz.
   a. In how many ways can a student select three of the questions to mark true?
   b. Complete the following calculation of the probability that exactly three of the questions on the quiz are true: C(______, ______) (______)——(______)—— = ______.
   c. Complete the probability distribution for the number of true answers:

Number of true	0	1	2	3	4	5
Probability	—	—	—	—	—	—

   d. Ms. Howe has a bias toward true answers, and so her questions have true answers about 70% of the time. Recalculate the probability distribution:

Number of true	0	1	2	3	4	5
Probability	___	___	___	___	___	___

3. Hale Ault is taking a 10-question true/false quiz on which the answers have equal chances of being true or false. Hale is doing the quiz by guessing and needs at least 6 correct in order to pass.
   a. Find the probability of exactly 6 correct answers.
   b. Find the probability of exactly 7 correct answers.
   c. Find the probability of exactly 8 correct answers.
   d. Find the probability of exactly 9 correct answers.
   e. Find the probability of exactly 10 correct answers.
   f. Hale will pass if he gets 6 right or if he gets 7 right or if he gets 8 right or if he gets 9 right or if he gets 10 right. What is the probability that Hale will pass the quiz?
4. Recall that the game of Chuk-a-Luk is played with three dice and that you win $1 for each time your number shows but lose $1 if your number doesn't show.
   a. Suppose you bet on the number 5. Complete the following calculation of the probability that 5 shows exactly once: C(______, ______)(______)______(______)______ = ______.
   b. Complete the probability distribution for the number of times that 5 shows:

Number of fives	0	1	2	3
Probability	___	___	___	___

   c. Complete the following calculation of your expected winnings: (Note that the distribution of winnings is different from the distribution of the number of fives because zero fives results in a loss of $1.)

   −1(______) + 1(______) + 2(______) + 3(______) = ______.

   d. If you play the game 100 times, about how much money should you expect to win or lose?
5. A list of people eligible for jury duty contains about 40% women. A judge is responsible for selecting six jurors from this list.
   a. If the judge's selection is made at random, what is the probability that three of the six jurors will be women?

b. Prepare a probability distribution table for the number of women among the six jurors.
c. The judge's selection includes only one woman. Do you think this is sufficient reason to suspect the judge of discrimination? Explain.

NEWS CLIP

Convictions in Jay Bias Homicide Reversed

Prosecutor Violated Equal Rights by Blocking Women From Jury

By Jon Jeter, WASHINGTON POST, April 28, 1993

The Maryland Court of Appeals reversed the convictions yesterday of two men in the killing of James S. "Jay" Bias, the brother of the late college basketball star, Len Bias, because the prosecutor blocked more than a dozen women from serving on the jury solely because of their sex.

The state's highest court ruled that a Prince George's County prosecutor who tried the murder case against Gerald Eiland and Jerry S. Tyler violated the state's equal rights law by barring the women from serving on the jury. The assistant state's attorney, Mark Foley, used 16 of his 20 peremptory strikes—the legal device by which attorneys from both sides may excuse potential jurors without explanation—to bar women from the jury.

The decision extends to women the protection from discrimination provided by a 1986 Supreme Court decision, which reaffirmed that black people cannot be barred from a jury solely because of their race. And it could end a long-standing practice by some prosecutors who eliminate women from juries based on the belief that they are more compassionate than men and less likely to convict, particularly when the accused is a young man.

6. Sickle cell anemia is a genetic disease that strikes an estimated 1 in 400 black children in this country. The disease causes red blood cells to have a crescent shape rather than the normal round shape, which inhibits their ability to carry oxygen. Victims suffer from severe pain and are susceptible to pneumonia and organ failure. Children of parents who both are carriers of the sickle cell gene are frequently stricken.

   a. Normal parents have two normal A genes, and carrier parents have one normal A gene and one sickle S gene. A victim of the disease has two S genes. A child inherits one gene independently from each parent. Complete the probability calculations in the tree diagram representing parents who both are carriers.

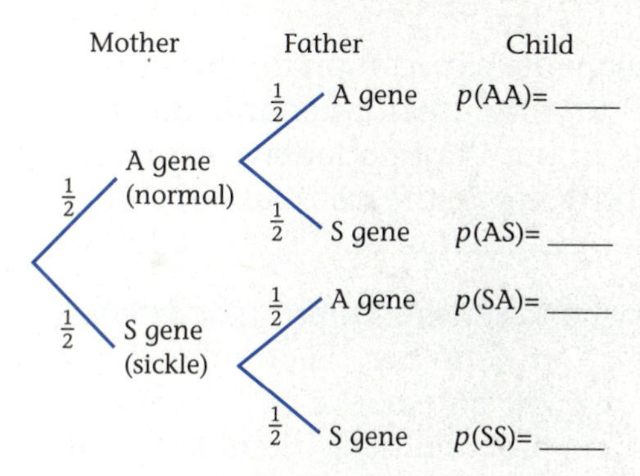

   b. What percentage of children of two carrier parents will have sickle cell anemia? What percentage are carriers? What percentage are normal?

   c. A couple who both are carriers have five children. Complete the following probability distribution for the number of children who have the disease:

Number of children	0	1	2	3	4	5
Probability	___	___	___	___	___	___

   d. Calculate the expectation. What does the expectation mean in this case?

7. A quality control engineer at a widget factory randomly selects three widgets each day for a thorough inspection. Suppose the assembly process begins producing 20% defective widgets.

a. Prepare a probability distribution for the number of defective widgets the engineer finds in the sample of three.
b. Do you think the engineer's quality control scheme is a reliable one? If not, suggest a way to improve it.

8. Suppose that a lottery ticket costs \$1 and requires you to select 6 numbers from the 44 available.
   a. The jackpot is \$27 million, and it is the only prize. Complete the following probability distribution for your expected winnings (see Exercise 9 of Lesson 6.3, p. 289):

Amount won	\$27 million	−\$1
Probability	—	—

   b. Calculate the expectation for this distribution.
   c. Recall that the International Lotto Fund purchased 5 million tickets in the Virginia lottery. Assume that the jackpot is the only prize, revise the distribution, and recalculate the expectation for the International Lotto Fund's winnings.

9. A country has a series of three radar defense systems that are independent of one another. The probability that an enemy plane escapes any one system is .2.
   a. Prepare a probability distribution for the number of radar systems that a plane will escape.
   b. What is the probability that an enemy plane will penetrate the country's defenses?

10. Recall the lottery that Hilary proposed in Lesson 6.1 on page 267. It requires selecting two numbers from the nine available. Hilary has proposed that the price of a ticket be \$1, that the prize for matching both numbers be \$20, and that a prize of \$1 be given to anyone who matches one of the two winning numbers.
   a. Prepare a probability distribution for the amount a player can expect to win.
   b. Calculate the player's expectation.
   c. Would the Central High council make money on this game? Explain. If not, suggest a revision of Hilary's plan for awarding prizes so that the council could make money.

11. A fair coin is tossed several times.
    a. Find the probability of obtaining exactly 5 heads in 10 tosses. Do not do the full probability distribution.
    b. Find the probability of obtaining exactly 10 heads in 20 tosses. Compare this with the previous answer.
    c. Prepare a partial distribution of the probabilities of obtaining 4, 5, or 6 heads in 10 tosses.
    d. Prepare a partial distribution of the probabilities of obtaining 8, 9, 10, 11, or 12 heads in 20 tosses.
    e. Are you more likely to obtain between 40% and 60% heads in 10 tosses or in 20?

12. Extrasensory perception (ESP) is the ability to communicate with another person without speaking. One common test for ESP requires that one person concentrate on a card selected at random from a special deck while the other person records the image that is received or felt.

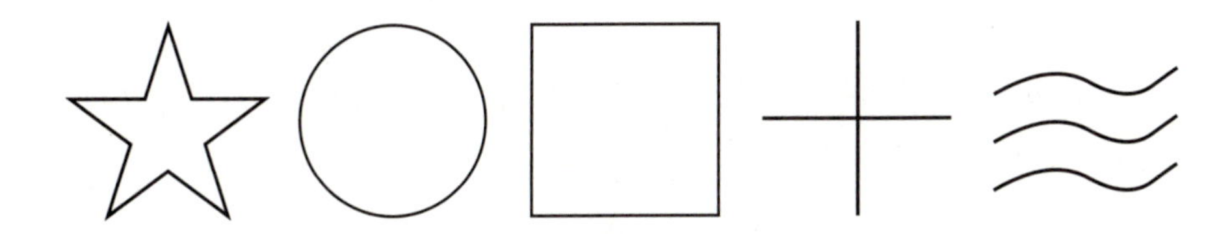

    a. If the deck consists of five of each of the cards shown above, what is the probability of guessing any one card correctly?
    b. Consider a test in which one person selects a card and concentrates while the other person records his or her impression. The card is placed back in the deck; the deck is shuffled; and the experiment is repeated a total of five times. Prepare a probability distribution for the number of cards the receiver can guess correctly.
    c. Suppose the receiver gets more than three correct. What is the probability of this happening by chance?

13. Recall the word game that Pierre proposed in the first lesson of this chapter on page 267. Suppose that the only two-letter words made from the letters of *Lions* that are considered legal are *in, is, on, no,* and *so*.

a. What is the probability that a player will draw a legal word from the letters recorded on the Ping-Pong balls?
b. Pierre proposes that the charge for playing the game be $0.50 and the prize for selecting a legal word be $1. Prepare a probability distribution for the winnings of someone who plays the game.
c. Calculate the player's expectation.
d. Would the council make money? If not, suggest a revision of Pierre's plan so that the council could make money.

14. Sara Swisher, the Central High Lions' star basketball player, has a field goal percentage of 65%.
    a. Sara attempts seven field goals in the first quarter of tonight's game. Prepare a probability distribution for the number of field goals that Sara makes.
    b. What assumption have you made? Do you think this is a realistic assumption?
    c. Calculate the expectation. What does it mean in this case?

15. *Pascal's triangle* is an array of numbers that you might have seen in an algebra class. To construct the triangle, begin with a row of two ones: 1 1. Each new row after this starts and ends with a one, and the other numbers are formed by adding the numbers above and on either side of them in this way:

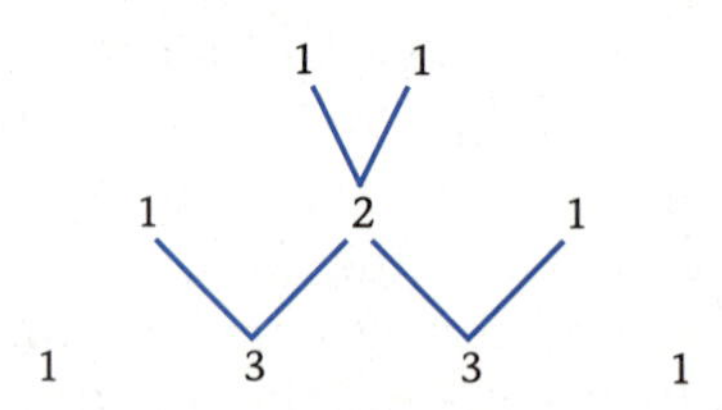

    a. Continue the triangle for three additional rows.
    b. Calculate all possible combinations of five things: $C(5, 0)$, $C(5, 1)$, $C(5,2)$, $C(5, 3)$, $C(5, 4)$, and $C(5, 5)$.
    c. Where do your answers in part b occur in the triangle?
    d. The tree diagram on the following page shows all possible ways of answering true/false quizzes with up to four questions. Fill in

the distributions of the number of trues for quizzes of one, two, three, and four questions by tracing the paths of the diagram. Compare these numbers with those in Pascal's triangle.

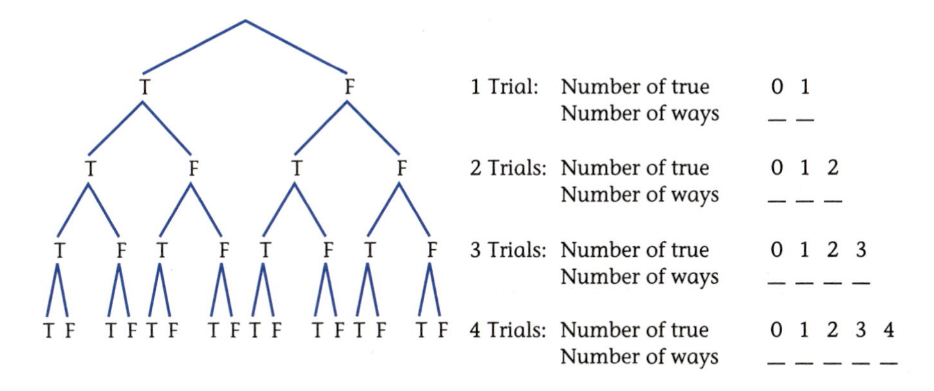

Projects

16. Pascal's triangle contains many patterns other than those in Exercise 15. Investigate and report on some of them.

17. Write a program for your computer or programmable calculator that does probability distributions of the type discussed in this lesson. The program should accept as input the probability of a single success and the number of trials. It should calculate and display each number of successes and the related probability.

1. The following table represents ownership of cats among professionals and nonprofessionals in a community:

	Own a Cat	Don't Own a Cat
Professionals	2,300	11,400
Nonprofessionals	2,600	27,600
Totals	7,900	39,000

a. If a person is selected at random from this group, what is the probability of selecting someone who owns a cat?
b. What is the probability of selecting a cat owner from the professionals?
c. What is the probability of selecting a professional?
d. What is the probability of selecting a person who owns a cat and is a professional?
e. What is the probability of selecting someone who owns a cat or is a professional?
f. Are the events selecting a cat owner and selecting a professional independent?
g. Are the events selecting a cat owner and selecting a professional disjoint?

2. Lesson 1.5 examined voting situations in which some voters received more votes than others. Recall that a coalition is a collection

of some of the voters. Suppose that there are five voters. In how many ways can you form a coalition of one, two, three, four, or five voters from this group?

3. Suppose that teams A and B are playing in the World Series and that each has a 50% chance of winning any game.
   a. What is the probability that team A will win the first four games?
   b. What is the probability that team B will win the first four games?
   c. The series will end in four games if either team A or B wins the first four games. What is the probability that the series will end in four games?

4. a. In how many ways can six books be arranged on a shelf?
   b. If two of the books are math books, how many times do you find the math books in the first two positions? (Hint: Draw six blanks and use the multiplication principle.)
   c. What is the probability that the math books will be in the first two positions?
   d. In how many arrangements are the math books next to each other?
   e. What is the probability that the math books will be next to each other?

5. You are playing a game in which you flip two coins. If both show heads, you will win \$2, if both show tails, you will win \$1, but if the coins don't match, you will lose \$1.
   a. Construct a probability distribution for the amount you could win on a single play of the game.
   b. Calculate your expectation.
   c. If you played the game 100 times, about how much could you expect to win or lose?

6. Being listed first on an election ballot is known to have a favorable effect on a candidate's chances. In order to minimize this effect, ballots can be printed by computer in such a way that each candidate is first on some ballots but not on all. There are three candidates for mayor and five candidates for city council in a local elec-

tion. How many different orderings are possible if the mayoral candidates must be listed before the council candidates? If 20,000 people vote, about how many would see each ballot?

7. Are you more likely to win the jackpot in a lottery that requires the selection of 5 numbers from 39 or in one that requires the selection of 6 numbers from 36? Suppose that 3 million tickets are sold in each lottery. About how many winning tickets can you expect in each?

8. Two cards are drawn from a standard deck, and the first card is not put back before the second card is drawn.
   a. What is the probability that the first card will be a heart?
   b. What is the probability that the second card will be a heart if the first card is a heart?
   c. What is the probability that the first card will be a heart and the second card will be a heart?
   d. The tree diagram below represents the two draws. Write the correct probabilities along the branches, and calculate the probabilities at the right.

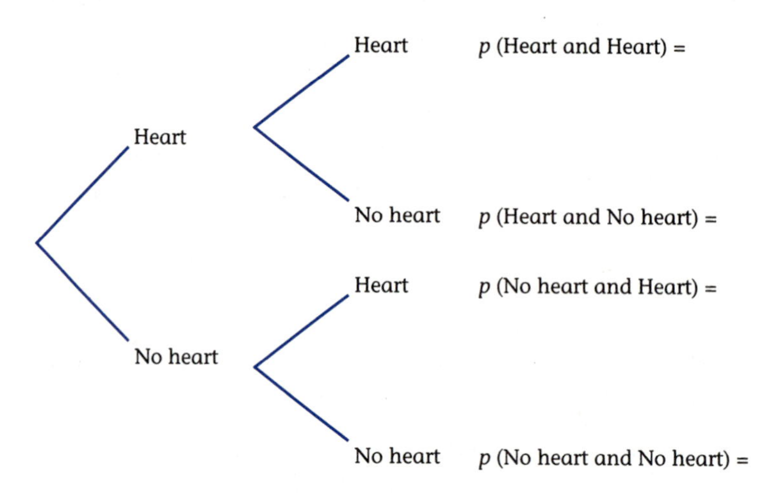

   e. Is the second draw independent of the first?

9. There are five boys and six girls in a group, and a committee of three is being selected.

a. In how many ways can the committee be formed?
b. How many committees consist of one boy and two girls?
c. What is the probability that the committee will consist of exactly one boy?
d. How many committees consist of one boy or one girl?
e. What is the probability that the committee will consist of one boy or one girl?

10. Ninety percent of all drivers are good; 5% of all good drivers get tickets; and 70% of all bad drivers get tickets.
a. Write the appropriate probabilities along each branch of the tree diagram below, and calculate the probabilities shown at the right.

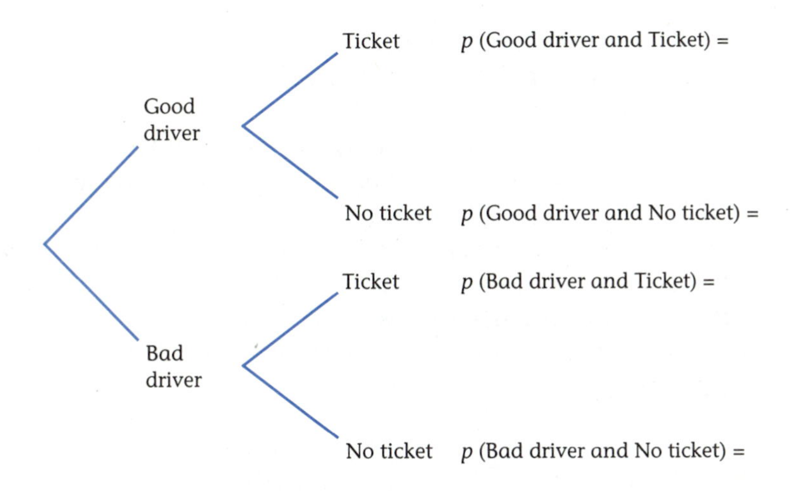

b. If a community has 50,000 drivers, about how many can be expected to get tickets?
c. How many of the people who get tickets are bad drivers?
d. What is the probability that a person who gets a ticket is a bad driver?

11. Suppose that you and three friends each choose a number between 1 and 10.
a. What is the probability that all three of your friends will pick the same number that you pick?

b. What is the probability that all three of your friends will pick a number different from yours?
c. What is the probability that all three of your friends will pick a number that is different from the number picked by any of the others?

12. Two different prizes are being awarded in a group of 10 people.
 a. In how many ways can this be done if the same person may win both prizes?
 b. In how many ways can this be done if each person may win no more than one prize?
 c. In how many ways can this be done if each person may win no more than one prize and both prizes are the same?

13. A fair die is rolled five times.
 a. What is the probability that the number of times a 6 will appear is two?
 b. What is the probability that a 6 will appear two or fewer times?
 c. What is the probability that a 6 will never appear?

14. Egbert fixes an omelette for breakfast every morning. Depending on what he has in his refrigerator, he adds one or more of the following ingredients to his eggs: mushrooms, green peppers, cheddar cheese, ham. How many different omelettes can Egbert make?

Bibliography

David, F. N. 1962. *Games, Gods and Gambling.* New York: Hafner.

Green, Thomas A., and Charles L. Hamburg. 1986. *Pascal's Triangle.* Palo Alto, CA: Dale Seymour.

Huff, Darrell. 1959. *How to Take a Chance.* New York: Norton.

Orkin, Mike. 1991. *Can You Win? The Real Odds for Casino Gambling, Sports Betting, and Lotteries.* New York: Freeman.

Paulos, John Allen. 1988. *Innumeracy: Mathematical Illiteracy and Its Consequences.* New York: Hill and Wang.

Weaver, Warren. 1963. *Lady Luck: The Theory of Probability.* Garden City, NY: Doubleday.

113030
LIVE HOGS
FDR CATTLE
EGGS
LUMBER
MILO

Matrices Revisited

The daily business activity that supplies us with the products and services we need generates large quantities of data that often must be organized into matrices in order to be understood. Proper organization of these data, as reflected in this board at the Chicago Commodity Exchange, is necessary not only for understanding but also for effective planning.

For example, how can a company that provides batteries for another company's compact disc players be sure that it will have enough batteries on hand to fill orders? How does a fast-food chain determine prices that will allow it to do as well as possible against a competitor? How can a meteorologist use data about recent weather activity to predict the weather for tomorrow or a week from now? Matrices demonstrate remarkable versatility in helping to solve these and other real-world problems.

The Leontief Input–Output Model, Part 1

The Leontief input–output model discussed in this lesson is used to analyze the flow of goods and services among sectors in an economy.

This model, which was developed by professor Wassily Leontief of Harvard University, can be applied to extremely complex economies with hundreds of production sectors, such as a nation (or even the world), or to a situation as small as a single company that produces only one product.

We begin our exploration of this model by looking at a simple case. Suppose, for example, the Best Battery Company of Lincoln, Nebraska, manufactures a particular type of battery that is used to power various kinds of electric motors. Not all of the batteries produced by the company, however, are available for sale outside the company. For every 100 batteries produced, 3 (3%) are used by various departments within the company. Thus, if the company produces 500 batteries during a week's time, 15 will be used within the company and 485 will be available for external sales. In general, for a total production of P batteries by this company, $0.03P$ batteries are used internally, and $P - 0.03P$ are available for external sales to customers.

In other words, the number of batteries available for sale outside the company equals the total production of batteries less 3% of that total production. If we let D represent the number of batteries available for external sales **(demand)** and P represent the total production of batteries, the following linear equation can be used to model this situation:

$$P - 0.03P = D. \qquad (1)$$

Suppose the company receives an order for 5,000 batteries. What must the total production be to satisfy this external demand for batteries? To find the total production necessary, we substitute 5,000 for D in the previous equation and solve for P, as follows:

$$P - 0.03P = 5{,}000 \qquad (2)$$

$$P(1 - 0.03) = 5{,}000 \qquad (3)$$

$$P = 5{,}155 \text{ batteries.} \qquad (4)$$

Hence, the company must produce a total of 5,155 batteries to satisfy an external demand for 5,000 batteries.

Notice that the total production of batteries equals the number of batteries used within the company during production plus the number of batteries necessary to fill the external demand.

Now let's look at a simple two-sector economy. Suppose that the Best Battery Company buys an electrical motor company and begins producing motors as well as batteries. The company's primary reason for this merger was that electrical motors are used to produce batteries. Batteries are also used to manufacture motors. The expanded company has two divisions, the battery division and the motor division. It also has changed its name to the Best Battery and Motor Company.

The production requirements of the newly expanded company's two divisions are as follows:

Battery Division

1. For the battery division to produce 100 batteries, it must use 3 (3%) of its own batteries.

2. For every 100 batteries produced, 1 motor (or 1% of the number of batteries produced) is required from the motor division.

Motor Division

1. For the motor division to produce 100 motors, it must use 4 (4%) of its own motors.

2. For every 100 motors produced, 8 batteries are required from the battery division. (Notice that the number of batteries required is 8% of the total number of motors produced.)

The production needs of this two-sector economy can be represented visually by using a weighted digraph, as shown in Figure 7.1. The digraph shows that if the company needs to produce a total of b batteries and m motors,

1. The battery division will require $0.03b$ batteries from its own division and $0.01b$ motors from the motor division.

2. The motor division will require $0.04m$ motors from its division and $0.08m$ batteries from the battery division.

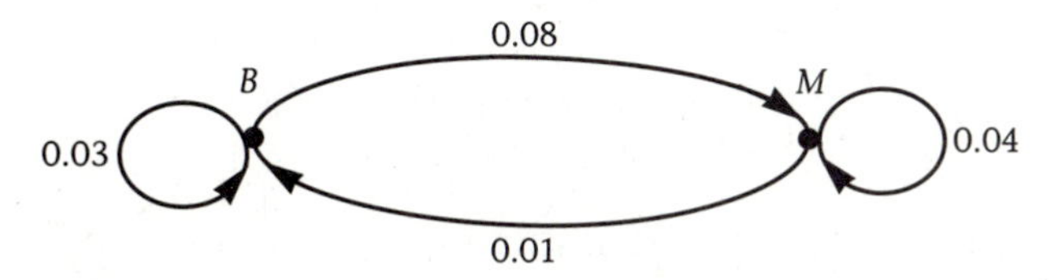

Figure 7.1 Weighted digraph showing the input required by each division.

NEWS CLIP

Raytheon Co. to Buy Business Jet Unit of British Aerospace for $391 Million

By David Stipp and Brian Coleman, *Staff Reporters of* THE WALL STREET JOURNAL, June 2, 1993

Raytheon Co., as expected, agreed to buy British Aerospace PLC's business jet unit for about $391 million.

The planned purchase would help Raytheon move beyond its core defense business and enable British Aerospace to reduce debt and better focus on its principal businesses, including automobiles and defense.

The purchase would nearly double Raytheon's corporate-jet market share and increase its overall aircraft sales to about $1.7 billion from $1.3 billion annually. Raytheon owns Beech Aircraft, a Wichita, Kan., maker of turboprop and jet aircraft, which has about 12% of the market for light to medium jets. The industry leader for business jets, Textron Inc.'s Cessna Aircraft unit, has an estimated 60% market share.

Max Bleck, Raytheon's president, said that although the small-plane market "has been in general malaise," Raytheon is positioning itself for an industry turnaround by buying when acquisition prices are low. Mr. Bleck reiterated Raytheon's goal of boosting to 50% from about 30% its portion of operating profit that is derived from non-defense business. He said Raytheon remains interested in possible acquisitions in the engineering, construction and appliance areas.

A third way to present the information regarding the company's production needs is with a matrix. This matrix, which shows the required input from each sector of the economy, is called the **technology matrix** for the economy.

$$\begin{array}{cc} & \text{To} \\ & \begin{array}{cc} \text{Battery} & \text{Motor} \end{array} \\ \text{From} \begin{array}{c} \text{Battery} \\ \text{Motor} \end{array} & \begin{bmatrix} 0.03 & 0.08 \\ 0.01 & 0.04 \end{bmatrix} \end{array}$$

As an example, to produce 200 batteries, the battery division will need (0.03)(200) = 6 batteries from the battery division and (0.01)(200) = 2 motors from the motor division.

Explore This

At the direction of your teacher, divide your class into groups of three to four people. Each group is to represent a production management team for the newly merged battery and motor company just described.

The company has given the battery division a daily total production quota of 1,000 batteries. The daily quota for the motor division is 200 motors. The production management team's problem is to determine how many batteries and motors each of the company's divisions will need during production and how many batteries and motors will be available for sales.

To solve the problem the management team must do the following tasks:

1. Find the number of batteries and motors needed by the battery division to meet its quota. Repeat for the motor division.

2. Find the number of batteries available for external sales if both divisions meet their daily quotas. Repeat for the number of motors available for sales outside the company.

3. Make up a formula for computing the number of batteries that the company will use internally to meet its production quotas. Repeat for the number of motors used internally. These formulas will be used to program a computer to do future calculations to save the management team some work. Let b represent the total number of batteries produced and m represent the total number of motors produced by the company.

4. Explore this problem from another point of view. Suppose the company needs to fill an order of 400 batteries and 100 motors. Estimate the total production needed from each division to fill these orders. Check your estimate, and revise it if necessary. Hint: Recall that the amount of a product available to fill an external order (outside demand) equals the total production minus the amount of that product consumed internally by the company.

5. If time permits, find a system of two equations in two unknowns that could be used to compute the total production for each division required in task 4.

After all of the groups have finished tasks 1 through 4, a spokesperson for each production management team should present the results of the team's discussion to the class.

Exercises

1. A utility company produces electrical energy. Suppose that 5% of the total production of electricity is used up within the company to operate equipment needed to produce the electricity. Complete the following production table for this one-sector economy:

Total Production Units	Units used Internally	Units for External Sales
500	.05(500) = —	500 − .05(500) = —
900	—	—
—	100	—
—	250	—
—	—	2,375
—	—	4,750
P	—	—

2. Suppose that for every dollar's worth of computer chips produced by a high-tech company, 2 cents' worth is used by the company in the manufacturing process.
   a. What percentage of the company's total production of computer chips is used within the company?
   b. What would the weighted digraph look like for this situation?
   c. Write an equation that represents the dollars' worth of computer chips available for external demands (D) in terms of the total production (P) of chips by the company.
   d. What must the total production of computer chips be in order for the company to meet an external demand for $20,000 worth of computer chips?

3. Suppose that the high-tech company described in Exercise 2 adds another division that produces computers. Each division within the expanded company uses some of the other division's product, as follows:

Computer Chip Division

Every dollar's worth of computer chips produced requires an input of 2 cents' worth of computer chips and 1 cent's worth of computers.

Computer Division

Every dollar's worth of computers produced requires an input of 20 cents' worth of computer chips and 3 cents' worth of computers.

a. Draw a weighted digraph that summarizes the production needs for this two-sector economy.
b. Construct the technology matrix for this economy.
c. Suppose the computer chip division produces $1,000 worth of chips. How much input does it need from itself and from the computer division?
d. Suppose the computer division produces $5,000 worth of computers. How much input does it need from itself and from the computer chip division?
e. What must the total production of computer chips be in order for the company to meet an external demand for $20,000 worth of computer chips?
f. What must the total production of computers be in order for the company to meet an external demand for $50,000 worth of computers?

4. A company has two departments: service and production. The needs of each department within this company (in cents per dollar's worth of output) are shown in the technology matrix below.

$$\begin{array}{l} \\ \text{Service} \\ \text{Production} \end{array} \begin{array}{c} \begin{array}{cc} \text{Service} & \text{Production} \end{array} \\ \begin{bmatrix} 0.05 & 0.20 \\ 0.04 & 0.01 \end{bmatrix} \end{array}$$

a. Draw a weighted digraph to show the flow of goods and services within this company.
b. Complete the following: For every dollar's worth of output, the service department requires ______ cents' worth of input from its own department and ______ cents' worth of input from the production department. For every dollar's worth of output, the production department requires ______ cents' worth of input from its own department and ______ cents' worth of input from the service department.
c. Suppose that the total output for the service department is $20 million over a certain period of time. How much of this amount is used within the service department? How much input is required from the production department?
d. The total output from the production department is $40 million over the same time period. How much of this total output is used within the production department? How much input is required from the service department?
e. Combine the information in parts c and d to find how much of the total output from the service and production departments will be available for sales demands outside the company.

5. The weighted digraph below represents the flow of goods and services in a two-sector economy involving transportation and agriculture. The numbers on the edges represent cents' worth of product (or service) used per dollar's worth of output.

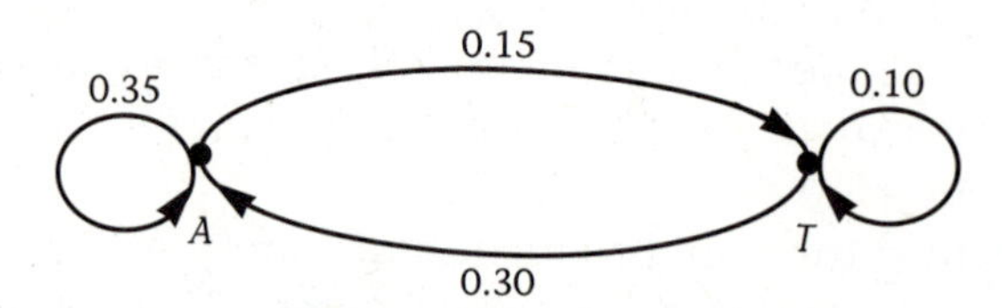

a. Construct the technology matrix for this situation.
b. Complete the following: For every dollar's worth of output the agriculture sector requires ______ cents' worth of input from its own sector and ______ cents' worth of input from the transportation sector. For every dollar's worth of output the transportation sector requires ______ cents' worth of input from its own sector and ______ cents' worth of input from the agriculture sector.

c. If the total output for the agriculture sector is $50 million and the total output for the transportation sector is $100 million over a certain period of time, find:
   i. The total amount of agricultural goods used internally by this two-sector economy.
   ii. The amount of agricultural goods available for external sales.
   iii. The total amount of transportation services used internally by this two-sector economy.
   iv. The amount of transportation services available for external sales.

d. Write an equation to represent the internal consumption of agricultural products for this economy. Let P_A represent the total production for agriculture and P_T represent the total output for transportation.

e. Write an equation to represent the internal consumption of transportation services. Use P_A and P_T, as in part d.

f. Using the information from parts b and c, write equations representing the amount of agricultural products and transportation available to fill external consumer demands. Recall: The amount of a product available to fill external demands is equal to the total production less the amount that is used internally.

g. Suppose that this economy has an external demand for $10 million worth of agricultural products and $15 million in transportation services. Write a system of equations that can be solved to find the total production of agriculture products and transportation services necessary to satisfy these demands.

6. Complete task 5 for the production management team of the Best Battery and Motor Company (see page 337).

The Leontief Input–Output Model, Part 2

As you can see from the exercises in Lesson 7.1, the Leontief model becomes quite complicated very quickly, even when dealing with only one- and two-sector economies. You are probably thinking that there has to be an easier way. You're right. There is an easier way.

Let's go back and look at the Best Battery and Motor Company example in Lesson 7.1 (page 335). The technology matrix (T) for that economy supplied us with the production needs of the two sectors:

$$T = \begin{array}{c} \\ \text{Battery} \\ \text{Motor} \end{array} \begin{array}{c} \begin{array}{cc} \text{Battery} & \text{Motor} \end{array} \\ \begin{bmatrix} 0.03 & 0.08 \\ 0.01 & 0.04 \end{bmatrix} \end{array}$$

In Exercise 6 (page 341) of that lesson we found a system of two equations in two unknowns that could be solved to find the total production of batteries (b) and motors (m) necessary to meet external sales demands of 400 batteries and 100 motors. To find two such equations, we used the fact that the total production for each product equaled the amount of the product used up by the two divisions during production plus the external demand for that product (sales outside the company).

For example, the total number of batteries produced (b) equals the number of batteries used within the battery division ($0.03b$) plus the number of batteries sent to the motor division ($0.08m$) plus the number of batteries required for sales outside the company (400). This gives us the equation $b = 0.03b + 0.08m + 400$.

Likewise, the total number of motors produced (m) equals the number of motors sent to the battery division ($0.01b$) plus the number of motors used within the motor division ($0.04m$) plus the number of motors required for sales outside the company (100). This gives us the equation $m = 0.01b + 0.04m + 100$. Thus, our system of two equations in two unknowns (unsimplified) is

$$b = 0.03b + 0.08m + 400$$

$$m = 0.01b + 0.04m + 100.$$

We know how to solve this system algebraically by combining terms and using linear combinations or some other technique. However, there is an easier way to do it that uses matrices and takes advantage of technology (calculators or computers) to do the work. Let's see, first, how this system of equations can be represented using matrices. We already have a technology matrix, T. If we let

$$P = \begin{bmatrix} b \\ m \end{bmatrix}$$

represent a total production matrix (P) and

$$D = \begin{bmatrix} 400 \\ 100 \end{bmatrix}$$

represent a demand matrix (D), we can represent our system of equations as the following matrix equation:

$$\begin{bmatrix} b \\ m \end{bmatrix} = \begin{bmatrix} 0.03 & 0.08 \\ 0.01 & 0.04 \end{bmatrix} \cdot \begin{bmatrix} b \\ m \end{bmatrix} + \begin{bmatrix} 400 \\ 100 \end{bmatrix}$$

Note: Perform the matrix multiplication shown here to show that the matrix equation does, indeed, represent the preceding two algebraic equations.

This matrix equation can also be written more simply as

$$P = TP + D$$

so that total production = internal consumption + external demand.

Notice that this matrix equation resembles the simple linear equation that we solved in Lesson 7.1 (page 334). Indeed, solving this matrix equation uses the same operations as in solving a linear equation. Let's see how this is so.

One-Sector Economy Linear Equation Ordinary Algebra	Two-Sector Economy Matrix Equation Matrix Algebra	Comments
$p = 0.03p + d$	$P = TP + D$	
$p - 0.03p = d$	$P - TP = D$	
$1p - 0.03p = d$	$IP - TP = D$	The identity matrix I times $P = P$.
$(1 - 0.03)p = d$	$(I - T)P = D$	
$\frac{1}{1 - 0.03}(1 - 0.03)p = \frac{1}{1 - 0.03}d$	$(I - T)^{-1}(I - T)P = (I - T)^{-1}D$	*
$1p = \frac{1}{1 - 0.03}d$	$IP = (I - T)^{-1}D$	Recall that $A^{-1}A = I$ (see Exercise 7 in Lesson 3.3, p. 127).
$P = \frac{1}{1 - 0.003}d$	$P = (I - T)^{-1} D$	

* Up to this point, the ordinary algebra operations and the matrix operations have been identical. In solving the linear equation, it would be natural to divide both sides of the equation by (1−0.03). But there is no division operation in matrix algebra as a counterpart. The thing to do, then, is to multiply both sides of the linear equation by the multiplicative inverse of (1 − 0.03). Multiplying both sides of the matrix equation by the inverse of $(I - T)$ is a valid matrix operation.

In summary, if we know the technology matrix (T) and the external demand matrix (D), we can find the total production matrix (P) using the matrix equation

$$P = (I - T)^{-1}D.$$

For the battery and motor problem, the solution is

$$\begin{bmatrix} b \\ m \end{bmatrix} = \left(\begin{bmatrix} 1 & 0 \\ 0 & 1 \end{bmatrix} - \begin{bmatrix} 0.03 & 0.08 \\ 0.01 & 0.04 \end{bmatrix} \right)^{-1} \cdot \begin{bmatrix} 400 \\ 100 \end{bmatrix}.$$

Now, using a calculator or computer to do the computations and rounding to the nearest whole number, we get

$$\begin{bmatrix} b \\ m \end{bmatrix} = \begin{bmatrix} 421 \\ 109 \end{bmatrix}.$$

Thus to fill an order for 400 barriers and 100 motors, the company must produce 421 batteries and 109 motors.

— Solving Systems of Linear Equations Using Matrices —

The matrix techniques used for solving systems of equations in this lesson can be used to solve any system of n independent equations in n unknowns. Look, for example, at the following system of two equations in two unknowns:

$$2x_1 + 3x_2 = 23$$
$$5x_1 - 2x_2 = 10.$$

This system can be written as a single matrix equation, as follows:

$$\begin{bmatrix} 2 & 3 \\ 5 & -2 \end{bmatrix} \cdot \begin{bmatrix} x_1 \\ x_2 \end{bmatrix} = \begin{bmatrix} 23 \\ 10 \end{bmatrix}.$$

If we let

$$A = \begin{bmatrix} 2 & 3 \\ 5 & -2 \end{bmatrix}, \quad X = \begin{bmatrix} x_1 \\ x_2 \end{bmatrix}, \quad \text{and} \quad B = \begin{bmatrix} 23 \\ 10 \end{bmatrix},$$

the matrix equation can be written in the form $AX = B$, which is similar to a simple linear equation such as $ax = b$.

One way to solve this linear equation is to multiply both sides of the equation by the inverse of a ($\frac{1}{a}$). The same strategy can be used to solve the matrix equation.

	Linear Equation Ordinary Algebra	Matrix Equation Matrix Algebra
Step 1:	$ax = b$	$AX = B$
Step 2:	$(\frac{1}{a})ax = (\frac{1}{a})b$	$A^{-1}AX = A^{-1}B$
Step 3:	$1x = (\frac{1}{a})b$	$IX = A^{-1}B$
Step 4:	$x = (\frac{1}{a})b$	$X = A^{-1}B$

Thus the solution for the above system of equations is

$$\begin{bmatrix} x_1 \\ x_2 \end{bmatrix} = \begin{bmatrix} 2 & 3 \\ 5 & -2 \end{bmatrix}^{-1} \cdot \begin{bmatrix} 23 \\ 10 \end{bmatrix}.$$

Using a calculator or computer to do the calculations, we find $x_1 = 4$ and $x_2 = 5$.

Exercises

Use either a calculator or computer software to do the matrix operations in the following exercises.

1. The total production for the high-tech company described in Exercises 2 and 3 in Lesson 7.1 (pages 338–339) over a period of time is \$40,000 worth of computer chips and \$50,000 worth of computers.
   a. Write the technology matrix, T, for this company. Label the rows and columns of your matrix.
   b. Write a production matrix, P, and label the rows and columns.
   c. Compute the matrix product TP to find the amount of each product that the company uses internally.
   d. Use the information from parts b and c and the matrix equation $D = P - TP$ to compute the amount of computer chips and computers available for sales outside the company (external demand).

e. The company has a order for $20,000 worth of computer chips and $70,000 worth of computers. Find the total production of computer chips and computers necessary to fill this order. Use the matrix equation $P = (I - T)^{-1}D$.

2. a. Use matrices to compute the results for parts c, d, and e of Exercise 4 in Lesson 7.1 (page 339).
   b. The company must meet external demands of $25 million in service and $50 million in products over a period of time. What must the total production be in service and products to meet this demand?

3. Use matrices to compute the results for parts c and g of Exercise 5 in Lesson 7.1 (page 340).

4. The techniques developed in this lesson using a two-sector economy can easily be extended to solve problems that involve economies of more than two sectors. For example, look at an economy that has three sectors—transportation, energy, and manufacturing. Each of these sectors uses some of its own products or services as well as some from each of the other sectors, as follows:

Transportation Sector

Every dollar's worth of transportation provided requires an input of 10 cents' worth of transportation services, 15 cents' worth of energy, and 25 cents' worth of manufactured goods.

Energy Sector

Every dollar's worth of energy produced requires an input of 25 cents of transportation, 10 cents' worth of energy, and 20 cents' worth of manufactured goods.

Manufacturing Sector

Every dollar's worth of manufactured goods produced requires an input of 20 cents' worth of transportation services, 20 cents' worth of energy, and 15 cents' worth of manufactured goods.

a. Draw a weighted digraph for this three-sector economy.
b. Construct a technology matrix (*T*) for this economy. Label the rows and columns of your matrix.
c. The total production over a period of time for this economy is \$150 million in transportation, \$200 million in energy, and \$160 million in manufactured goods. Write a production matrix (*P*) for this economy. Label the rows and columns.
d. Compute the matrix product *TP* to find the amount of each product that is used internally by the economy. Write your answer as a matrix and label the rows and columns.
e. Use the information from parts c and d and the matrix equation $D = P - TP$ to find the amount of goods available for external demand (sales outside the three sectors described here).
f. The estimated consumer demand for transportation, energy, and manufactured goods and services in millions of dollars is 100, 95, and 110, respectively. Find the total production necessary to fulfill these demands. Use the matrix equation $P = (I - T)^{-1}D$.

5. An economy consisting of three sectors (services, manufacturing, and agriculture) has the following technology matrix:

$$T = \begin{array}{l} \text{Services} \\ \text{Manufacturing} \\ \text{Agriculture} \end{array} \overset{\begin{array}{ccc} \text{Serv.} & \text{Manu.} & \text{Agri.} \end{array}}{\begin{bmatrix} 0.1 & 0.3 & 0.2 \\ 0.2 & 0.3 & 0.1 \\ 0.2 & 0.1 & 0.2 \end{bmatrix}}.$$

a. Draw a weighted digraph for this economy.
b. On what sector of the economy is manufacturing the most dependent? The least dependent?
c. If the services sector has an output of \$40 million, what is the input in dollars from manufacturing? From agriculture?
d. The production matrix, *P*, in millions of dollars is shown below. Find the internal consumption matrix, *TP*, and the external demand matrix, *D*.

$$P = \begin{array}{l} \text{Services} \\ \text{Manufacturing} \\ \text{Agriculture} \end{array} \begin{bmatrix} 20 \\ 25 \\ 15 \end{bmatrix}$$

e. The external demand matrix, D, in millions of dollars is shown below. How much must be produced by each sector to meet this demand?

$$D = \begin{matrix} \text{Services} \\ \text{Manufacturing} \\ \text{Agriculture} \end{matrix} \begin{bmatrix} 4.6 \\ 5.0 \\ 4.0 \end{bmatrix}.$$

6. An economy consisting of four sectors (transportation, manufacturing, agriculture, and services) has the following technology matrix (in millions of dollars' worth of products):

$$T = \begin{matrix} \\ \text{Transportation} \\ \text{Manufacturing} \\ \text{Agriculture} \\ \text{Services} \end{matrix} \begin{matrix} \begin{matrix} \text{Trans.} & \text{Manu.} & \text{Agri.} & \text{Serv.} \end{matrix} \\ \begin{bmatrix} 0.25 & 0.28 & 0.22 & 0.20 \\ 0.15 & 0.15 & 0.17 & 0.23 \\ 0.19 & 0.20 & 0.21 & 0.15 \\ 0.20 & 0.24 & 0.19 & 0.25 \end{bmatrix} \end{matrix}.$$

a. Draw a weighted digraph for this economy.
b. On what sector of the economy are services the most dependent? The least dependent?
c. If the manufacturing sector has an output of $20 million, what is the input in dollars from services? From transportation?
d. The production matrix, P, in millions of dollars, is shown below. Find the internal consumption matrix, TP, and the external demand matrix, D.

$$P = \begin{matrix} \text{Transportation} \\ \text{Manufacturing} \\ \text{Agriculture} \\ \text{Services} \end{matrix} \begin{bmatrix} 50 \\ 40 \\ 45 \\ 50 \end{bmatrix}.$$

e. The external demand matrix, D, in millions of dollars, is shown below. How much must be produced by each sector to meet this demand?

$$D = \begin{matrix} \text{Transportation} \\ \text{Manufacturing} \\ \text{Agriculture} \\ \text{Services} \end{matrix} \begin{bmatrix} 10 \\ 12 \\ 10 \\ 15 \end{bmatrix}.$$

7. A two-industry system consisting of services and manufacturing has the following input–output matrix, T:

$$T = \begin{array}{l} \text{Services} \\ \text{Manufacturing} \end{array} \overset{\begin{array}{cc} \text{Services} & \text{Manufacturing} \end{array}}{\begin{bmatrix} 0.5 & 0.5 \\ 0.2 & 0.3 \end{bmatrix}}$$

   a. Can this system satisfy a consumer demand for 15 units of services and 25 units of manufacturing? Explain your answer.
   b. The consumer demand is for 30 units of services and 50 units of manufacturing. Find the production needed to fill these demands.
   c. Based on the results in parts a and b, predict the total production of services and goods for a consumer demand for 45 units of service and 75 units of manufacturing. Check your prediction.

8. A company has two divisions: service and production. The flow of goods and services within this company is described by the input–output matrix below.

$$T = \begin{array}{l} \text{Services} \\ \text{Production} \end{array} \overset{\begin{array}{cc} \text{Services} & \text{Production} \end{array}}{\begin{bmatrix} 0.10 & 0.25 \\ 0.05 & 0.10 \end{bmatrix}}$$

   a. Draw a weighted digraph for this situation.
   b. The total output for the company during 1 year is \$50 million in services and \$75 million in products. How much of the total output is used internally by each of the company's divisions?
   c. What total output is needed to meet an external consumer demand of \$15 million in services and \$25 million in products?
   d. If the consumer demand increases to \$22 million for services and to \$30 million for products, what will be the effect on the total production of goods and services?

Project 1

Research and report to the class on the following:

   a. What computer software is available in your school for solving systems of equations?

b. What is the largest number of variables that the software can handle?
c. How long does it take the computer to solve a system with the largest number of variables possible using this software?
d. It took Professor Leontief 56 hours to solve a system of 42 equations in 42 unknowns using the Mark II in the 1940s and 3 hours to solve a system of 81 equations in 81 unknowns in the 1960s using the IBM 7090. If the software available in your school can handle systems of 42 and 81 equations, find out
    a. How long it takes to solve a system of 42 independent equations in 42 unknowns.
    b. How long it takes to solve a system of 81 independent equations in 81 unknowns.
e. Investigate parts a and b for computer software such as Mathematica! or Theorist that your school may not own.

Project 2

Research and report to the class about the size, cost, and capability of the Mark II and IBM 7090 computers. In your report, compare the information you find out about these computers with similar information about the personal computers that are available to students in your school.

Markov Chains

A **Markov chain** (named after the Russian mathematician A. A. Markov, 1856–1922) is a process that arises naturally in problems that involve a finite number of events or states that change over time. In this lesson, we examine a situation that illustrates the significant characteristics of a Markov chain.

Consider the following: Students at Central High have two choices for lunch. They can either eat in the cafeteria or eat elsewhere. The director of food service wants to be able to predict how many students can be expected to eat in the cafeteria over the long run. She has asked the discrete mathematics class to help her out by conducting a survey of the student body during the first two weeks of school. The results of the survey show that if a student eats in the cafeteria on a given day, the probability that he or she will eat there again the next day is 70% and the probability that he or she will not eat there is 30%. If a student does not eat in the cafeteria on a given day, the probability that he or she will eat in the cafeteria the next day is 40% and the probability that he or she will not eat there is 60%. On Monday, 75% of the students ate in the cafeteria, and 25% did not. What can be expected to happen on Tuesday?

A good way to organize all of these statistics is with a tree diagram similar to the one in which you organized probabilities in Chapter 6 (see Figure 7.2).

The director wants to know what portion of the students can be expected to eat in the cafeteria on Tuesday. Look at the tree diagram and notice that this happens if a student eats in the cafeteria on both Monday and Tuesday, or if a student eats elsewhere on Monday and in the cafeteria on Tuesday. The correct probability is .525 + .100 = .625. Similarly, the portion of students who will eat elsewhere on Tuesday is .225 + .150 = .375. Note that this could also be calculated by subtracting .625 from 1.

The tree diagram model is fine if only two stages are required to reach a solution. The director, however, is interested in continuing this process for many days. Because the number of branches of the tree diagram doubles with each additional day, the model soon becomes impractical, and so an alternative is needed.

The Monday student data are called the **initial distribution** of the student body and can be represented by a row (or **initial state**) vector, D_0, where

$$\begin{array}{ccc} & C & E \\ D_0 = & [.75 & .25] \end{array} \quad \begin{array}{l} C = \text{eat in Cafeteria} \\ E = \text{eat Elsewhere.} \end{array}$$

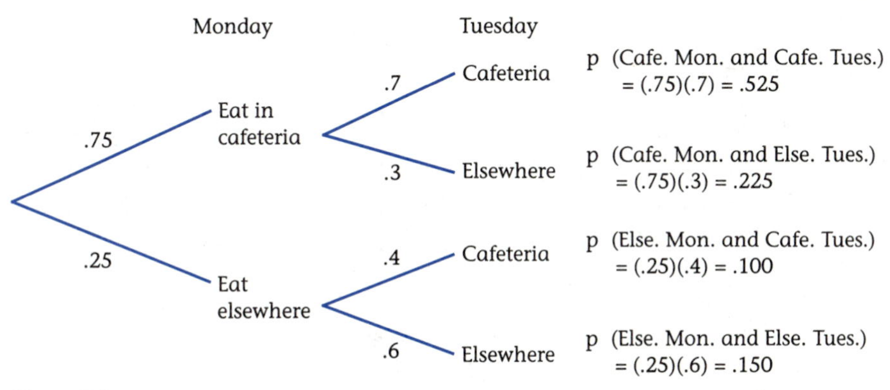

Figure 7.2 Cafeteria statistics organized in a tree diagram.

Movement from one state to another is often called a **transition,** so the data on where students choose to eat from one day to the next are written in a matrix called a **transition matrix,** T, where

$$T = \begin{array}{c} \\ C \\ E \end{array}\begin{array}{c} \begin{array}{cc} C & E \end{array} \\ \begin{bmatrix} .7 & .3 \\ .4 & .6 \end{bmatrix} \end{array}.$$

Notice that the entries of a transition matrix are probabilities, values between 0 and 1 inclusive. Also notice that the transition matrix is a square matrix and that the sum of the probabilities in any row is 1.

Now calculate the product of matrix D_0 and matrix T:

$$D_0T = [.75 \quad .25]\begin{bmatrix} .7 & .3 \\ .4 & .6 \end{bmatrix}$$
$$= [.75(.7) + .25(.4) \quad .25(.3) + .25(.6)] = [.625 \quad .375].$$

Compare these calculations with those made in the tree diagram model. The values in the resulting row vector can be interpreted as the portion of students that eat in the cafeteria and eat elsewhere on Tuesday. This row vector is called D_1 to indicate that it occurs one day after the initial day. To see what happens on Wednesday, it is necessary to repeat the process using D_1 in place of D_0:

$$D_1T = [.625 \quad .375]\begin{bmatrix} .7 & .3 \\ .4 & .6 \end{bmatrix}$$
$$= [.625(.7) + .375(.4) \quad .625(.3) + .375(.6)]$$
$$= [.5875 \quad .4125].$$ (Therefore, approximately 59% of the students will eat in the cafeteria on Wednesday and 41% will eat elsewhere.)

This row vector is called D_2 to indicate that it occurs two days after the initial day.

Consider how D_2 was calculated:

$$D_2 = D_1T, \text{ but } D_1 = D_0T, \text{ so by substitution, } D_2 = (D_0T)(T).$$

Because matrix multiplication is associative,

$$D_2 = (D_0T)(T) = D_0(T^2).$$

This means that the calculation of the distribution of students on Wednesday can be completed by multiplying the initial state matrix times the square of the transition matrix.

This observation simplifies additional calculations. If, for example, you want to know the distribution on Friday, four days from Monday, calculate $D_4 = D_0(T^4)$ on a calculator that has matrix features or on a computer equipped with matrix software:

$$D_0T^4 = [.75 \quad .25]\begin{bmatrix} .7 & .3 \\ .4 & .6 \end{bmatrix}^4 = [.572875 \quad .427125].$$

About 57% of the students can be expected to eat in the cafeteria on Friday. For a school with 1,000 students, about 573 of them can be expected in the cafeteria on that day.

Exercises

1. a. Find the distribution of students eating and not eating in the cafeteria each day for the first week of school using the initial distribution $D_0 = [.75 \quad .25]$ and the transition matrix, T, of this lesson.
   b. Find the distribution of students eating and not eating in the cafeteria after 2 weeks (10 school days) have passed. Repeat for 3 weeks (15 days).
   c. Choose any other initial distribution of students and repeat parts a and b.
   d. Compare the results of parts b and c. Does the initial distribution appear to make a difference in the long run?
   e. Calculate the fifteenth power of matrix T. Compare the entries with the distribution after the fifteenth day.

2. When successive applications of a Markov process are made and the rows of powers of the transition matrix converge to a single vector, this common vector is called the **stable-state vector** for the Markov chain. A sufficient condition, which we will not prove in this text, for a Markov chain to have a stable-state vector is that some power of its transition matrix has only positive entries. Since all of the entries in the transition matrix, T, are nonzero probabilities, this condition is clearly met for the cafeteria Markov chain.
   a. What is the stable-state vector for the transition matrix T?
   b. Make a conjecture about the relationship between the distribution of students in the long run and the stable-state vector of the transition matrix.

3. The entire student body eats in the cafeteria on the first day of school. The initial distribution in this case is $D_0 = [1 \quad 0]$. Repeat parts a and b of Exercise 1 for this distribution. After several weeks, what percentage of students will be eating in the cafeteria?

4. a. Which of the matrices below could be Markov transition matrices?
   b. For the matrices that cannot be transition matrices, explain why not.

a. $\begin{bmatrix} .7 & .3 \\ .6 & .6 \end{bmatrix}$ b. $\begin{bmatrix} .1 & .4 & .5 \\ .2 & .6 & .2 \end{bmatrix}$

c. $\begin{bmatrix} 1.2 & -.4 \\ 1 & 0 \end{bmatrix}$ d. $\begin{bmatrix} .6 & .3 & .1 \\ .3 & .3 & .3 \end{bmatrix}$

e. $\begin{bmatrix} .75 & .25 \\ 1 & 0 \end{bmatrix}$ f. $\begin{bmatrix} .4 & .6 \\ 0 & 1 \end{bmatrix}$

5. There is a 60% chance of rain today. Suppose we know that tomorrow's weather depends on today's, as shown in the tree diagram below.

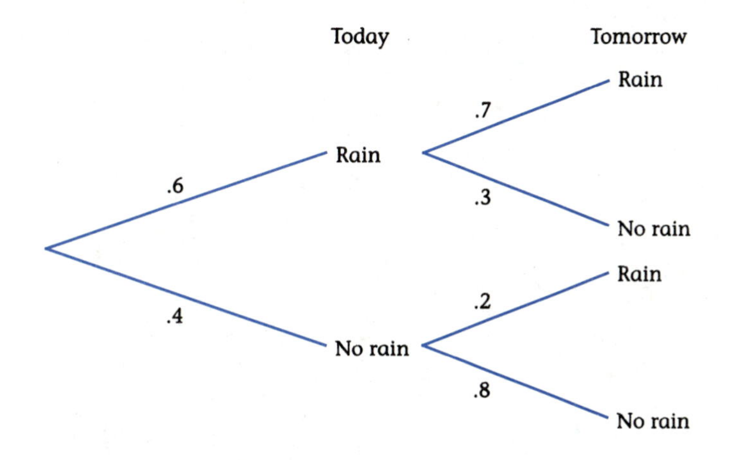

   a. What is the probability that it will rain tomorrow if it rains today?
   b. What is the probability that it will rain tomorrow if it doesn't rain today?
   c. Write an initial-state matrix that represents the weather forecast for today.
   d. Write a transition matrix that represents the transition probabilities shown in the tree diagram.
   e. Calculate the forecast for 1 week (7 days) from now.
   f. In the long run, on what percentage of days does it rain?

6. A taxi company has divided the city into three districts—Westmarket, Oldmarket, and Eastmarket. By keeping track of

NEWS CLIP

Chicago, Suburban Forecast

CHICAGO TRIBUNE, October 14, 1991

Chicago and Vicinity: Partly cloudy to cloudy Monday with morning showers likely, possible thundershowers. South wind turns southwest 25 to 30 m.p.h. and gusty. Mostly cloudy Monday night with rain likely. The low will be 36, with a west wind 12 to 20 m.p.h. Cold and mostly cloudy Tuesday with rain mixed with snow likely. The high will be mid to upper 40s.

Monday

Morning showers likely, possible thundershowers, high 58/low 36

Tuesday

Rain mixed with snow likely, cloudy and cold, high 48/low 35

Wednesday

Fair to partly cloudy, high 55/low 42

Thursday

Increasingly cloudy, warmer, high 70/low 49

pickups and deliveries, the company found that of the fares picked up in the Westmarket district, only 10% were dropped off in that district, 50% were taken to the Oldmarket district, and 40% went to the Eastmarket district. Of the fares picked up in the Oldmarket district, 20% were taken to the Westmarket district, 30% stayed in the Oldmarket district, and 50% were dropped off in the Eastmarket district. Of the fares picked up in the Eastmarket district, 30% were delivered to each of the Westmarket and Oldmarket districts, and 40% stayed in the Eastmarket district.

a. Draw a transition digraph for this Markov chain, as shown on p. 361.
b. Construct a transition matrix for these data.
c. Write an initial-state matrix for a taxi that begins by picking

up a fare in the Oldmarket district. What is the probability that it will end up in the Oldmarket district after three additional fares?

d. Find and interpret the stable-state vector for this Markov process.

7. Jim, Nan, and Maria are tossing a football around. Jim always tosses to Nan, and Nan always tosses to Maria, but Maria is equally likely to toss the ball to either Jim or Nan.
   a. Draw a transition digraph to represent this information.
   b. Represent this information as the transition matrix of a Markov chain.
   c. What is the probability that Jim will have the ball after three tosses if he is the first one to throw it to one of the others?
   d. Find and interpret the stable-state vector for this Markov chain.
   e. Explain why there are no zeros in the stable-state matrix, even though there are several zeros in the transition matrix.

8. A discrete math student observes a bug crawling from vertex to vertex along the edges of a tetrahedron model on the teacher's desk (see the figure below). From any vertex the bug is equally likely to go to any other vertex.

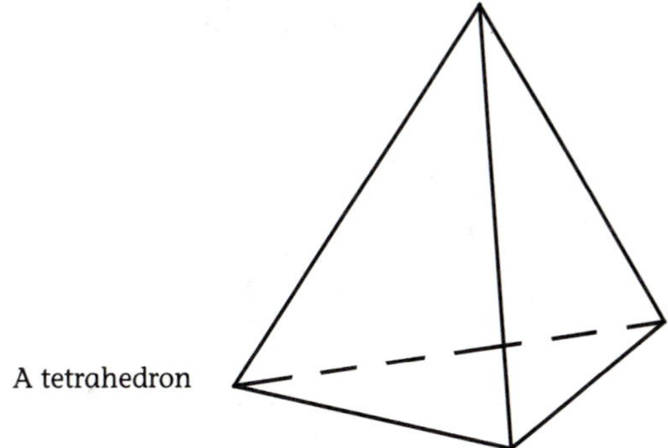

A tetrahedron

   a. Set up a transition matrix for this situation.
   b. Determine the probabilities for the location of the bug when the passing bell rings if it moves to a different vertex about 20 times during the class period. (Note: To minimize roundoff errors, approximate one-third to at least four decimal places when you enter the data into your calculator or computer program.)

9. Dick's old hound dog, Max, spends much of his time during the day running from corner to corner along the fence surrounding his square-shaped yard. There is a 0.5 probability that Max will turn in either direction at a corner. The corners of Max's yard point north, east, south, and west.
   a. Draw an initial-state diagram that represents Max's movement.
   b. Construct a transition matrix for this situation.
   c. Look at the behavior of successive powers of the transition matrix. Notice the oscillation of the transition probabilities between the states represented by the rows of the matrices. Does this system appear to stabilize in some way? Explain your answer.
   d. Approximately what percentage of the time will Max spend at each of the corners of his yard? (Note: You need to halve the entries in the matrix to account for the oscillating pattern.)
   e. Max changes his routine one day, and the pattern of his new movements is represented by the transition matrix below. Answer part d for this situation.

$$\begin{array}{c} \\ N \\ E \\ S \\ W \end{array}\begin{array}{c} \begin{array}{cccc} N & E & S & W \end{array} \\ \begin{bmatrix} 0 & \frac{1}{2} & 0 & \frac{1}{2} \\ \frac{3}{8} & 0 & \frac{5}{8} & 0 \\ 0 & \frac{3}{8} & 0 & \frac{5}{8} \\ \frac{3}{4} & 0 & \frac{1}{4} & 0 \end{bmatrix} \end{array}$$

10. A prison places its inmates in one of three types of confinement: minimum security, maximum security, and death row. A prisoner in one type of confinement is left there or placed in another type according to the probabilities shown in the following transition matrix:

	Min. security	Max. security	Death row
Min. security	.8	.2	0
Max. security	.1	.6	.3
Death row	0	0	1

   a. Write an initial-state vector for a prisoner who enters the prison by being placed in minimum security.

b. If prisoners are reassigned every week, predict the prisoner's future after one month of confinement.
c. Predict the prisoner's future in the long run.
d. This Markov chain has a state that is called an **absorbing state.** Which state do you think it is? Why?

11. A hospital categorizes its patients as well (in which case they are discharged), good, critical, or deceased. Data show that the hospital's patients move from one category to another according to the probabilities shown in this transition matrix:

$$\begin{array}{l} \\ \text{Well} \\ \text{Good} \\ \text{Critical} \\ \text{Dead} \end{array} \begin{array}{c} \begin{array}{cccc} \text{Well} & \text{Good} & \text{Critical} & \text{Dead} \end{array} \\ \begin{bmatrix} 1 & 0 & 0 & 0 \\ .5 & .3 & .2 & 0 \\ 0 & .3 & .6 & .1 \\ 0 & 0 & 0 & 1 \end{bmatrix} \end{array}$$

a. Write an initial-state matrix for a patient who enters the hospital in critical condition.
b. If patients are reclassified daily, predict a patient's future after 1 week in the hospital.
c. Predict the future of any patient in the long run.
d. Does this Markov chain have any absorbing states? Which states do you think are absorbing? Why?

A Transition Digraph

The movement of students from one state to another can also be shown with a weighted digraph called a **transition digraph** or **state diagram** (see Figure 7.3).

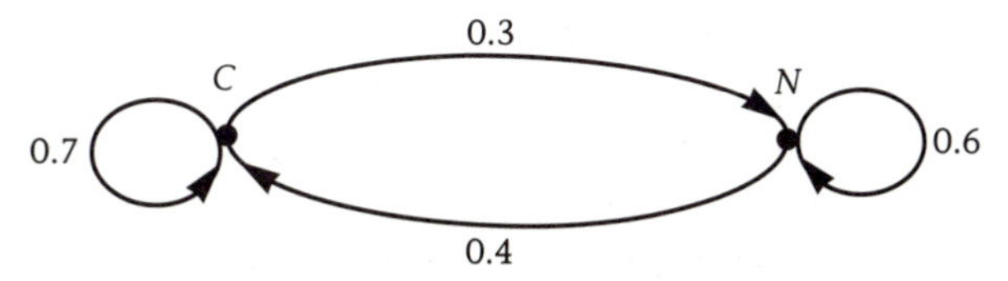

Figure 7.3 Transition digraph for the cafeteria statistics.

Game Theory, Part 1

The basic ideas of game theory were first researched by John Von Neumann in the 1920s, but it was not until the 1940s that game theory was recognized as a legitimate branch of mathematics. Thus, most of the work in this area has been done in the last 50 years.

You probably tend to think of games as being fun or relaxing ways to spend your time. There are, however, many decision-making situations in fields such as economics or politics that can also be thought of as games. Such games have two or more players with conflicting interests. These players may be individuals, teams of people, whole countries, or even forces of nature. Each player (or side) has a set of alternative courses of action called **strategies** that can be used in making decisions. Mathematical game theory involves selecting the best strategies for a player to follow in order to achieve the most favorable outcomes.

In this lesson, we consider some examples of games with two players and use matrices to determine the best strategy for each player to choose. As the first example, consider a simple coin-matching game that Sol and Tina are playing. Each conceals a penny with either heads or tails turned upward. They display their pennies simultaneously. Sol will win three pennies from Tina if both are heads. Tina will win two pennies from Sol if both are tails, and one penny from Sol if the coins don't match. What is the best strategy for each player?

If you think carefully about the game, you will probably decide that it isn't such a good deal for Sol. So long as Tina displays tails, she cannot lose. If Sol knows that Tina is going to play tails, he should display heads because he will lose more if he doesn't.

You probably think this is a rather boring game. In a sense it is, because both players will do the same thing every time. A game in which both players should pursue the same strategy every time is called **strictly determined.** Although strictly determined games are fairly boring, there are situations in life in which they can't be avoided, and knowing how to analyze them properly can be beneficial. Although strictly determined games are often very simple, they can be difficult to analyze without an organizational scheme. Matrices offer a way of doing this.

The matrix below presents Sol's view of the game. It is customary to write a game matrix from the viewpoint of the player associated with the matrix rows rather than the player associated with the columns. Such a matrix is called a **payoff matrix.** The entries are the payoffs to Sol for each outcome of the game.

$$\begin{array}{cc} & \text{Tina} \\ & \begin{array}{cc}\text{Heads} & \text{Tails}\end{array} \\ \text{Sol}\ \begin{array}{c}\text{Heads}\\ \text{Tails}\end{array} & \begin{bmatrix} 3 & -1 \\ -1 & -2 \end{bmatrix} \end{array} \quad \begin{array}{l}\text{Payoff matrix from}\\ \text{Sol's point of view.}\end{array}$$

This matrix is easy to follow if you are Sol, but the entries are just the opposite if you are Tina. If you find it difficult to think of all the numbers as their opposites, you may find it preferable to write a second matrix from Tina's point of view:

$$\begin{array}{cc} & \text{Tina} \\ & \begin{array}{cc}\text{Heads} & \text{Tails}\end{array} \\ \text{Sol}\ \begin{array}{c}\text{Heads}\\ \text{Tails}\end{array} & \begin{bmatrix} -3 & 1 \\ 1 & 2 \end{bmatrix} \end{array} \quad \begin{array}{l}\text{Payoff matrix from}\\ \text{Tina's point of view.}\end{array}$$

Consider the game from Sol's point of view. Sol does not want to lose any more money than necessary, so he analyzes his strategies from the standpoint of his losses. If he displays heads, the worst he can do is lose 1 cent. If he displays tails, the worst he can do is lose 2 cents. Because it is better to lose 1 cent than 2 cents, Sol decides to display heads.

Sol's analysis can be related to the payoff matrix by writing the worst possible outcome of each strategy to the right of the row that represents it. The worst possible outcome of each strategy is the smallest value of each row, often referred to as the **row minimum.** Sol's best strategy is to select the option that produces the largest of these minimums or, in other words, to select the "best of the worst." Because this value is the largest of the smallest row values, it is called the **maximin** (the maximum of the row minimums).

		Tina Heads	Tina Tails	Row Minimums
Sol	Heads	3	−1	(−1)
	Tails	−1	−2	−2

In general, the best strategy for the row player in a strictly determined game is to select the strategy associated with the largest of the row minimums.

Because Tina's point of view is exactly the opposite of Sol's, she views the minimums as maximums and vice versa. Therefore, her best strategy is the one associated with the smallest of the largest column values, the **minimax** (the minimum of the column maximums).

		Tina Heads	Tina Tails
Sol	Heads	3	−1
	Tails	−1	−2
Column Maximums		3	(−1)

In general, the best strategy for the column player in a strictly determined game is to select the strategy associated with the smallest of the column maximums.

Remember, if you find it confusing to reverse your thinking when analyzing the columns, change the sign of all matrix entries and use the same thinking you used for the rows.

In this game, the value selected by both Sol and Tina is the same one, that is, the -1 that appears in the upper right-hand corner of the matrix. This is the identifying characteristic of strictly determined games. If the value selected by the two players is not the same, the game is not strictly determined and is much less boring. Games that are not strictly determined are considered in the next lesson.

A **strictly determined** game is one in which the *maximin* (maximum of the row *minimums*) and the *minimax* (minimum of the column *maximums*) are the same value. This value is called the **saddle point** of the game. The saddle point can be interpreted as the amount won per play by the row player.

When players have more than two strategies, a game is somewhat harder to analyze. It often is helpful to eliminate strategies that are **dominated** by other strategies. For example, in a competition between two pizza restaurants, Dino's and Sal's both are considering four strategies: running no special, offering a free minipizza with the purchase of a large pizza, offering a free medium pizza with the purchase of a large one, and offering a free drink with any pizza purchase.

A market study estimates the gain or loss in dollars per week to Dino's over Sal's according to the following matrix:

		Sal's			
		No special	Mini	Medium	Drink
Dino's	No special	200	−400	−300	−600
	Mini	500	100	200	600
	Medium	400	−100	−200	−300
	Drink	300	0	400	−200

What should the managers of Dino's and Sal's do?

Suppose you are the manager of Dino's and examine the first two rows carefully. You notice that no matter what your competitor does, you always achieve a larger payoff by offering the free mini. It would make no sense, therefore, to offer no special. The first row of the matrix is dominated by the second and can be eliminated by drawing a line through it. Similarly, the second row also dominates the third, and so the third can be eliminated.

		Sal's No special	Sal's Mini	Sal's Medium	Sal's Drink
Dino's	No special	200	−400	−300	−600
	Mini	500	100	200	600
	Medium	400	−100	−200	−300
	Drink	300	0	400	−200

Now think of the matrix from the point of view of Sal's manager. Because all the payoffs to Sal's are opposites of the payoffs to Dino's, a column is dominated if all its entries are larger, rather than smaller, than those of another column. Notice that all the values in the first column are larger than the corresponding values in the second column. Because the first column is dominated by the second, it is unwise for Sal's to run no special, and so this strategy can be eliminated. Similarly, the second column dominates the third, and so the third can be eliminated.

		Sal's No special	Sal's Mini	Sal's Medium	Sal's Drink
Dino's	No special	200	−400	−300	−600
	Mini	500	100	200	600
	Medium	400	−100	−200	−300
	Drink	300	0	400	−200

Once these strategies are eliminated, the game is easier to examine for a minimax and a maximin:

		Sal's No special	Sal's Mini	Sal's Medium	Sal's Drink	Row min
Dino's	No special	200	−400	−300	−600	
	Mini	500	100	200	600	100
	Medium	400	−100	−200	−300	
	Drink	300	0	400	−200	−200
Column max			100		600	

The game is strictly determined with a saddle point of 100. Dino's best strategy is to offer the free mini, and Sal's best strategy is to do the same. By pursuing this strategy, Dino's will gain about \$100 a week over Sal's.

Exercises

1. Each of the following matrices represents a payoff matrix for a game. Determine the best strategies for the row and the column players. If the game is strictly determined, find the saddle point of the game.

   a. $\begin{bmatrix} 16 & 8 \\ 12 & 4 \end{bmatrix}$ b. $\begin{bmatrix} 0 & 4 \\ -1 & 2 \end{bmatrix}$ c. $\begin{bmatrix} 2 & -3 \\ -3 & 4 \end{bmatrix}$

   d. $\begin{bmatrix} 0 & 1 & 2 \\ 3 & -2 & 0 \end{bmatrix}$ e. $\begin{bmatrix} 0 & -6 & 1 \\ -4 & 8 & 2 \\ 6 & 5 & 4 \end{bmatrix}$ f. $\begin{bmatrix} 0 & 3 & 1 \\ -3 & 0 & 2 \\ -1 & -4 & 0 \end{bmatrix}$

2. a. For the game defined by the following matrix, determine the best strategies for the row and the column players and the saddle point of the game.

   $$\begin{bmatrix} -4 & 2 \\ 5 & 3 \end{bmatrix}$$

   b. Add 4 to each element in the matrix given in part a. How does this affect the best strategies and the saddle point of the game?
   c. Multiply each element in the matrix in part a by 2. How does this affect the saddle point of the game and the best strategies?
   d. Make a conjecture based on the results of parts b and c.

3. Discuss what would happen in the game discussed in this lesson if Sol decided to depart from his best strategy. Suppose he switches to displaying tails occasionally. Do you think Tina should still play tails every time? Explain your answer.

4. Use the concept of dominance to solve each of the following games. Give the best row and column strategies and the saddle point of the game.

a.
$$\begin{array}{c} \\ A \\ B \\ C \\ D \end{array}\begin{array}{c} \begin{array}{ccc} E & F & G \end{array} \\ \begin{bmatrix} 3 & 1 & 7 \\ 0 & 1 & 3 \\ 4 & 3 & 4 \\ 1 & 3 & 6 \end{bmatrix} \end{array}$$

b.
$$\begin{array}{c} \\ A \\ B \\ C \\ D \end{array}\begin{array}{c} \begin{array}{ccc} E & F & G \end{array} \\ \begin{bmatrix} 4 & -1 & -2 \\ 0 & 1 & 1 \\ 0 & -2 & 5 \\ 3 & 2 & 4 \end{bmatrix} \end{array}$$

5. The Democrats and Republicans are engaged in a political campaign for mayor of a small midwestern community. Both parties are planning their strategies for winning votes for their candidate in the final days. The Democrats have settled on two strategies, A and B, and the Republicans plan to counter with strategies C and D. A local newspaper got wind of their plans and conducted a survey of eligible voters. The results of the survey show that if the Democrats choose plan A and the Republicans choose plan C, then the Democrats will gain 150 votes. If the Democrats choose A and the Republicans choose D, the Democrats will lose 50 votes. If the Democrats choose B and the Republicans choose C, the Democrats will gain 200 votes. If the Democrats choose B and the Republicans choose D, the Democrats will lose 75 votes. Write this information as a matrix game. Find the best strategies and the saddle point of the game.

6. Two major discount companies, Salemart and Bestdeal, are planning to locate stores in Nebraska. If Salemart locates in city A and Bestdeal in city B, then Salemart can expect an annual profit of $50,000 more than Bestdeal's annual profit. If both locate in city A, they can expect to have equal profits. If Salemart locates in city B and Bestdeal in city A, then Bestdeal's profits will exceed Salemart's by $25,000. If both companies locate in city B, then Salemart's profits will exceed Bestdeal's by $10,000. What are the best strategies in this situation, and what is the saddle point of the game?

7. Carol and Bill each have three dimes. They both hold one, two, or three coins in a clenched fist and open their fists together. If they both are holding the same number of coins, Carol will take the coins that Bill is holding. If they are holding a different number of coins, Bill will take the coins that Carol is holding.

a. Write the payoff matrix from Carol's point of view.
b. Does this game have a saddle point? If so, what are the best strategies for Carol and Bill?

8. Mike is going over to see his girlfriend, Nancy, after track practice, when he suddenly remembers that today may be a special anniversary for him and Nancy, and he always brings her a single red rose on this occasion. But he's not sure. Maybe the anniversary is next week. What should he do? If it is their anniversary and he doesn't bring a rose, he'll be in trouble. On a scale from 0 to 10, he'd score a −10. If he doesn't bring a rose and it isn't their anniversary, Nancy won't know anything about his frustration, and he'll score a 0. If he brings a rose and it is not their anniversary, then Nancy will be suspicious that something funny is going on, but he'll score about a 2. If it is their special anniversary and he brings a rose, then Nancy will be expecting it, and he'll score a 5. Write a payoff matrix for this situation. What is Mike's best strategy?

9. School Board and Teacher Education Association representatives are meeting to negotiate a contract. Each side can either threaten (reduction in staff or strike), refuse to negotiate, or negotiate willingly. Each side decides on its strategy before coming to the negotiating table. The payoff matrix below gives the percentage pay increases for the teachers that would result from each combination of strategies. Find the best strategies for each side.

		School Board		
		Threaten	Refuse	Negotiate
Teachers	Threaten	5	4	3
	Refuse	3	0	2
	Negotiate	4	3	2

Game Theory, Part 2

The games considered in the previous lesson were strictly determined. In this lesson, we examine games for which there is not a single best strategy for each player.

Look again at the game of the previous lesson. Suppose that Sol, knowing that he can lose only if Tina plays rationally, proposes changing the game. He now will win a penny if both coins are tails but will lose two pennies if the coins don't match. The new payoff matrix is shown below:

$$\begin{array}{cccc} & & \multicolumn{2}{c}{\text{Tina}} \\ & & \text{Heads} & \text{Tails} \\ \text{Sol} & \begin{array}{c}\text{Heads}\\ \text{Tails}\end{array} & \multicolumn{2}{c}{\left[\begin{array}{cc} 3 & -2 \\ -2 & 1 \end{array}\right]} \end{array}$$

Here is the same matrix with the row minimums and column maximums:

$$\begin{array}{ccccc} & & \multicolumn{2}{c}{\text{Tina}} & \\ & & \text{Heads} & \text{Tails} & \text{Row minimums} \\ \text{Sol} & \text{Heads} & 3 & -2 & -2 \\ & \text{Tails} & -2 & 1 & -2 \\ \multicolumn{2}{r}{\text{Column maximums}} & 3 & 1 & \end{array}$$

There is a tie for the maximin, and the minimax is 1. Neither maximin agrees with the minimax, so the game is not strictly determined. The best strategy for either player is to display a mixture of heads and tails and to keep the other player guessing. One way to do this would be to flip the coin and allow it to appear heads or tails at random. But such a strategy would cause heads and tails to appear in roughly equal portions, and it is not clear that this would be best. Sol could, for example, roll a die and show heads if one, two, three, or four spots appear, and tails otherwise. He might reason that this would benefit him because he would show heads two-thirds of the time and he would win more if two heads appeared.

Consider what will happen if Sol and Tina each decide to flip their coins. The probability of heads is .5, as is the probability of tails. Because Sol's flip and Tina's flip are made independently, the probability of both showing heads is $.5 \times .5 = .25$. The probability of both showing tails is also .25. The probability of one head and one tail is $2 \times .5 \times .5 = .5$ because there are two ways this can happen. The table shows a probability distribution for Sol's winnings.

Outcome	HH	TT	HT or TH
Amount won	3	1	−2
Probability	.25	.25	.5

Sol's expectation is $3(.25) + 1(.25) - 2(.5) = 0$.

Because Tina's payoffs are the opposite of Sol's, her expectation is $-3(.25) - 1(.25) + 2(.5) = 0$. If both players display heads and tails in equal proportions, the game is fair because their expectations are equal.

But suppose that Tina decides to play heads 40% of the time and Sol continues flipping his coin. The probability of both heads is now $.4 \times .5 = .2$, and the probability of both tails is $.6 \times .5 = .3$. The probability of one of each can be determined by subtracting the previous two from 1: $1 - (.2 + .3) = .5$. The distribution for Sol's winnings in this situation is shown at the top of the next page.

Sol's expectation is now $3(.2) + 1(.3) - 2(.5) = -.1$, which means he will lose about 0.1 pennies per play or 1 penny every 10 plays. Tina has an advantage!

Outcome	HH	TT	HT or TH
Amount won	3	1	−2
Probability	.2	.3	.5

You have seen that Tina can gain an advantage over Sol if she knows he will display heads and tails in equal proportions. She does not know that Sol is going to do this, however, so how can she decide the best mixture of strategies? How can Sol decide what is best for him?

Reconsider the game from Sol's point of view, and suppose that Tina plays heads every time while Sol continues to flip his coin. Sol's expectation is now $3(.5) - 2(.5) = .5$. If Tina decides to play tails each time while Sol continues to flip his coin, Sol's expectation will be $-2(.5) + 1(.5) = -.5$. A simple way to make these calculations is to write the probabilities of Sol's displaying heads and tails in a row vector and to multiply it by the payoff matrix:

$$[.5 \quad .5]\begin{bmatrix} 3 & -2 \\ -2 & 1 \end{bmatrix} = [.5 \quad -.5]$$

Suppose Sol switches to displaying heads 60% of the time. The matrix product is

$$[.6 \quad .4]\begin{bmatrix} 3 & -2 \\ -2 & 1 \end{bmatrix} = [.1 \quad -.8]$$

This means that if Sol displays heads 60% of the time, he will gain 0.1 pennies per play if Tina always displays heads and lose .8 pennies per play if Tina always displays tails.

In general, if the probability that Sol will display heads is p, his expected winnings per play if Tina displays all heads or all tails are

$$[p \quad 1-p]\begin{bmatrix} 3 & -2 \\ -2 & 1 \end{bmatrix} = [3p - 2(1-p) \quad -2p + 1(1-p)]$$

Because it is not very likely that Tina will display all heads or all tails, Sol's best strategy is to act in such a way that the two expectations are balanced or equalized. To find the value of p that does this, set the two expectations equal to each other and solve the resulting equation:

$$\begin{aligned} 3p - 2(1 - p) &= -2p + 1(1 - p) \\ 3p - 2 + 2p &= -2p + 1 - p \\ 5p - 2 &= 1 - 3p \\ 8p &= 3 \\ p &= \tfrac{3}{8}. \end{aligned}$$

Sol's best strategy is to display heads three-eighths of the time and tails five-eighths of the time. One way he could accomplish this is to generate a random number on a calculator and display heads if it is less than $\frac{3}{8} = 0.375$.

Tina's best strategy can be determined in a similar way. Call the probability that she displays heads q. Because she is the column player, multiply the payoff matrix by a column matrix to obtain her expectations if Sol plays either all heads or all tails:

$$\begin{bmatrix} 3 & -2 \\ -2 & 1 \end{bmatrix} \begin{bmatrix} q \\ 1 - q \end{bmatrix} = \begin{bmatrix} 3q - 2(1 - q) \\ -2q + 1(1 - q) \end{bmatrix}.$$

Equate the two entries in the resulting matrix and solve. In this case, Tina's best strategy is also to display heads three-eighths of the time.

If both players pursue these strategies, a pair of heads will appear $(\frac{3}{8})(\frac{3}{8}) = \frac{9}{64}$ of the time; a pair of tails will appear $(\frac{5}{8})(\frac{5}{8}) = \frac{25}{64}$ of the time; and one of each will appear $1 - (\frac{9}{64} + \frac{25}{64}) = \frac{30}{64}$ of the time. The resulting probability distribution from Sol's point of view is

Amount won	3	1	−2
Probability	$\frac{9}{64}$	$\frac{25}{64}$	$\frac{30}{64}$

Sol's expectation is $3(\frac{9}{64}) + 1(\frac{25}{64}) - 2(\frac{30}{64}) = -\frac{8}{64}$. If both players pursue their best strategy, the game will be in Tina's favor, and so she can expect to win about $\frac{8}{64} = \frac{1}{8}$ of a penny per play from Sol.

Exercises

1. Suppose that in the example of this lesson, Sol decides to play heads 70% of the time while Tina continues to pursue her best strategy of playing heads three-eighths of the time.

a. What is the probability that both coins will appear heads?
b. What is the probability that both coins will appear tails?
c. What is the probability that one of each will appear?
d. Write a probability distribution for Sol's winnings.
e. Calculate Sol's expectation.
f. Compare Sol's expectation to the negative one-eighth that he achieves by displaying heads three-eighths of the time.

2. Suppose that Sol and Tina change their game so that the payoffs to Sol are as follows:

$$\begin{array}{cc} & \text{Tina} \\ & \begin{array}{cc}\text{Heads} & \text{Tails}\end{array} \\ \text{Sol} \begin{array}{c}\text{Heads} \\ \text{Tails}\end{array} & \begin{bmatrix} 4 & -2 \\ -3 & 1 \end{bmatrix} \end{array}$$

a. Use the row matrix $[p \quad 1 - p]$ and the resulting equation to find Sol's best strategy.
b. Use the column matrix

$$\begin{bmatrix} q \\ 1 - q \end{bmatrix}$$

to find Tina's best strategy.
c. Use the results of parts a and b to prepare a probability distribution for Sol's winnings.
d. Find Sol's expectation.
e. If the game were played 100 times, about how much would you expect Sol to win or lose?

3. The procedure outlined in this lesson is designed to determine the best mixture of strategies when a game is not strictly determined. Therefore, you should always inspect a game to see whether it is strictly determined and apply the saddle point technique of the previous lesson if it is. It is, however, easy to forget to do this. To see what will happen if you attempt to determine a mixture of strategies for a game that is strictly determined, apply the techniques of this lesson to the strictly determined game that Sol and Tina were playing in the last lesson, and try to find the best mixture of strategies for each of them. The game is reprinted on the next page.

$$\begin{array}{cc} & \text{Tina} \\ & \begin{array}{cc} \text{Heads} & \text{Tails} \end{array} \\ \text{Sol} \begin{array}{c} \text{Heads} \\ \text{Tails} \end{array} & \begin{bmatrix} 3 & -1 \\ -1 & -2 \end{bmatrix} \end{array}$$

4. a. For the game defined by the following matrix, determine the best strategies for both players:

$$\begin{bmatrix} 1 & 3 \\ 4 & 2 \end{bmatrix}$$

   b. Add 5 to each element in the matrix given in part a. How does this affect the best strategies?
   c. Multiply each element in the matrix in part a by 3. How does this affect the best strategies?
   d. Make a conjecture based on the results of parts b and c.

5. In a game known as Two-finger Morra, two players simultaneously hold up either one or two fingers. If they hold up the same number of fingers, player 1 will win the sum (in pennies) of the digits from player 2. If they hold up different numbers, then player 2 will win the sum from player 1. Write the payoff matrix for this game. Find the best strategy for each player and the expectation for the row player. Is this a fair game? Explain your answer.

6. In another version of the game in Exercise 5, if the sum of the fingers held out by each player is even, player 1 will win 5 cents. If the sum is odd, player 2 will win 5 cents. Write the payoff matrix for this version. Find the best strategy for each player and the expectation for the row player. Is this a fair game? Explain your answer.

7. A group of parents in a small town are upset about a new social studies program that the school district has adopted. They are seeking to have the program removed from the curriculum. A second group of parents believe the new program is a solid choice and are organizing in favor of keeping it. In order to bring the issue before the voters in the town, the opposing group must collect 400 supporting signatures from registered voters. Both sides are campaigning vigorously by making telephone calls, sending out mail-

ings, and going door to door to contact voters. The local newspaper has estimated the number of signatures that the opposing group is expected to collect with each combination of strategies. What are the best strategies for both groups of parents? If both follow their best strategies, can the opposing group expect to gather enough signatures to get the issue on the ballot? (Hint: Use the concept of dominance to eliminate a row and column.)

		Group in favor		
		Phone	Mail	Door
Group against	Phone	150	75	100
	Mail	350	300	200
	Door	500	100	400

8. Two rival TV networks compete for prime time audiences by showing comedy, drama, and sports. The following matrix gives the payoffs for network A in terms of the percentages of regular viewers who watch its channel for various combinations of programs. Find the best strategy for each network and the expectation for network A.

		Network B		
		Comedy	Drama	Sports
Network A	Comedy	10	50	20
	Drama	40	30	50
	Sports	30	20	60

9. In a campaign for student council president at North High, the top two candidates, Betty and Bob, each are making two promises about what they will do if they are elected. The payoff matrix in terms of the number of votes that Betty will gain is shown below. What are the best strategies for each candidate, and what is Betty's expectation?

		Bob	
		A	B
Betty	1	200	100
	2	50	180

Project

The games you studied in this and the previous lesson are known as *zero-sum games,* because one person's loss is the other's gain. If, for example, Sol wins $2, Tina will lose the same amount. In some games, a particular outcome may be worth 2 to one player, but not −2 to the other. Examples of such games include prisoner's dilemma and arms races between countries. Research and report on games that are not zero-sum.

A Visual Explanation for Sol's Best Strategy

Sol's search for a best strategy can be visualized in the following way: Draw a horizontal line to represent the probability of Sol's displaying heads. It can be drawn from 0 to 1 or 1 to 0 and should be scaled in tenths (see Figure 7.4a).

Draw two vertical lines at each end of the horizontal line, and scale them from the minimum amount Sol can win (−2 in this case) to the maximum (3 in this case) (see Figure 7.4b).

Draw a diagonal line to represent what will happen if Tina always displays heads. To do this, notice that if Tina always displays heads and Sol always displays heads, Sol will win 3 cents. So, above the probability of Sol's always displaying heads (the 1 on the horizontal line), place a dot at 3 on the vertical line. Similarly, notice that if Sol always displays tails when Tina displays heads, he will lose 2 cents, so place a dot at −2 on the vertical line above the 0 on the horizontal line. Then connect the two dots with a diagonal line. Sol's winnings for his various strategies when Tina always displays heads can be read from the line (see Figure 7.4c).

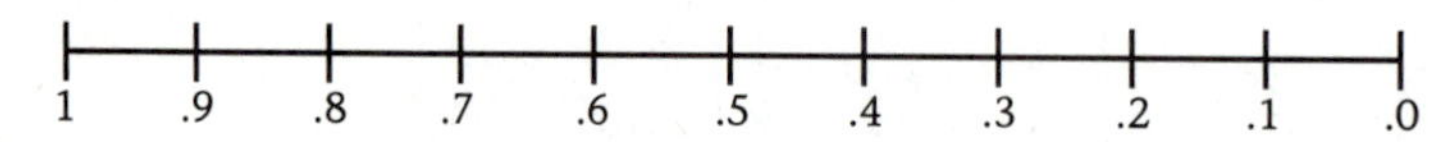

Figure 7.4a

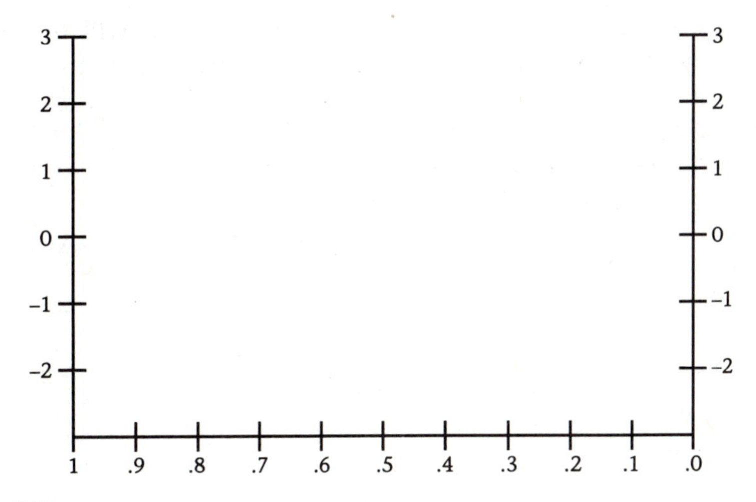

Figure 7.4b

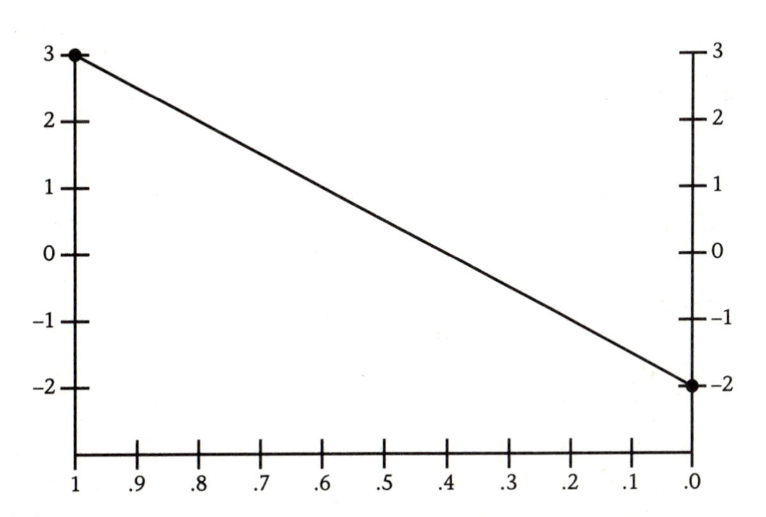

Figure 7.4c

Repeat the previous procedure to represent what will happen if Tina always displays tails. Place a dot at −2 above the 1 on the horizontal line because Sol will lose 2 cents if he always displays heads when Tina displays tails. Similarly, place a dot at 1 above the 0 on the horizontal line and connect the two dots. The intersection of the two lines is directly above Sol's best strategy (see Figure 7.4d).

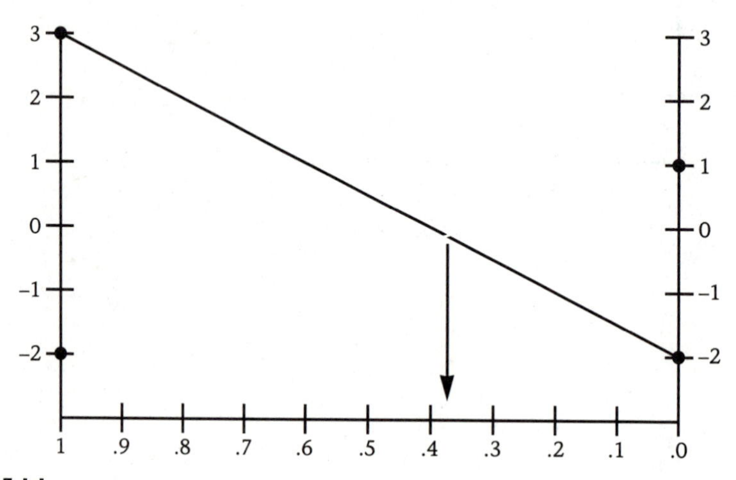

Figure 7.4d

1. Suppose that a three-sector economy has the following technology matrix:

$$T = \begin{array}{c} \\ A \\ B \\ C \end{array} \begin{array}{c} \begin{array}{ccc} A & B & C \end{array} \\ \begin{bmatrix} 0.1 & 0.2 & 0.3 \\ 0.1 & 0.3 & 0.2 \\ 0.2 & 0.1 & 0.2 \end{bmatrix} \end{array}$$

a. Draw a weighted digraph for this economy.
b. A production matrix, P, is shown below. Find the internal consumption matrix, TP, and the external demand matrix, D, where $D = P - TP$.

$$P = \begin{bmatrix} 8 \\ 12 \\ 15 \end{bmatrix}$$

c. An external demand matrix, D, is shown below. Find the production matrix, P, for this economy. Recall that $P = (I - T)^{-1}D$.

$$D = \begin{bmatrix} 6 \\ 8 \\ 12 \end{bmatrix}$$

2. Dan and Sarah are playing poker for pennies to kill time during lunch. Dan is holding a very poor hand and is considering bluffing or not bluffing. Sarah can either call or not call the bluff. The payoff matrix for this situation is shown below. Over the course of several

games in which Dan comes up with a poor hand, what should his strategy be?

$$\text{Dan}\quad\begin{array}{l} \text{Bluff} \\ \text{Not bluff} \end{array}\overset{\begin{array}{c}\text{Sarah}\\ \begin{array}{cc}\text{Call} & \text{Not call}\end{array}\end{array}}{\begin{bmatrix} -10 & 10 \\ -2 & 0 \end{bmatrix}}$$

3. The discrete mathematics teacher has three class starter activities, one of which she uses to begin class every day: a pop quiz, a quickie review, and a small-group problem-solving activity. She never uses the same activity two days in a row. If she gave a pop quiz yesterday, she will toss a coin and do a quickie review if it comes up heads. If she used a review, she will toss two coins and switch to problem solving if two heads come up. If she did a problem-solving activity, she will toss three coins, and if three heads come up, she will give a pop quiz again. The transition matrix for this scheme is shown below.

$$\begin{array}{c} \\ Q \\ R \\ P \end{array}\begin{array}{c} \begin{array}{ccc} Q & R & P \end{array} \\ \begin{bmatrix} 0 & \frac{1}{2} & \frac{1}{2} \\ \frac{3}{4} & 0 & \frac{1}{4} \\ \frac{1}{8} & \frac{7}{8} & 0 \end{bmatrix} \end{array}$$

a. If the teacher gives a quiz on Monday, what is the probability that she will give another quiz on Friday?
b. In the long run, how often should the students expect that the teacher will start class with a quiz?
c. What activity will the teacher use most often to start class, and how frequently will she use it?

4. The Super X sells three kinds of sandwiches that many of the students at Jackson High especially like for lunch—Super X Original, Italian Special, and Barbecue Beef. The Super X clerk observed that the same students were coming in for sandwiches for lunch every school day and that the kind of sandwich that each student purchased depended on what he or she had ordered on the previous visit. He conducted a survey and found that of the students who ordered the Original on their last visit, 20% ordered it again the next time, whereas 25% switched to Italian and 55% switched to

Barbecue Beef. Of the students who ordered the Italian sandwich the last time, 35% did so again the next time, but 45% switched to the Original, and 20% switched to the Barbecue Beef. Of the students who got the Barbecue Beef the last time, 55% ordered it the next time, 20% switched to the Original, and 25% switched to Italian.

a. Set up the transition matrix for this Markov chain.
b. If the same students tend to buy Super X sandwiches for lunch every day, what is the probability that a student who buys the Italian sandwich on Monday will have it again on Wednesday?
c. In the long run, what percentage of the orders will be for the Original? For the Italian? For the Barbecue Beef?
d. How will access to this information help the Super X clerk?

5. A certain economy consists of three industries: transportation, petroleum, and agriculture. The production of $1 million worth of transportation requires an internal consumption of $0.2 million worth of transportation, $0.4 million worth of petroleum, and no agriculture. The production of $1 million worth of petroleum requires an internal consumption of $0.3 million worth of transportation, $0.2 million worth of petroleum, and $0.3 million worth of agriculture. The production of $1 million worth of agriculture requires an internal consumption of $0.3 million worth of transportation, $0.2 million worth of petroleum, and $0.25 million worth of agriculture.

a. Draw a weighted digraph for this economy.
b. Write a technology matrix, T, representing this information.
c. On what sector of the economy is transportation the most dependent? The least dependent?
d. If the agriculture sector has an output of $5.4 million, what is the input in dollars for petroleum? From agriculture?
e. A production matrix, P, in millions of dollars, is shown below. Find the internal consumption matrix, TP, and the external demand matrix, D.

$$P = \begin{matrix} \text{Trans.} \\ \text{Petr.} \\ \text{Agri.} \end{matrix} \begin{bmatrix} 20 \\ 25 \\ 15 \end{bmatrix}$$

f. An external demand matrix, D, in millions of dollars, is shown below. How much must each sector produce in order to meet this demand?

$$D = \begin{matrix} \text{Trans.} \\ \text{Petr.} \\ \text{Agri.} \end{matrix} \begin{bmatrix} 4.6 \\ 5.2 \\ 3.0 \end{bmatrix}$$

6. Two computer companies (1 and 2) are competing for sales in two large school districts (A and B). The payoff matrix below shows the differences in sales for companies 1 and 2 in hundreds of thousands of dollars if they focus their full sales force on either school district. Find the best strategy for each company.

$$\text{Computer company 1} \quad \begin{matrix} \\ A \\ B \end{matrix} \begin{matrix} \text{Computer company 2} \\ \begin{matrix} A & B \end{matrix} \\ \begin{bmatrix} 3 & 7 \\ -7 & -3 \end{bmatrix} \end{matrix}$$

7. Suppose that in the final days of a political campaign for mayor of a small midwestern city, the Democrats and Republicans are planning their strategies for winning undecided voters to their political camps. The Democrats have decided on two strategies, plan A and plan B. The Republicans plan to counter with plans C and D. The matrix below gives the payoff for the Democrats of the various combinations of strategies. The numbers represent the percentage of undecided voters joining the Democrats in each case. Find the best strategy for both parties and the expectation for the Democrats.

$$\text{Democrats} \quad \begin{matrix} \\ \text{Plan A} \\ \text{Plan B} \end{matrix} \begin{matrix} \text{Republicans} \\ \begin{matrix} \text{Plan C} & \text{Plan D} \end{matrix} \\ \begin{bmatrix} 30 & 60 \\ 50 & 40 \end{bmatrix} \end{matrix}$$

8. A manufacturing company has divisions in Massachusetts, Nebraska, and California. The company divisions use goods and services from one another, as shown in the input–output matrix, T, below.

$$T = \begin{array}{r} \\ \text{Mass.} \\ \text{Neb.} \\ \text{Calif.} \end{array} \begin{array}{c} \begin{array}{ccc} \text{Mass.} & \text{Neb.} & \text{Calif.} \end{array} \\ \begin{bmatrix} 0.04 & 0.02 & 0.03 \\ 0.03 & 0.01 & 0.05 \\ 0.01 & 0.02 & 0.04 \end{bmatrix} \end{array}$$

a. Draw a weighted digraph for this situation.

b. Find the total output needed to meet a final consumer demand of \$50,000 from Massachusetts, \$30,000 from Nebraska, and \$40,000 from California.

c. What will the internal consumption be for each division to meet the demands in part b?

d. Suppose there is an increase in consumer demand of \$10,000 from Massachusetts, \$8,000 from Nebraska, and \$12,000 from California. What will be the change in internal consumption and in the total production of goods and services for each division?

9. Two competing dairy stores choose daily strategies of raising, not changing, or lowering their milk prices. The following payoff matrix shows the percentage of customers who go from store A to store B for each combination of strategies. What should each store do?

		Store B		
		Raise	No change	Lower
Store A	Raise	4	−1	−4
	No change	2	1	−2
	Lower	5	2	3

Bibliography

Bittinger, M. L., and J. C. Crown. *Finite Mathematics.* Reading, MA: Addison-Wesley.

Bogart, Kenneth P. 1988. *Discrete Mathematics.* Lexington, MA: Heath.

Cozzens, M. B., and R. D. Porter. 1987. *Mathematics and Its Applications.* Lexington, MA: Heath.

Keller, M. K. 1983a. *Food Service Management and Applications of Matrix Methods.* Lexington, MA: COMAP.

Keller, M. K. 1983b. *Markov Chains and Applications of Matrix Methods: Fixed Point and Absorbing Markov Chains.* Lexington, MA: COMAP.

Kemeny, J. G., J. L. Snell, and G. L. Thompson. 1957. *Finite Mathematics.* Englewood Cliffs, NJ: Prentice Hall.

Mauer, S. B., and A. Ralston. 1991. *Discrete Algorithmic Mathematics.* Reading, MA: Addison-Wesley.

National Council of Teachers of Mathematics. 1988. *Discrete Mathematics across the Curriculum K-12.* Reston, VA: NCTM.

North Carolina School of Science and Mathematics. *New Topics for Secondary School Mathematics: Matrices.* 1988. Reston, VA: National Council of Teachers of Mathematics.

Ross, K. A., and C. R. B. Wright. 1985. *Discrete Mathematics.* Englewood Cliffs, NJ: Prentice-Hall.

Tuchinsky, P. M. 1989. *Management of a Buffalo Herd.* Lexington, MA: COMAP.

Wheeler, R. E., and W. D. Peebles. 1987. *Finite Mathematics with Applications to Business and the Social Sciences.* Monterey, CA: Brooks/Cole.

Williams, J. D. 1982. *The Compleat Strategyst*. New York: Dover Publications.

Zagare, F. C. 1985. *The Mathematics of Conflict*. Lexington, MA: COMAP.

Recursion

Many patterns can be found in nature and in the man-made world. Patterns not only repeat themselves, but they also often contain themselves. Each of the quadrilaterals formed by the arms and legs of these sky divers is contained in several other larger quadrilaterals, which in turn are contained in still larger quadrilaterals, or would be if the figure continued indefinitely.

How are the patterns that occur in such figures related to patterns in the numbers that count them? What mathematical models can we create to represent these patterns? How can the models we create to represent the patterns in figures be adapted to help us plan for financial security? The mathematics of recurring patterns provides the answers to these and many other important questions.

Introduction to Recursive Thinking

Reconsider a problem introduced in Lesson 2.6. Luis and Britt were examining the number of handshakes that occur when every person in a group shakes hands with every other person. A table of data is reprinted below.

Number of People in the Group	Number of Handshakes
1	0
2	1
3	3
4	6
5	10

When a new person entered a group in which everyone had shaken hands, the new person had to shake hands with each of the people who were already in the group. Thus, the number of handshakes in a group

of n people is $n - 1$ more than the number of handshakes in a group of $n - 1$ people. If H_n represents the number of handshakes in a group of n people, this recurrence relation can be expressed symbolically as $H_n = H_{n-1} + (n - 1)$.

Your work with this recurrence relation included writing a formula called a **solution to the recurrence relation** and using mathematical induction to prove the formula correct. In the case of the handshake problem, the solution, also called a **closed-form solution,** is $H_n = \dfrac{n(n-1)}{2}$.

The counting techniques discussed in Chapter 6 can also be applied to certain kinds of recurrence relations to find the closed-form solution. Imagine a handshake between two people in a group who were selected at random. There are $C(n, 2)$ ways of selecting two people from a group of n people, and so there are $C(n, 2) = \frac{n!}{(n-2)!2!}$ handshakes in a group of n people. But $n! = n(n-1)(n-2)!$, so the counting solution is equivalent to the solution you hypothesized and proved in Chapter 2.

The advantage to the closed-form solution is that it can be used to determine the number of handshakes in a large group of people without consideration of a smaller group. If, for example, you want to know the number of handshakes in a group of 100 people, the closed-form solution gives $\frac{100 \times 99}{2} = 4{,}950$. To find the same value using a recurrence relation requires a table of 100 rows. Closed-form solutions for recurrence relations, however, can be difficult to find.

You can create a table of data for the handshake problem by calculating and recording each piece of information individually, but this method is time-consuming and tedious if the table is large. Computers and programmable calculators offer more efficient ways of doing this.

A computer *spreadsheet* is really a matrix consisting of columns labeled with the letters A, B, C, . . . and rows labeled with the numbers 1, 2, 3, ... A particular location in the spreadsheet is called a *cell* and is denoted by its column letter and row number, such as A1 or C5. Cells may contain verbal labels, numeric values, or formulas based on references to other cells. Spreadsheets include copy features that allow formulas to be copied quickly into other cells so that tables can be generated rapidly. (See "Using a Computer Spreadsheet to Build a Table" at the end of this lesson.)

Another way to build a table is by programming a computer or calculator. To do so requires that an appropriate algorithm be adapted to the computing device's particular language. The following is an algorithm for the handshake problem that can be adapted to a computer or calculator. N represents the number of people in the group, and H represents the number of handshakes.

1. Store the number 1 for variable N and the number 0 for variable H.

2. Display N and H.

3. Add 1 to N and store this value as the new value of N.

4. Add $N - 1$ to H and store this value as the new value of H.
5. Repeat Steps 2 through 4.

A BASIC implementation is given in "Building a Table with a BASIC Algorithm" at the end of this lesson. Step 4 of this algorithm uses a recurrence relation to calculate the appropriate value of H. The closed-form may also be used. To do so, replace Step 4 with the following: Store $N(N - 1)/2$ as the new value of H. In the following exercises, generate tables by any means you feel appropriate. But be sure to use a spreadsheet or program for at least some of the problems.

Exercises

1. Consider a variation of this lesson's handshake problem. Suppose there are an equal number of men and women in Luis's and Britt's group and that each person shakes hands with all the people of the opposite sex.
   a. Draw a graph for a group of four couples, in which the vertices represent the men and women in the group and the edges represent the handshakes. Recall your work in graph theory. What kind of a graph is this?
   b. If there are only one man and one woman in the group, how many handshakes will be made? With two couples? With three couples?
   c. Complete the following table to investigate the number of handshakes that are made:

Number of Couples	Number of Handshakes	Recurrence Relation
1		
2		
3		
4		
5		

d. Assume that there are H_{n-1} handshakes for $n - 1$ couples and that another couple joins the party. How many additional handshakes are now possible?
e. Write a recurrence relation between the number of handshakes (H_n) for n couples and the number of handshakes (H_{n-1}) for $n - 1$ couples.

2. Consider another variation of the handshake problem, in which each man shakes hands with each of the women *except* his date.
   a. Make a table showing the number of handshakes that occur when there is one couple. Two couples. Three couples. Four couples.
   b. Assuming you know the number of handshakes with $n - 1$ couples, how many additional handshakes are made when the nth couple arrives?
   c. Write a recurrence relation for the total number of handshakes (H_n) when there are n couples.

3. a. Write a recurrence relation to describe each of the following patterns. (Note: No closed-form formulas are allowed.)
      i. 1, 4, 7, 10, 13, ...
      ii. 1, 2, 4, 8, 16, 32, ...
      iii. 1, 3, 6, 10, 15, 21, 28, ...
      iv. 1, 2, 6, 24, 120, 720, ...
   b. What is the next term in each one of the patterns?

4. The terms of a recurrence relation cannot be listed without an initial value. For example, $t_n = 2t_{n-1} - 3$ has terms 5, 7, 11, 19 . . . if the initial value t_1 is 5. But if the initial value is 6, then the terms are 6, 9, 15, 27, ... Notice, however, that if the initial value is 3, then all of the terms of the recurrence relation are 3. An initial value for which all of the terms of the recurrence relation are the same is called a **fixed point.**

   Find the fixed point for each of the following recurrence relations if one exists:
   a. $t_n = 2t_{n-1} - 4$.
   b. $t_n = 3t_{n-1} + 2$.
   c. $t_n = 2t_{n-1}$.
   d. $t_n = t_{n-1} + 3$.

5. If two rays have a common endpoint, one angle is formed. If a third ray is added, three angles are formed. See the figure below.

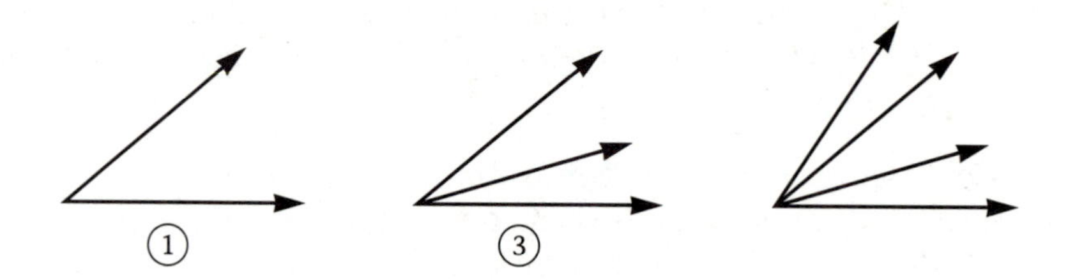

   a. How many angles are formed if a fourth ray is added? A fifth ray?
   b. Write a recurrence relation for the number of angles formed with n rays.
   c. Write a closed-form solution.
   d. Use your closed-form solution to find the number of angles formed by 10 rays.

6. For the original handshake problem in which everyone shakes hands with everyone else, construct a table for eight people in the following manner:

   - First column: term number.
   - Second column: number of handshakes.
   - Third column: differences of successive numbers from column 2.
   - Fourth column: differences of successive numbers from column 3.

   a. What do you notice about the last column?
   b. What degree is the polynomial that was obtained for the closed-form solution of the handshake problem? How many difference columns are there?

7. Consider the closed-form polynomial $S_n = 4n^3 - 3n + 2$.
   a. Make a table, as in Exercise 6, for $n = 1 \ldots 8$. Include the difference columns until all the numbers in the difference column are the same.

b. How many difference columns were needed?
c. What is the degree of the original generating polynomial?

8. Let V_n be the number of vertices in a complete binary tree of n levels. A binary tree is complete if each vertex of the tree has either two or no children. Level 0 is the root of the tree. The first three trees are shown below.

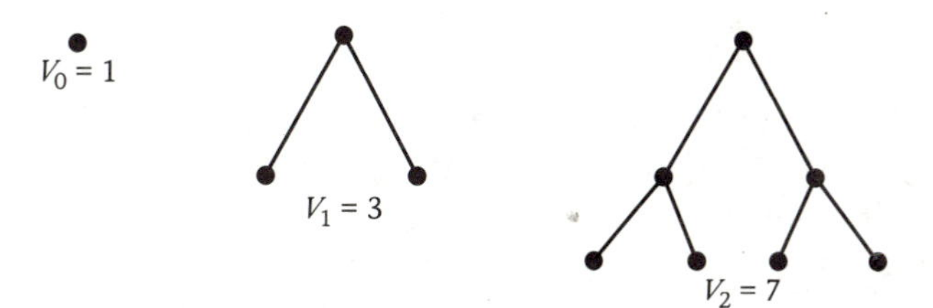

a. Make a table for $V_0 \ldots V_6$.
b. Find a recurrence relation to describe V_n.

In the previous exercise it was convenient to begin the table with V_0 because the root of the tree is considered level 0. There are other cases in which the initial value of the recurrence relation is labeled with the subscript 0. It is especially helpful when working with time intervals, as in Exercises 9 and 10.

9. The number of African bees in Texas in 1987 was estimated to be 5,000. It is also estimated that this number will increase by about 12% each year. Let B_n be the number of African bees in Texas each year, where $n = 0$ corresponds to the beginning of the year 1987. Then B_1 would indicate "the end of year 1."
   a. Make a table with entries for $B_0 \ldots B_4$.
   b. Write a recurrence relation for B_n.
   c. Use a spreadsheet or calculator to determine when the population of African bees exceeds 100,000. In what year will this occur?

10. Susie put $500 in a bank that offers 7% interest per year, compounded yearly. Let n be the number of years she leaves the money in the bank; let A_0 be the initial amount of money ($500); and let A_n be the amount of money in the bank at the end of n years.

a. By creating a table, find out how many years it will take for Susie's money to double in value.
b. Write a recurrence relation for this problem.

Using a Computer Spreadsheet to Build a Table

A spreadsheet is created by typing optional labels in the first row. Initial values of 1 and 0 for the number of people and the number of handshakes are typed in cells A2 and B2 of the second row. Cell A3, directly below cell A2, is given the formula A2 + 1 because its value is 1 more than the value in cell A2. Cell B3, directly below cell B2, contains the recurrence relation $H_n = H_{n-1} + (n - 1)$, which becomes B2 + A2 because cell B2 contains the previous number of handshakes and cell A2 contains the previous number of people. The remaining rows are filled with the appropriate formulas, by copying the formulas in row 3. Note that most spreadsheets require that the initial character of a formula be either "+" or "=."

A spreadsheet can also be built by using a closed-form solution. In this case, the formula in cell B3 becomes A3 * (A3 − 1)/2. All the other steps described in the previous paragraph remain unchanged. The figures shown below demonstrate both methods. The first contains the formulas for the recursive method in the second column and those for the closed-form method in the third. The second figure contains the values that result from the formulas.

	A	B	C
1	Number of people	Number of handshakes	Closed form
2	1	0	= A2*(A2 − 1)/2
3	= A2 + 1	= A2 + B2	= A3*(A3 − 1)/2
4	= A3 + 1	= A3 + B3	= A4*(A4 − 1)/2
5	= A4 + 1	= A4 + B4	= A5*(A5 − 1)/2
6	= A5 + 1	= A5 + B5	= A6*(A6 − 1)/2
7	= A6 + 1	= A6 + B6	= A7*(A7 − 1)/2
8	= A7 + 1	= A7 + B7	= A8*(A8 − 1)/2
9	= A8 + 1	= A8 + B8	= A9*(A9 − 1)/2
10	= A9 + 1	= A9 + B9	= A10*(A10 − 1)/2

Continued

	A	B	C
1	Number of people	Number of handshakes	Closed form
2	1	0	0
3	2	1	1
4	3	3	3
5	4	6	6
6	5	10	10
7	6	15	15
8	7	21	21
9	8	28	28
10	9	36	36

Building a Table with a BASIC Algorithm

The following is a BASIC algorithm that generates a table for the handshake problem:

```
10 N = 1:H = 0
20 PRINT N,H
30 N = N + 1
40 H = H + N − 1
50 GO TO 20
```

Because this algorithm does not end, a statement can be added between 40 and 50 to terminate the table at some value. If, for example, the number of handshakes for no more than 10 people is needed, the statement 45 could be IF N > 10 THEN STOP. To modify the algorithm to use the closed-form method, change line 40 to H = N * (N − 1)/2.

Finite Differences

In the last lesson we discussed a closed-form solution to the handshake problem. In this lesson we consider another way to find a closed-form solution for this and other problems.

Recall the recurrence relation for the handshake problem: $H_1 = 0$, $H_n = H_{n-1} + (n - 1)$. Below is a table generated by this recurrence relation that is similar to the one in Exercise 6 of the last lesson. The

Number of People	Number of Handshakes	Differences First	Differences Second
1	0	—	—
2	1	1	—
3	3	2	1
4	6	3	1
5	10	4	1
6	15	5	1
7	21	6	1
8	28	7	1

third column contains the differences between successive values in the second column, and the fourth column contains the differences between successive values in the third column. (See "Including Differences in Your Spreadsheet" at the end of this lesson.)

The constant second differences lead us to suspect that the closed-form solution for this recurrence relation is a second-degree polynomial. The general form of a second-degree polynomial is $an^2 + bn + c$.

Consider what happens when a second-degree polynomial is evaluated for successive values of n and differences of successive values of the polynomial are found. The table below summarizes this.

		Differences	
Value of n	Value of Polynomial	First	Second
1	$a + b + c$	—	—
2	$4a + 2b + c$	$3a + b$	—
3	$9a + 3b + c$	$5a + b$	$2a$
4	$16a + 4b + c$	$7a + b$	$2a$
5	$25a + 5b + c$	$9a + b$	$2a$

Notice that not only are the second differences constant for any second-degree polynomial but that also the value of the difference is always twice the value of the coefficient of n^2. This means that in the case of the handshake problem, the constant difference of 1 indicates that one of the terms of the closed-form solution is $\frac{1}{2}n^2$.

The remaining terms of the closed-form solution can be found by substituting values from the table into the polynomial $H_n = \frac{1}{2}n^2 + bn + c$. This method of completing the problem can be tedious, particularly when the degree of the polynomial is higher than second.

A quicker way to find the solution once it is known to be second degree is to form a system of equations and apply the matrix techniques given in the last chapter (see p. 345). Reconsider the problem. You know the solution is second degree: $H_n = an^2 + bn + c$. You must determine the values of a, b, c. To do so, select any three pairs of values from your table. Any three will work, but the first three are the most convenient. Form three equations by substituting these three pairs into the general second-degree polynomial $H_n = an^2 + bn + c$.

$$\text{When } n = 1, \quad 0 = a + b + c.$$

$$\text{When } n = 2, \quad 1 = 4a + 2b + c.$$

$$\text{When } n = 3, \quad 3 = 9a + 3b + c.$$

Solve this system by using the matrix techniques of the last chapter:

$$\begin{bmatrix} 1 & 1 & 1 \\ 4 & 2 & 1 \\ 9 & 3 & 1 \end{bmatrix}^{-1} \times \begin{bmatrix} 0 \\ 1 \\ 3 \end{bmatrix} = \begin{bmatrix} 0.5 \\ -0.5 \\ 0 \end{bmatrix}$$

This method for finding closed-form solutions is called the **method of finite differences.** It can be used whenever the differences in values of the recurrence relation become constant in a finite number of columns. The number of columns that this requires indicates the degree of the polynomial solution. The number of equations in the system needed to find the solution is one more than its degree.

Example

Consider a stack of cannonballs at Fort Recurrence (see Figure 8.1). The following table displays the number of cannonballs in a pyramid of n layers:

Number of Layers	Number of Cannonballs	Differences First	Differences Second	Differences Third
1	1	—	—	—
2	5	4	—	—
3	14	9	5	—
4	30	16	7	2
5	55	25	9	2
6	91	36	11	2

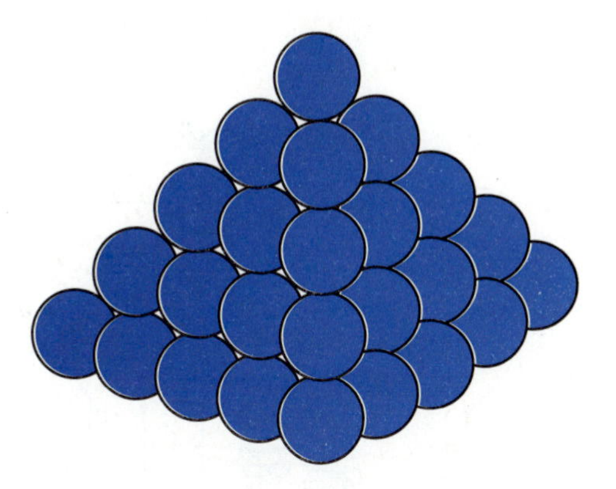

Figure 8.1 Cannonballs at Fort Recurrence.

The recurrence relation that describes the number of cannonballs in a stack of n layers is $C_n = C_{n-1} + n^2$. The constant differences in the third column indicate that the solution is third-degree: $C_n = an^3 + bn^2 + cn + d$. The system created by this general third-degree polynomial and the first four values in the table is as follows:

When $n = 1$, $1 = a + b + c + d$.

When $n = 2$, $5 = 8a + 4b + 2c + d$.

When $n = 3$, $14 = 27a + 9b + 3c + d$.

When $n = 4$, $30 = 64a + 16b + 4c + d$.

The matrix solution is

$$\begin{bmatrix} 1 & 1 & 1 & 1 \\ 8 & 4 & 2 & 1 \\ 27 & 9 & 3 & 1 \\ 64 & 16 & 4 & 1 \end{bmatrix}^{-1} \times \begin{bmatrix} 1 \\ 5 \\ 14 \\ 30 \end{bmatrix} = \begin{bmatrix} 0.3333 \\ 0.5 \\ 0.1667 \\ 0 \end{bmatrix}$$

The closed-form solution, therefore, is $C_n = \frac{1}{3}n^3 + \frac{1}{2}n^2 + \frac{1}{6}n$. Note that unlike the case in which the solution is second degree, the coefficient of the first term of the solution is not one-half of the constant difference.

Unfortunately, the finite difference technique does not apply to recurrence relations for which the differences never become constant. In such cases, other methods that are described in later lessons of this chapter are often successful. The following exercises investigate several recurrence relations and apply the techniques of this lesson to situations in which the differences eventually become constant.

Exercises

1. Using finite differences, determine the degree of the closed-form formula that was used to generate the given sequence:

a. −3, −2, 3, 12, 25, 42, 63, 88, 117, 150, 187, 228, 273, 322, ...
b. 0.29, 0.52, 0.75, 0.98, 1.21, 1.44, 1.67, 1.90, 2.13, 2.36, 2.59, ...
c. 0, −2, −2, 0, 4, 10, 18, 28, 40, 54, 70, 88, 108, 130, 154, ...
d. 1,3,9,27,81,243,729,2187,6561,19683,59049,111147, ...

2. For each part of Exercise 1, determine the closed-form formula that will generate the sequence.

3. a. Write a recurrence relation for the number of edges T_n in a complete graph, K_n.
   b. For your recurrence relation in part a, what is the initial condition? (Hint: How many edges are in a graph with one vertex, $T_1 = ?$)
   c. Using finite difference techniques, determine a closed-form formula for the number of edges in a K_n graph.

4. $a_1 = 1$ and $a_n = 3a_{n-1} - 5$.
   a. For this recurrence relation, find the first 6 to 8 terms.
   b. Find the fixed point for this recurrence relation. (Hint: When a recurrence relation has a fixed point, all of the terms are the same. Replace a_n and a_{n-1} with a single variable such as X, and then solve. Check your solution by using it as an initial value in the recurrence relation.)

5. A triangle has no diagonal; a quadrilateral has two diagonals; and a pentagon has five diagonals.
   a. Write a recurrence relation for the number of diagonals in an n-sided polygon.
   b. Use finite difference techniques to find a closed-form formula for the number of diagonals in an n-sided polygon.

6. An auditorium has 24 seats in the front row. Each successive row, moving toward the back of the auditorium, has 2 additional seats. The last row has 96 seats.
   a. Create a table with a column for the number of the row and a column for the number of seats in that row. Complete at least the first six entries in the table.
   b. Write a recurrence relation for the number of seats in the nth row.

NEWS CLIP

Battle Lines Drawn, as Deer and People Collide

IRONDEQUOIT JOURNAL, December 28, 1992

The pastoral sight of deer feeding around the Durand-Eastman County Park has not necessarily been a welcome one to residents here.

The deer, having stripped the park of greenery in several areas, have taken to munching on expensive landscaping in the yards of houses nearby. And as they cross the roadways looking for food, the deer are involved in more than 100 car accidents a year in this town of about 50,000 residents, the police say.

After more than 10 years of debate and frustration, a committee of town and county lawmakers, members of the County Parks Department and officials from the State Department of Environmental Conservation came up with a plan they thought would solve the problem: lure the deer to feeding sites in the 900-acre park, which borders Lake Ontario, and have sharpshooters thin the herd.

Conservation officials have estimated that as many as 300 or 400 deer live in and around the park, but no one is exactly sure.

Two sharpshooters from local police agencies would be assigned to each site. The shooting would probably be unannounced and the park closed off to prevent animal-rights advocates from disrupting the effort.

Under the plan, Mr. Eckert said, approximately 40 deer would be killed in January and 40 in March. He said that the effort would be repeated annually.

c. Find a closed-form solution for the number of seats in the nth row. (One way to do this is to use finite differences.)
d. How many rows are in the auditorium?
e. Add a third column, "Total seats," to your table from part a. Complete at least the first six sums in this column.
f. Write a recurrence relation for the number of seats in the first n rows of the auditorium.
g. Write a closed-form formula for the number of seats in the first n rows of the auditorium. (One way to do this is to use finite differences.)

7. A house purchased in 1980 increased in value at the rate of 8% per year.
   a. If the original cost of the house was $38,000, calculate the value of the house each year from 1980 to 1993. (A spreadsheet might be nice to use here.)
   b. Write a recurrence relation for the value of the house at the end of the nth year since 1980.
   c. Calculate the finite differences for your numbers in part a. Do you eventually obtain constant differences?

8. In 1993, a herd of 50 deer is increasing at the rate of approximately 4% per year.
   a. Make a table that gives the number of deer at the end of each year ($T_0 = 50$).
   b. If the grazing area can support only 325 deer, in what year will there not be enough food?
   c. Write a recurrence relation for the number of deer at the end of the nth year.
   d. Calculate the finite differences for your table in part a. Do you eventually obtain constant differences?

Including Differences in Your Spreadsheet

If your spreadsheet contains the number of people and the number of handshakes in columns A and B, then adding columns to include differences requires very little additional effort. It is possible to type just one additional formula and to copy it into as many additional rows and columns as you desire. If, for example, your spreadsheet for the handshake problem contains the number of handshakes for a group of one in cell B2, for a group of two in cell B3, and so forth, the first difference is placed in cell C3 by typing the formula B3-B2. This formula is then copied into the remaining cells of column C. These are not constant, so the same formula is copied into the cells of column D starting in cell D4. Because these cells are constant, the process is terminated in column D.

The figures shown below demonstrate this method. The first contains the formulas, and the second contains the resulting values.

	A	B	C	D
1	Number of people	Number of handshakes	First differences	Second differences
2	1	0		
3	= A2 + 1	= B2 + A3	= B3 − B2	
4	= A3 + 1	= B3 + A4	= B4 − B3	= C4 − C3
5	= A4 + 1	= B4 + A5	= B5 − B4	= C5 − C4
6	= A5 + 1	= B5 + A6	= B6 − B5	= C6 − C5
7	= A6 + 1	= B6 + A7	= B7 − B6	= C7 − C6
8	= A7 + 1	= B7 + A8	= B8 − B7	= C8 − C7
9	= A8 + 1	= B8 + A9	= B9 − B8	= C9 − C8
10	= A9 + 1	= B9 + A10	= B10 − B9	= C10 − C9
11				

	A	B	C	D
1	Number of people	Number of handshakes	First differences	Second differences
2	1	0		
3	2	2	2	
4	3	5	3	1
5	4	9	4	1
6	5	14	5	1
7	6	20	6	1
8	7	27	7	1
9	8	35	8	1
10	9	44	9	1
11				

If you generated your table of data with a calculator or computer program, you might find it easier to calculate and record the differences rather than to incorporate the differences into the program.

LESSON 8.3

Arithmetic and Geometric Recursion

In our discussion of recurrence relations in the previous two lessons, we saw that in certain recurrence relations, a term is generated by adding a constant to the previous term. In other types, a term is generated by multiplying the previous term by a constant. Those in which the terms are generated by adding a constant are called **arithmetic,** and the constant that is added is called the **common difference.** Those in which the terms are generated by multiplying by a constant are called **geometric,** and the constant by which a term is multiplied is called the **common ratio.** This lesson considers arithmetic and geometric recurrence relations.

As examples of arithmetic and geometric recurrence relations, consider a job that employs you for 30 days and in which you have a choice of two methods of payment. Method 1 pays $5,000 the first day and includes a $10,000 raise each day after that. Method 2 pays only $0.01 the first day, but doubles the amount you are paid each successive day. Which salary is better?

The first method of payment is arithmetic, and the common difference is $10,000. The second method is geometric, and the common

Thomas Malthus (1766–1834), a nineteenth-century economist, based a well-known prediction that the increase in food supplies would eventually be unable to match increases in population on his perception that the growth of the human population was geometric and the rate of growth of food supplies was arithmetic.

ratio is 2. If P_n represents the amount paid on the nth day, a recurrence relation for the first method of payment is $P_n = P_{n-1} + 10{,}000$, and a recurrence relation for the second is $P_n = 2P_{n-1}$. Another important issue other than the pay on the nth day is the total pay after n days. In both methods of payment, if T_n represents the total amount paid after n days, a recurrence relation that describes T_n is $T_n = T_{n-1} + P_n$. (See "Including Sums in Spreadsheets and Programs" on pages 413–414.)

It is easy to find a closed-form solution for the recurrence relation that describes the pay on the nth day. In Method 1, the pay on the first day is \$5,000; the pay on the second is \$5,000 + \$10,000 = \$15,000; and on the third day it is \$5,000 + \$10,000 + \$10,000 = \$25,000. The pay on the nth day includes \$5,000 plus $n - 1$ raises of \$10,000. Thus, the pay on the nth day is \$5,000 + $(n - 1)$\$10,000.

In general, the nth term of an arithmetic recurrence relation is found by adding $n - 1$ common differences to the first term: $t_n = t_1 + (n - 1)d$.

In Method 2, the pay on the first day is \$0.01; the pay on the second day is \$0.01(2) = \$0.02; and the pay on the third day is \$0.01(2)(2) =

\$0.04. The pay on the nth day is \$0.01 doubled $n - 1$ times. Thus, the pay on the nth day is $\$0.01(2^{n-1})$.

In general, the nth term of a geometric recurrence relation is found by multiplying the first term by the common ratio $n - 1$ times: $t_n = t_1(r^{n-1})$.

The most important issue in this problem is the total pay after n days, T_n. To examine the total pay in the first method, consider the general arithmetic recurrence relation with first term t_1 and common difference d.

Term Number	Term	Sum of First n Terms	Differences: First	Differences: Second
1	t_1	t_1	—	—
2	$t_1 + d$	$t_1 + (t_1 + d) = 2t_1 + d$	$t_1 + d$	—
3	$t_1 + 2d$	$(t_1 + 2d) + (2t_1 + d) = 3t_1 + 3d$	$t_1 + 2d$	d
4	$t_1 + 3d$	$(t_1 + 3d) + (3t_1 + 3d) = (4t_1 + 6d)$	$t_1 + 3d$	d

The constant second differences indicate that the closed-form solution is a second-degree polynomial, $t_n = an^2 + bn + c$. The related system of equations created from the first three pairs in the table is

$$\text{When } n = 1, \quad t_1 = a + b + c.$$

$$\text{When } n = 2, \quad 2t_1 + d = 4a + 2b + c.$$

$$\text{When } n = 3, \quad 3t_1 + 3d = 9a + 3b + c$$

This system can be solved using matrices:

$$\begin{bmatrix} 1 & 1 & 1 \\ 4 & 2 & 1 \\ 9 & 3 & 1 \end{bmatrix}^{-1} \times \begin{bmatrix} t_1 \\ 2t_1 + d \\ 3t_1 + 3d \end{bmatrix} = \begin{bmatrix} 1 & 1 & 1 \\ 4 & 2 & 1 \\ 9 & 3 & 1 \end{bmatrix}^{-1} \times \begin{bmatrix} 1 & 0 \\ 2 & 1 \\ 3 & 3 \end{bmatrix} \times \begin{bmatrix} t_1 \\ d \end{bmatrix}$$

$$= \begin{bmatrix} 0 & 0.5 \\ 1 & -0.5 \\ 0 & 0 \end{bmatrix} \times \begin{bmatrix} t_1 \\ d \end{bmatrix} = \begin{bmatrix} 0.5d \\ t_1 - 0.5d \\ 0 \end{bmatrix}.$$

In general, the sum of the first n terms of an arithmetic recurrence relation is $0.5dn^2 + (t_1 - 0.5d)n$.

The sums formed by the second method of payment do not generate constant differences in a finite number of steps. Thus, the closed-form solution is not a polynomial and cannot be found by the finite difference method. It can, however, be derived by algebraic means.

Consider the general geometric recurrence relation with first term t_1 and common ratio r. The sum of the first n terms, S_n, is

$$S_n = t_1 + t_1r + t_1r^2 + \cdots + t_1r^{n-1}.$$

Write this equation twice:

$$S_n = t_1 + t_1r + t_1r^2 + \cdots + t_1r^{n-1}$$

$$S_n = t_1 + t_1r + t_1r^2 + \cdots + t_1r^{n-1}.$$

Multiply the second equation by r:

$$S_n = t_1 + t_1r + t_1r^2 + \cdots + t_1r^{n-1}$$

$$rS_n = r(t_1 + t_1r + t_1r^2 + \cdots + t_1r^{n-1}).$$

Distribute the r on the right side of the second equation, and subtract the second equation from the first:

$$S_n = t_1 + t_1r + t_1r^2 + \cdots + t_1r^{n-1}$$

$$rS_n = t_1r + t_1r^2 + t_1r^3 + \cdots + t_1r^n$$

$$S_n - rS_n = t_1 - t_1r^n.$$

Now factor both sides and divide to obtain the solution:

$$S_n(1 - r) = t_1(1 - r^n) \text{ or } S_n = t_1\frac{1 - r^n}{1 - r}$$

In general, the sum, S_n, of the first n terms of a geometric recurrence relation is $t_1\dfrac{(1 - r^n)}{(1 - r)}$.

The closed-form formulas for the terms and sums of arithmetic and geometric recurrence relations can be used to compare quickly the two methods of payment. The results are summarized in the table on the next page.

	Method 1 (Arithmetic)	Method 2 (Geometric)
Pay on 30th day	$P_{30} = 5{,}000 + (30 - 1)10{,}000$ $= 295{,}000$	$P_{30} = 0.01(2^{30} - 1)$ $= 5{,}368{,}709.12$
Total pay after 30 days	$S_{30} = 0.5(10{,}000)30^2 + (5{,}000 - 0.5(10{,}000))30$ $= 4{,}500{,}000$	$S_{30} = 0.01\,\dfrac{1 - 2^{30}}{1 - 2}$ $= 10{,}737{,}418.23$

The total wages for the second method of payment are over twice those of the first!

Including Sums in Spreadsheets and Programs

If your spreadsheet includes the term numbers and the terms of a recurrence relation in columns A and B, then adding a column to include the sum requires entering a formula and copying it into as many cells as you need. If, for example, the first term of the recurrence relation is in cell B1, the second in B2, and so forth, type the simple formula B1 in cell C1 and the formula C1 + B2 in cell C2. Copy the formula in cell C2 into cells C3, C4, and so on.

The algorithm for generating the terms of a recurrence relation and the corresponding BASIC translation that were given in Lesson 8.1 are reprinted below. Recall that the algorithm is for the recurrence relation $H_n = H_{n-1} + (n - 1)$ with first term 0.

1. Store the number 1 for variable *N* and the first term for variable *H*.	10 N = 1:H = 0
2. Display *N* and *H*.	20 PRINT N,H
3. Add 1 to *N* and store this value as the new value of *N*.	30 N = N + 1
4. Add *N* − 1 to *H* and store this value as the new value of *H*.	40 H = H + N − 1
5. Repeat Steps 2 through 4.	50 GO TO 20

The algorithm and BASIC translation adapted to include sums are shown below. The variable S represents the sum of the first n terms of the recurrence relation.

Algorithm	BASIC
1. Store the number 1 for variable N, the first term for variable H, and H for variable S.	10 N = 1:H = 0:S = H
2. Display N, H, and S.	20 PRINT N, H, S
3. Add 1 to N and store this value as the new value of N.	30 N = N + 1
4. Add $N - 1$ to H and store this value as the new value of H.	40 H = H + N − 1
5. Add the value of H to S and store this as the new value of S.	50 S = S + H
6. Repeat Steps 2 through 5.	60 GO TO 20

Exercises

1. Consider the following sequences:
   i. 2, 5, 8, 11, 14, ...
   ii. 64, 32, 16, 8, 4, 2, 1, ...
   iii. 10, 12, 14.4, 17.28, 20.736, ...
   iv. 2, 3, 5, 8, 13, 21, ...
   v. 3/10, 3/100, 3/1,000, 3/10,000, ...
   vi. 3, 4, 6, 9, 13, 18, ...
   a. Which of the sequences are geometric, which are arithmetic, and which are neither?
   b. Write a recurrence relation for each sequence.
   c. For those sequences that are arithmetic or geometric, write a closed-form formula for the nth term.

2. Consider the general recurrence relation for a geometric sequence, $t_n = rt_{n-1}$. Find the fixed point for this recurrence relation if one exists. (Recall the hint in Exercise 4b, p. 405, of the previous lesson.)

3. Consider the general recurrence relation for an arithmetic sequence, $t_n = t_{n-1} + d$. Find the fixed point for this recurrence relation if one exists.

4. A teen chat line is a 900 telephone number line that charges \$2.00 for the first minute and \$0.95 for each additional minute. The purpose of the line is to let teens talk to many other teens around the nation, but in reality it is a money-making scheme for the telephone company and the company running the chat line.
   a. What recurrence relation does the statement of the problem suggest? (Make a table if necessary.)
   b. Find a closed-form formula to describe the cost of the call.
   c. Assume that 5,000 teens use the line each week and talk for an average of 15 minutes. How much income is produced? Sup-

pose also that the bulk of the cost to the company is the cost of the long distance line, which averages $3.00 for a 15-minute call. How much profit does the company make each week?

5. George deposits $5,000 in the bank at an interest rate of 8.4% compounded yearly.
   a. Write a recurrence relation for the amount of money in the account at the end of n years.
   b. Write a closed-form formula for the amount of money in the account at the end of n years.
   c. How much money will be in George's account at the end of 3 years?
   d. Suppose the bank decides to become competitive with other banking institutions in the city by offering an interest rate of 8.4% compounded monthly. What monthly interest rate will George receive?
   e. Write a recurrence relation for the amount of money in the account at the end of n months.
   f. Write a closed-form formula for the amount of money in the account at the end of n months.
   g. Find the amount of money in George's account at the end of 3 years (36 months).
   h. Compare the amount of money in George's account at the end of 3 years when the interest is compounded yearly with the amount at the end of 3 years when the interest is compounded monthly.

6. Suppose George is given the choice of investing his $5,000 at 7.2% compounded monthly or at 8.0% compounded yearly. Compare the two methods of investment at the end of 1 year. At the end of 2 years. At the end of 3 years.

7. Find the fifteenth term and the sum of the first 15 terms for the following sequences:
   a. 4, 9, 14, 19, ...
   b. 45.75, 47, 48.25, 49.5, ...
   c. 3,650, 3,623, 3,596, 3,569, 3,542, ...

8. The number of deer on Fawn Island is currently estimated to be 500 and is increasing at a rate of 8% per year. At the present time,

Because of the demise of natural predators, deer populations sometimes threaten to multiply out of control. Discrete mathematical models play an important role in managing deer populations.

the island can support 4,000 deer, but acid rain is destroying the vegetation on the island, and the number of deer that can be fed is decreasing by 100 per year. In how many years will there not be enough food for the deer on Fawn Island?

9. The cost of n shirts selling for \$14.95 each is given by $C_n = 14.95n$.
   a. What is the equivalent recurrence relation for this problem?
   b. Is this arithmetic, geometric, or neither? If it is arithmetic or geometric, what is the first term and the common difference/ratio?

10. At 8.5% compounded yearly, what amount must Bill's parents invest for Bill when he's age 10 if they want Bill to be a millionaire when he reaches age 50?

11. In order to double your investment at the end of 11 years, what annual interest rate do you need to get?

12. The following table gives data relating temperature in degrees Fahrenheit to the number of chirps per minute for a cricket. Although the data are not quite a perfect arithmetic sequence, they are close to one.

Temp (°F)	50	52	55	58	60	64	68
Chirps/min.	40	48	60	73	80	98	114

   a. Assuming that the data are an arithmetic sequence, with the temperature as the term number and the chirps per minute as the term value, what is the common difference per degree rise in temperature?
   b. What is the temperature when the cricket is chirping at 110 chirps per minute?
   c. At what temperature does a cricket stop chirping?

13. A ball is dropped from 8 feet high and rebounds on each bounce to 75% of its height on the previous bounce.
   a. How high does the ball reach after the sixth bounce?
   b. What is the total distance that the ball has traveled just before the seventh bounce?

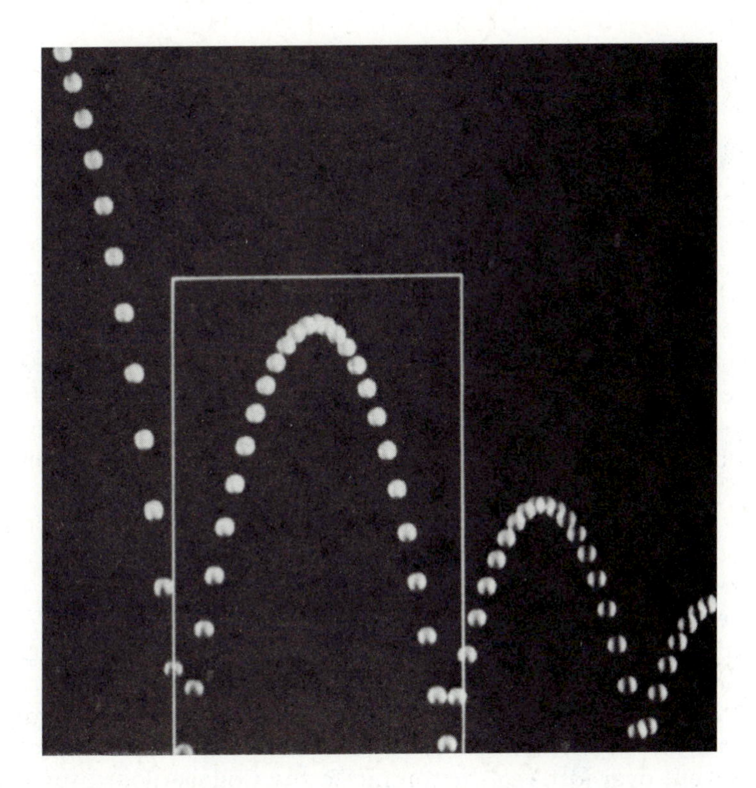

A rubber ball rebounds to a fraction of the height of the previous bounce, as shown in this time-lapse photograph.

14. The tenth term of an arithmetic sequence is 4 and the twenty-fifth term is 20. Find the first term and the common difference of this arithmetic sequence.

15. The average cost of a year of college education is currently $8,750. The cost is rising an average of 8% per year.
    a. What will be the cost of a year of college education in 30 years?
    b. Suppose the best interest rate you can get 10 years from now when you start saving is 9% compounded annually. What amount would you have to put in the bank in 10 years in order to pay for a year of college 30 years from now?

NEWS CLIP

College Tuitions, Fees Rise Average 7–10 Percent, Study Finds

By Anthony Flint, BOSTON GLOBE, October 14, 1992

The cost of attending private colleges and universities rose an average 7 percent this year, while tuition and fees at public colleges and universities shot up an average 10 percent, according to the College Board's annual survey of the price of higher education.

For the private institutions, where price tags for tuition, room and board are now approaching $25,000 a year, the increase is relatively modest compared to the double-digit increases of the 1980s. But the public institutions, strained by fiscal crises in many states during the past few years, continue to rely on bigger student charges to keep them afloat.

At private four-year colleges and universities, the average price for tuition and fees is $10,498 this year, an increase of 7 percent over last year, according to the College Board survey. That is about the same level of increase from the previous year. At private two-year colleges, tuition and fees average $5,621, an increase of 6 percent over last year.

Tuition and fees at public four-year colleges and universities average $2,315, while the average price at public two-year institutions is $1,292. In both cases, that is a 10 percent increase over last year. The year before, average tuition and fees rose 13 percent.

16. Over time the number of bacteria in a culture increases in a geometric sequence. There are 600 bacteria at $t = 0$ and 950 bacteria at $t = 3$ hours.
    a. What is the approximate common ratio for this sequence?
    b. Approximately how many bacteria will there be in 10 hours?
    c. At what time will there first be 50,000 bacteria in the culture?

17. A 50-meter-long pool is constructed with the shallow end 0.85 meters deep. For each meter, starting at the shallow end, the pool deepens by 0.06 meters.
   a. How deep is the deepest part of the pool?
   b. A rope is to be placed across the pool when the pool depth is 1.6 meters, in order to mark the end of the shallow section. How far from the shallow end of the pool should this rope be placed?

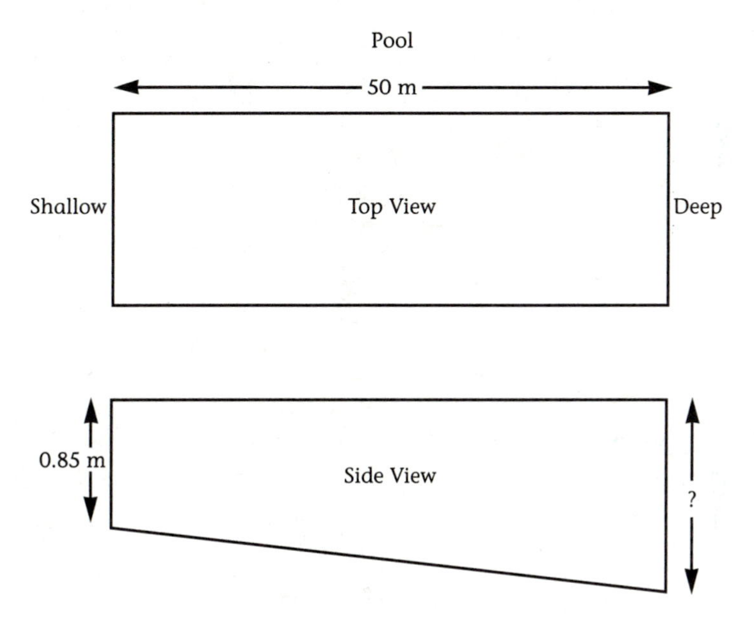

Mixed Recursion, Part 1

The previous lesson looked at arithmetic and geometric recurrence relations. This lesson examines recurrence relations that have both a geometric and an arithmetic component.

Consider the game called Towers of Hanoi. It consists of three pegs and disks of varying sizes stacked from largest to smallest on one of the pegs (see Figure 8.2).

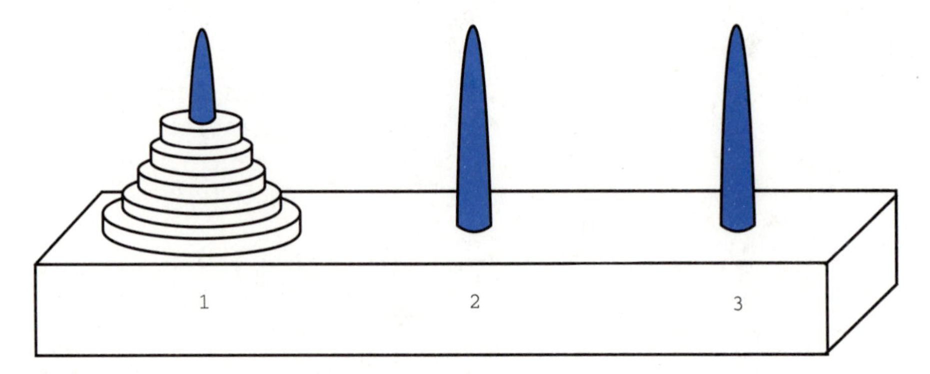

Figure 8.2 Towers of Hanoi.

The object of the game is to move the disks from the peg they are on to the third peg. Disks may be placed temporarily on the middle peg or moved back to the first peg, but only one disk may be moved at a time, and a disk may never be placed on top of one that is smaller than it.

As an experiment, cut several round or square pieces of different sizes from paper or posterboard. Use the pieces to represent the disks. Number three spots on a piece of paper to represent the pegs, and stack several of the pieces of paper from largest to smallest on spot number one. You now have a rudimentary version of the Towers of Hanoi puzzle.

At the direction of your instructor, team up with another member of your class and play the game. You should try to complete it in as few moves as possible. Keep a record of the fewest moves in which the puzzles with different numbers of disks can be completed in a table like the one below.

Number of Disks	Fewest Moves to Complete Puzzle
1	1
2	
3	
4	
5	
etc.	

As you progress, look for patterns in the data as well as in the way you are completing the puzzle. If M_n represents the fewest moves in which a puzzle of n disks can be completed, try to find a recurrence relation that describes M_n. Before reading further try to find a closed-form solution for your recurrence relation.

As you investigated the Towers of Hanoi puzzle, you should have noticed that neither an arithmetic nor a geometric model works. You cannot move from one term to the next by adding a single number or by multiplying by a single number. This recurrence relation requires both.

A **mixed recurrence relation** is one in which both multiplication by a constant and addition of a constant are required. The general form of a mixed recurrence relation is $t_n = at_{n-1} + b$. Many common situations can be modeled with mixed recurrence relations. Examples are financial applications such as annuities and loan repayment, the way a warm object such as a cup of coffee cools, and the way in which diseases spread in a population.

It is important to be able to recognize that a given set of data can be modeled by a mixed recurrence relation. Consider, for example, a table of data generated by the mixed recurrence relation $t_n = 2t_{n-1} + 3$ with $t_1 = 4$.

n	t_n	Differences
1	4	—
2	2(4) + 3 = 11	11 − 4 = 7
3	2(11) + 3 = 25	25 − 11 = 14
4	2(25) + 3 = 53	53 − 25 = 28
5	2(53) + 3 = 109	109 − 53 = 56

Note that the values in the difference column grow by a factor equal to the value of $a = 2$. This characteristic of data generated by a mixed recurrence relation makes it possible to identify situations in which a mixed recurrence relation is an appropriate model and to find the values of a and b.

When data are generated by a mixed recurrence relation, the ratio of any difference to the one preceding it is the same as the value of a in $t_n = at_{n-1} + b$.

For example, the increase in value of a house over several years is as indicated in the table below.

Year	Value	Differences
0	50,000	—
1	56,000	6,000
2	62,600	6,600
3	69,860	7,260

The ratio of the second difference to the first is $\frac{6,600}{6,000} = 1.1$. The ratio of the third difference to the second is $\frac{7,260}{6,000} = 1.1$. A mixed recurrence relation is an appropriate model, and the value of a is 1.1.

The recurrence relation for the value of the house after n years must be $V_n = 1.1\ V_{n-1} + b$. Completion of the model requires determining the value of b, which can be done by substituting into this equation any two successive values from the second column of the table. If the first pair is used, the result is $56,000 = 1.1(50,000) + b$. This gives $56,000 - 1.1(50,000) = b$, or $b = 1,000$. The completed model for the value of the house after n years is $V_n = 1.1\ V_{n-1} + 1,000$.

In previous lessons in this chapter, we completed several exercises involving savings accounts. A common type of account that can be modeled with a mixed recurrence relation is an **annuity.** An annuity is an account to which you make regular (usually monthly) additions. Often the purpose of such an account is to save for retirement and/or future expenses such as a child's college education.

Example

An account pays 6% annual interest compounded monthly, to which monthly additions of $200 are made. Note that if the interest is compounded monthly, the monthly rate is $0.06/12 = 0.005$. The following table shows the growth of the account during the first few months:

Month	Balance
0	200
1	$1.005(200) + 200 = 401$
2	$1.005(401) + 200 = 603.01$
3	$1.005(603.01) + 200 = 806.03$

In general, if B_n represents the balance at the end of the nth month, then $B_n = 1.005B_{n-1} + 200$. This is a mixed recurrence relation with $a = 1.005$ and $b = 200$. It can be used to create a table that charts the growth of the account over many months or years.

Annuities on Spreadsheets

An annuity can easily be modeled on a spreadsheet using techniques you have already learned. However, it is useful to create the spreadsheet in a way that allows monthly payments and interest rates to be easily changed. Here is a way to do this.

Create a spreadsheet in which a monthly payment of $200 and an annual interest rate of 0.06 are stored at the top of the spreadsheet in cells A1 and B1. Type column headings of "Month" and "Balance" in the next row. In cell A3 type 0 and in cell B3 type the formula A1 so that the initial balance will be the same as the first payment. In cell A4 type the formula A3 + 1. In cell B4 type the formula (1 + $B$1/12)*B3 + $A$1. This formula is the key to the spreadsheet model. The purpose of the dollar signs is to "fix" the cells that contain the interest rate and the monthly payment so that the spreadsheet does not change these cells when the formula is copied.

	A	B
1	200	0.06
2	Month	Balance
3	0	= A1
4	= A3 + 1	= (1 + $B$1/12)*B3 + $A$1
5		
6		

The formulas in row 4 can be copied to as many rows as you desire.

Exercises

1. Use the annuity example of this lesson with a monthly addition of $200 and an annual rate of 6% compounded monthly.
   a. Use a spreadsheet or program to extend the table to 20 years (240 months). Do not record the entire table, only the balance at the end of 20 years.
   b. Determine the portion of the balance after 20 years that was paid into the account by its owner.
   c. Determine the portion of the balance after 20 years that was paid in interest.
   d. Extend the account for another 10 years, and compare the amount in the account after 30 years with the amount after 20.
   e. Change the interest rate to 8% compounded monthly, and determine the amount in the account after 20 years. How does it compare with the balance after 20 years when the interest is 6%?
   f. Change the monthly addition to $300, and determine the amount in the account paying 6% after 20 years. How does the balance compare with the balance after 20 years in the 6% account with $200 monthly additions?

2. Below is a table similar to the one you made for the Towers of Hanoi puzzle of this lesson.

Number of Disks	Number of Moves
1	1
2	3
3	7
4	15
5	31

The recurrence relation that describes these data is $M_n = 2M_{n-1} + 1$. The closed-form solution to this recurrence relation is

$M_n = 2^n - 1$. The closed form is easy to detect if you notice that the number of moves is always 1 smaller than a power of 2.

a. There is an old legend that at the beginning of time God created a temple in which a group of priests are working tirelessly together to complete a 64-disk version of the puzzle. Find the number of moves necessary to complete the 64-disk version.

b. The legend states that the priests move 1 disk a second and work nonstop in shifts. When the puzzle is complete, God will end the world. Find the number of years from the beginning of creation until the end of the world according to this legend.

3. Marty had $2,000 in her savings account when she finished college and began work. Her goal was to deposit $100 each month in her account that earns 6.0% interest compounded monthly.

a. Complete this table.

n (in Months)	t_n
0	2,000
1	$1.005(2{,}000) + 100 = 2{,}110$
2	
3	

b. Find the recurrence relation for the amount of money at the end of n months.

c. Using your recurrence relation, find the amount in Marty's account at the end of the fourth month. At the end of the fifth month.

4. The Gingerichs have borrowed $12,000 to buy a boat. The yearly interest on the loan is 10.8% and the monthly payment is $306.00.

a. Complete the table on the next page.

t (in Months)	t_n
0	12,000
1	1.009(12,000) − 306 = 11,802
2	
3	

b. Write a recurrence relation for the amount left on the loan at the end of n months.

c. Use a spreadsheet, calculator, or computer program to complete the table in part a until the loan is paid off. The values that you are calculating form what is known as an *amortization schedule.*

d. Explore what will happen to the Gingerichs' loan if their interest rate and monthly payment remain the same but if they had borrowed $34,000. What will be the amount in the loan at the end of 1 month? At the end of 2 months? At the end of n months? What is the $34,000 called?

5. Newton's law of cooling says that over a fixed time period, the change in temperature of an object is proportional to the difference of the temperature of the object and its surrounding environment. If t is the temperature of the object, s is the temperature of the surroundings, and a is the constant of proportion, then $t_n - t_{n-1} = a(t_{n-1} - s)$.

   A cup of hot chocolate was brewed to a temperature of 170°F. When set in a room whose temperature is 70°F for one minute, the temperature of the hot chocolate dropped to 162°F.

   a. Write the recurrence relation for the temperature of the hot chocolate.

   b. Simplify the recurrence relation in part a to the form $t_n = at_{n-1} + b$.

   c. Use the recurrence relation to find the temperature of the hot chocolate after 2 minutes.

   d. Find the fixed point for this recurrence relation. What is the significance of this value?

6. Sam took out a college loan of $5,000. At the end of the first year, there was a $4,500 balance, and at the end of the second year, $3,900 remained. The amount of money left at the end of n years can be modeled by the mixed recurrence relation $t_n = at_{n-1} + b$.
   a. The information stated above is summarized in the following table:

n	t_n
0	5,000
1	4,500
2	3,900

   Use the general form of recurrence relation and the data in the table to write a system of equations. Solve for a and b. What is the recurrence relation for the amount of money in Sam's account after n years?
   b. What will be the balance owed on the loan at the end of the third year? At the end of the fourth year?
   c. What is the rate of interest on this loan?
   d. Find the fixed point for this recurrence relation. What is the significance of this amount of money?

7. Suppose the cost of college tuition over the past 3 years has risen from $8,000 to $8,700 to the present cost of $9,435. Use a mixed recurrence relation of the form $t_n = at_{n-1} + b$ to predict the cost of next year's tuition.

8. Chicken pox is spreading through Central High. In the following table, n represents a given period of time, and t_n represents the number of people exposed to the disease at the end of the time period.

n	t_n
5	500
6	750
7	900
8	990
9	1,044

a. Write a recurrence relation for these data.
b. Using the recurrence relation, find the number of people exposed to the disease during time period 10. During time period 4.

9. The exposure to disease in Exercise 8 can be modeled by a mixed recurrence relation, by assuming that the number of people exposed during a given time interval is directly proportional to the number of people not yet exposed at the beginning of the time interval. If t_n represents the number of people exposed during the time period n, P represents the total population, and k is the constant of proportion, then $t_n - t_{n-1} = k(P - t_{n-1})$.
   a. Using your recurrence relation and data from Exercise 8, find the total population of Central High.
   b. What is the fixed point for your recurrence relation from Exercise 8? What is the significance of the fixed point for this recurrence relation?

Computer/Calculator Explorations

10. Eric is opening an annuity account that pays 8% annual interest compounded monthly. How much should he deposit each month if he wants to have $10,000 in the account at the end of 5 years? (Hint: One possible way to do this is to experiment with various amounts in your spreadsheet until you reach $10,000 after 5 years.)

LESSON 8.5

Mixed Recursion, Part 2

In the last lesson, we worked with several situations that were modeled with mixed recurrence relations. The long duration of some applications such as annuities required creating tables of several hundred rows. Although calculators and computers help make this reasonable, a closed-form solution for mixed recurrence relations would further decrease the time necessary to complete some calculations. This lesson develops the closed form for mixed recurrence relations.

Recall that a mixed recurrence relation is of the form $t_n = at_{n-1} + b$. As an example, consider $t_n = 3t_{n-1} - 4$ with $t_1 = 5$. The first four terms are 5, 11, 29, 83. The fixed point for this recurrence relation can be found by equating t_n and t_{n-1}. The equation $x = 3x - 4$ gives a fixed point of $x = 2$. To see that this is correct, write the first few terms of the recurrence relation $t_n = 3t_{n-1} - 4$, with $t_1 = 2$: 2, $3(2) - 4 = 2$, and so forth. Because all terms are 2, the closed form of this recurrence relation is extremely simple: $t_n = 2$.

Now consider the original recurrence relation $t_n = 3t_{n-1} - 4$ with $t_1 = 5$. If this recurrence relation did not include the subtraction of 4, it would be geometric, and the closed form would be obtained from the general closed form for geometric recurrence relations $t_n = t_1a^{n-1}$, which in this case gives $t_n = 5(3^{n-1})$. The fixed point of this geometric recurrence relation is 0, and when 0 is the first term, the general closed-form formula for geometric recurrence relations becomes $t_n = 0(3^{n-1})$.

The closed form for $t_n = 3t_{n-1} - 4$ behaves in a similar way. That is, because the recurrence relation involves multiplication by 3, the solution involves powers of 3, and this part of the solution becomes 0 when the fixed point is used for the first term.

One way to do this is to modify the geometric closed form $t_n = 5(3^{n-1})$ by subtracting the fixed point from the first term: $t_n = (5 - 2)(3^{n-1})$. If the first term is 2, this becomes $t_n = (2 - 2)(3^{n-1}) = 0$. If the first term is 2, however, the solution must be $t_n = 2$ rather than $t_n = 0$. This problem can be remedied by adding the fixed point to the solution: $t_n = (5 - 2)(3^{n-1}) + 2$.

Recall that the first four terms of $t_n = 3t_{n-1} - 4$ with $t_1 = 5$ are 5, 11, 29, and 83. Our formula gives

$$t_1 = (5 - 2)(3^{1-1}) + 2 = 3(1) + 2 = 5$$

$$t_2 = (5 - 2)(3^{2-1}) + 2 = 3(3) + 2 = 11$$

$$t_3 = (5 - 2)(3^{3-1}) + 2 = 3(9) + 2 = 29$$

$$t_4 = (5 - 2)(3^{4-1}) + 2 = 3(27) + 2 = 83.$$

All are correct.

The closed form looks good, but it was derived by a bit of guessing. Will it continue to work? One way to be certain is to use mathematical induction to verify the formula.

Suppose that $t_n = (5 - 2)(3^{n-1}) + 2$ generates the nth term of the recurrence relation. It must be shown that $t_{n+1} = (5 - 2)(3^{(n+1)-1}) + 2 = (3)(3^n) + 2$ generates the $n + 1$st term. The term t_{n+1} is generated by multiplying the previous term by 3 and subtracting 4: $t_{n+1} = 3t_n - 4$, but $t_n = (5 - 2)(3^{n-1}) + 2$, so by substitution $t_{n+1} = 3((5 - 2)(3^{n-1}) + 2) - 4$. Simplify this expression:

$$3((5 - 2)(3^{n-1}) + 2) - 4 =$$

$$3((3)(3^{n-1}) + 2) - 4 =$$

$$3((3^n) + 2) - 4 =$$

$$3(3^n) + 6 - 4 =$$

$$3(3^n) + 2.$$

Mathematical induction guarantees that the closed form $t_n = (5 - 2)(3^{n-1}) + 2$ will always generate the terms of $t_n = 3t_{n-1} - 4$ with $t_1 = 5$.

> In general, the closed-form solution for the mixed recurrence relation $t_n = at_{n-1} + b$ is $t_n = (t_1 - p)(a^{n-1}) + p$, where p is the fixed point $b/(1 - a)$.

If $a = 1$, there is no fixed point, but the recurrence relation is arithmetic, and the solution given in Lesson 8.3 (see p. 410) can be applied.

The previous mathematical induction proof guarantees this solution only for the specific mixed recurrence relation $t_n = 3t_{n-1} - 4$ with $t_1 = 5$. Verifying the general closed form given above requires a general proof that we consider in the exercises.

Example

Consider the annuity problem that was posed in the last lesson. The account paid 6% annual interest compounded monthly and included monthly additions of \$200. The recurrence relation for the balance at the end of the nth month is $B_n = \left(1 + \frac{0.06}{12}\right)B_{n-1} + 200$, or $B_n = 1.005\ B_{n-1} + 200$. The account balances for the first few months are given in table below.

Month	Balance
0	200
1	1.005(200) + 200 = 401
2	1.005(401) + 200 = 603.01
3	1.005(603.01) + 200 = 806.03

Note that $t_0 = 200$, $t_1 = 401$, $a = 1.005$, and $b = 200$. The fixed point is $\frac{200}{1 - 1.005} = -40{,}000$. Substituting for t_1, a, and b in the general closed form $t_n = (t_1 - p)(a^{n-1}) + p$ gives $t_n = (401 + 40{,}000)(1.005^{n-1}) - 40{,}000$ or $t_n = 40{,}401(1.005^{n-1}) - 40{,}000$. Determining the amount in the account after 20 years requires evaluating this for $n = 240$: $40{,}401(1.005^{240-1}) - 40{,}000 = 93{,}070.22$.

Suppose that you want to know how long it will take this account to grow to \$200,000. You need to determine when $B_n = 200{,}000$ or when $40{,}401(1.005^{n-1}) - 40{,}000 = 200{,}000$. This requires a little algebra:

$$40{,}401(1.005^{n-1}) - 40{,}000 = 200{,}000$$

$$40{,}401(1.005^{n-1}) = 200{,}000 + 40{,}000$$

$$40{,}401(1.005^{n-1}) = 240{,}000$$

$$(1.005^{n-1}) = 240{,}000/40{,}401$$

$$(n - 1)\log(1.005) = \log(240{,}000/40{,}401)$$

$$n - 1) = \log(240{,}000/40{,}401)/\log(1.005)$$

$$n = \log(240{,}000/40{,}401)/\log(1.005) + 1.$$

The expression on the right can be evaluated on a scientific calculator or on a computer spreadsheet to obtain about 358 months, or just a little less than 30 years.

In the following exercises, the general closed form for mixed recurrence relations is applied to situations similar to those examined in the last lesson. Try to solve the problems by using the closed form rather than the recurrence relation, but use the recurrence relation with your spreadsheet or calculator program to check your work.

Exercises

1. Find the fixed point and the closed form for each of the following recurrence relations. Use the closed form to find the one-hundredth term.
   a. $t_1 = 1$, $t_n = 2t_{n-1} + 3$.
   b. $t_1 = 5$, $t_n = 3t_{n-1} - 7$.
   c. $t_1 = 2$, $t_n = 4t_{n-1} - 5$.

2. An annuity pays an annual rate of 8% compounded monthly and includes monthly additions of \$150.
   a. Write the recurrence relation for B_n, the amount in the account at the end of the nth month.
   b. Use your recurrence relation to build a table showing B_n for $n = 0, 1, 2, 3, 4$.
   c. Find the fixed point for the recurrence relation.
   d. Use the fixed point and the value of B_1 to write the closed form.
   e. Use the closed form to find the account balance at the end of 30 years.
   f. Use the closed form to determine the amount of time that it will take the account to grow to \$500,000.
   g. Suppose the owner of the account would like it to reach \$500,000 in 30 years. What monthly additions are required? (Hint: You must find the value of b in $B_n = (1 + 0.08/12)B_{n-1} + b$. Calculate the fixed point, leaving b as an unknown. Write the appropriate closed form; set it equal to 500,000 when $n = 360$; and solve for b.)

3. Jilian borrowed \$10,000 to buy a car. The annual interest rate is 12% compounded monthly, and the monthly payments are \$220.
   a. Write a recurrence relation for the unpaid balance at the end of the nth month.
   b. Use the recurrence relation to tabulate the unpaid balance at the end of months 0, 1, 2, 3, and 4.
   c. Find the fixed point for the recurrence relation.
   d. Find the closed form.
   e. Use the closed form to determine the unpaid balance at the end of 2 years.
   f. Use the closed form to determine the amount of time needed to repay the loan. That is, set the closed form equal to 0, and solve for n. Round your answer to the nearest whole number of months.
   g. Multiply your previous answer by the monthly payment, and determine the amount Jilian really paid for her car. What is the total amount of interest that she paid?
   h. Suppose Jilian wanted her payments to run for 3 years. What would her monthly payment be? (See the hint in part g of the previous exercise.)

NEWS CLIP

Your Wheels

Deciding on the *Other* Car Options

Ralph Vartabedian, LOS ANGELES TIMES, June 24, 1992

Many consumers never gave up on cash as a way to buy a car.

[Peter] Levy, who authors the "Complete Cost Car Guide," said that current interest rates and inflation rates favor paying cash. His book provides a useful work sheet for the calculation.

Let's say you buy a new car for $20,000, using a 7.5% loan offered by the manufacturer. You would be paying nearly $484 monthly for 48 months, and over the loan's term you would pay out $23,212.

Alternatively, let's say you select the option of taking a $1,000 cash rebate offered by the same auto maker and scrape together your savings to pay the $19,000 balance. Then, you set up a saving program to replace the $19,000 over the next four years.

Each month, assume you can earn a generous 8%, amounting to 5.2% after you pay income taxes. You will have to save $357 per month over the next four years, amounting to an outlay $17,136.

Thus, over the four-year period, it looks like you are $6,076 ahead by paying cash.

But there is a hitch. You obviously would have earned interest on the $19,000 if you hadn't used it to buy a car.

4. In this lesson the closed-form solution for $t_n = at_{n-1} + b$ was given as $t_n = (t_1 - p)(a^{n-1}) + p$, where p is the fixed point $\frac{b}{1-a}$. It was verified for only one specific case. In this exercise we use mathematical induction to verify that this closed form is correct for all mixed recurrence relations.
   a. To begin, verify that the solution works for t_1. Replace n with 1 in $(t_1 - p)(a^{n-1}) + p$, and show that this really is t_1.
   b. The next step is to show that whenever the closed form generates the correct value of t_n, it will also generate the correct value of t_{n+1}. To begin this process, write the closed form for t_n and for t_{n+1}.
   c. The general form for this type of recurrence relation says that a term is generated by multiplying the previous term by a and adding b. Generate the $(n + 1)$st term by multiplying the closed form for the nth term by a and adding b.
   d. Algebraically simplify the previous expression until it matches the closed form for the $(n + 1)$st term. (Hint: You might find it helpful to replace the second occurrence of the fixed point with $\frac{b}{1-a}$, but not the first.)

5. In Exercise 5 of the previous lesson (p. 429) you applied Newton's law of cooling to a cup of hot chocolate that dropped from 170° to 162° in 1 minute in a room whose temperature was 70°.
   a. Rewrite the recurrence relation, and recalculate the fixed point for the recurrence relation.
   b. Write the closed form for the temperature of the hot chocolate after n minutes.
   c. Use the closed form to find the temperature of the object after 5 minutes.
   d. Use the recurrence relation to create a table showing the temperature each minute through the first 5 minutes. Compare the last entry of the table with your answer to part c.
   e. Would it make sense to use the closed form to determine the amount of time needed for the temperature of the hot chocolate to freeze? Try to do so. What happens?

6. The following data represent the number of people at Central High who have heard a rumor:

Number of Hours after Rumor Began	Number of People Who Have Heard It
1	80
2	240
3	320
4	360

a. Write a recurrence relation for the number of people who have heard the rumor after n hours.
b. Find the fixed point for the recurrence relation.
c. Use the fixed point to write the closed form.
d. Use the closed form to determine the number of people who have heard the rumor after 10 hours.
e. The way in which rumors and other information spread through a population is similar to the way in which disease spreads through a population. Compare this exercise with Exercises 8 and 9 of the previous lesson (pp. 430–431) and explain the significance of the fixed point in this exercise.

Cobweb Diagrams

The process of finding successive terms of a recurrence relation can be visualized with the aid of graph paper. Consider the recurrence relation $t_n = 2t_{n-1} - 1$ with $t_1 = 3$. The table shows the first four terms.

n	t_n
1	3
2	5
3	9
4	17

Although it may seem obvious, it is important to understand that each term fills the roles of both t_n and t_{n-1}. The first term 3, for example, is t_n if you are thinking of n as 1, but it is t_{n-1} if you are thinking of n as 2. As you build a table of values, each term first takes on the role of t_n, then the role of t_{n-1}, then fades into history something like an officer of an organization who serves a term as president, then a term as past president, and finally disappears from office. This succession of terms can be visualized by thinking of the recurrence relation $t_n = 2t_{n-1} - 1$ as separate functions $y = t_n$ and $y = 2t_{n-1} - 1$, or $y = x$ and $y = 2x - 1$.

The process of plotting the activity that occurs when the terms of the recurrence relation are found goes like this:

1. Graph the lines $y = x$ and $y = 2x - 1$.

2. Locate the first term of the recurrence relation on the x-axis.

3. Draw a vertical line from the first term to the line $y = 2x - 1$. This represents substituting the first term into $2t_{n-1} - 1$ to find the second term.

4. The second term $2(3) - 1 = 5$ becomes the current value of t_n. Because 5 now changes its role from that of t_n to t_{n-1}, draw a horizontal line from $y = 2x - 1$ (i.e., $2t_{n-1} - 1$) to $y = x$ (i.e., t_n). Now 5 changes its role from that of t_n to t_{n-1}, so draw a vertical line from $y = x$ to $y = 2x - 1$ to represent finding the third term. The value of the third term can be read from the y-axis.

5. Continue this process for as many terms as desired.

The graph that results is sometimes called a **cobweb diagram** (see Figure 8.3). Note that if the first term of the recurrence relation $t_n = 2t_{n-1} - 1$ is 1, the first vertical segment will reach the point of intersection of the two lines and the process can go nowhere from there, because 1 is the fixed point.

The process of constructing a cobweb diagram can be adapted to a graphing calculator with a short program. Here is an algorithm that you can adapt to your graphing calculator:

1. Set a suitable graphing range.

2. Graph the line $y = x$ and the line whose equation is found by replacing t_n with y and t_{n-1} with x in the recurrence relation. Give these functions suitable names such as Y_1 and Y_2.

3. Input the value of the initial point for variable A.

4. Replace X with the value of A.

5. Draw a line from the point $(X, 0)$ to (X, Y_2).

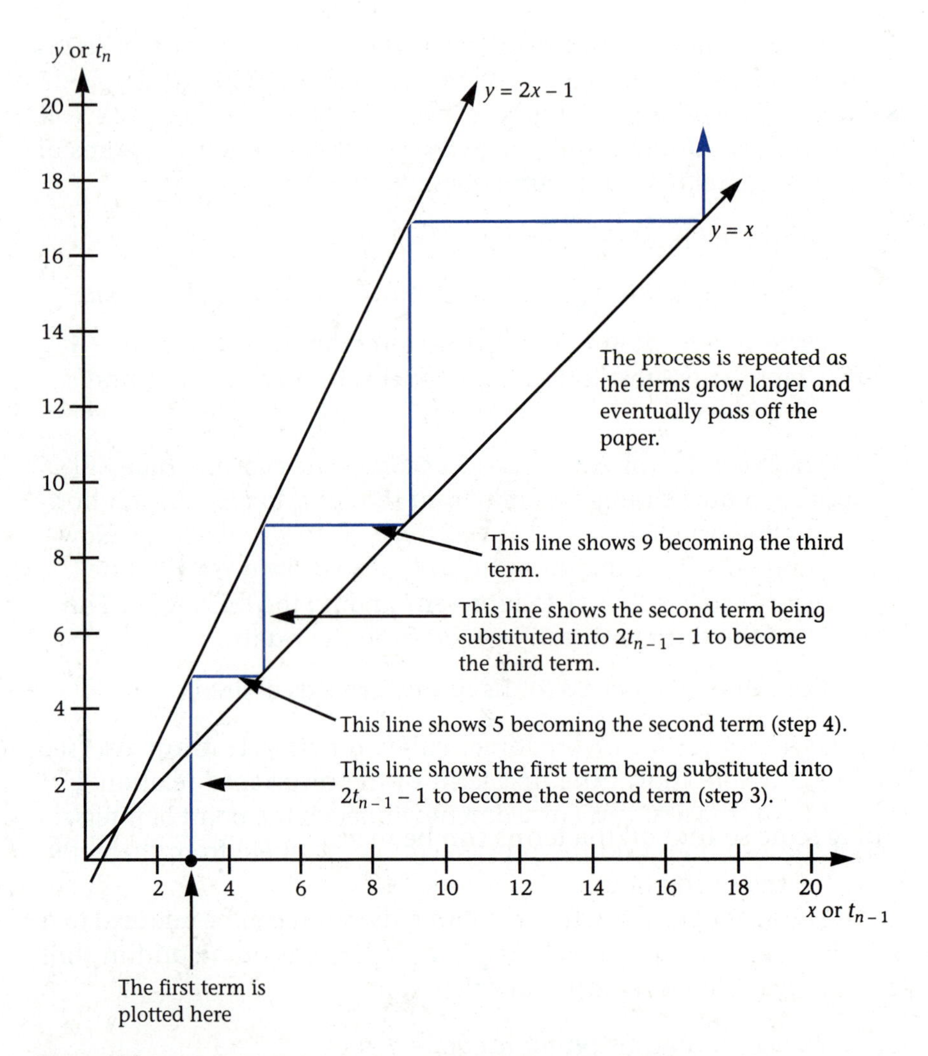

Figure 8.3 Cobweb diagram for the recurrence relation $t_n = 2t_{n-1} - 1$ with $t_1 = 3$.

6. Pause and wait for the user to press ENTER.
7. Draw a line from (X, Y_2) to (Y_2, Y_2).
8. Pause and wait for the user to press ENTER.
9. Replace X with the current value of Y_2.

10. Draw a line from (X, X) to (X, Y_2).

11. Pause and wait for the user to press ENTER.

12. Go back to Step 7.

The first three steps can be included in the program or performed separately before the program is run. The exact commands that implement the steps depend on the model of your calculator. Consult your calculator manual or talk with someone who is experienced in the graphing and programming features of your calculator if you are unsure of what to do.

A cobweb diagram is a useful way to visualize what happens when successive terms of a recurrence relation are calculated. The behavior exhibited in the example of this lesson is not the only kind of behavior that can occur. The following exercises explore some different types of behavior.

Exercises

1. Construct a cobweb diagram for the indicated number of terms of each recurrence relation. Find all the terms first, then choose a suitable scale so that all the terms can be seen, and draw the diagram on a piece of graph paper.
   a. $t_n = 3t_{n-1} - 8$ with $t_1 = 5$, four terms.
   b. $t_n = 5 - t_{n-1}$ with $t_1 = 3$, four terms.
   c. $t_n = 9 - 0.5t_{n-1}$ with $t_1 = 2$, four terms.
   d. $t_n = 0.5t_{n-1} + 6$ with $t_1 = 2$, four terms.

2. Find the fixed point for each of the recurrence relations in Exercise 1 and mark the fixed point on your cobweb diagram. Fixed points are often categorized in a variety of ways. Some, for example, are *repelling,* and others are *attracting.* Which fixed points in Exercise 1 do you think are attracting. Which are repelling? Which appear to be neither?

3. Experiment with mixed recurrence relations $t_n = at_{n-1} + b$. Try various values of a and b. When is the fixed point attracting, and when is it repelling? When is it neither?

4. The figure below is a cobweb diagram for a recurrence relation.

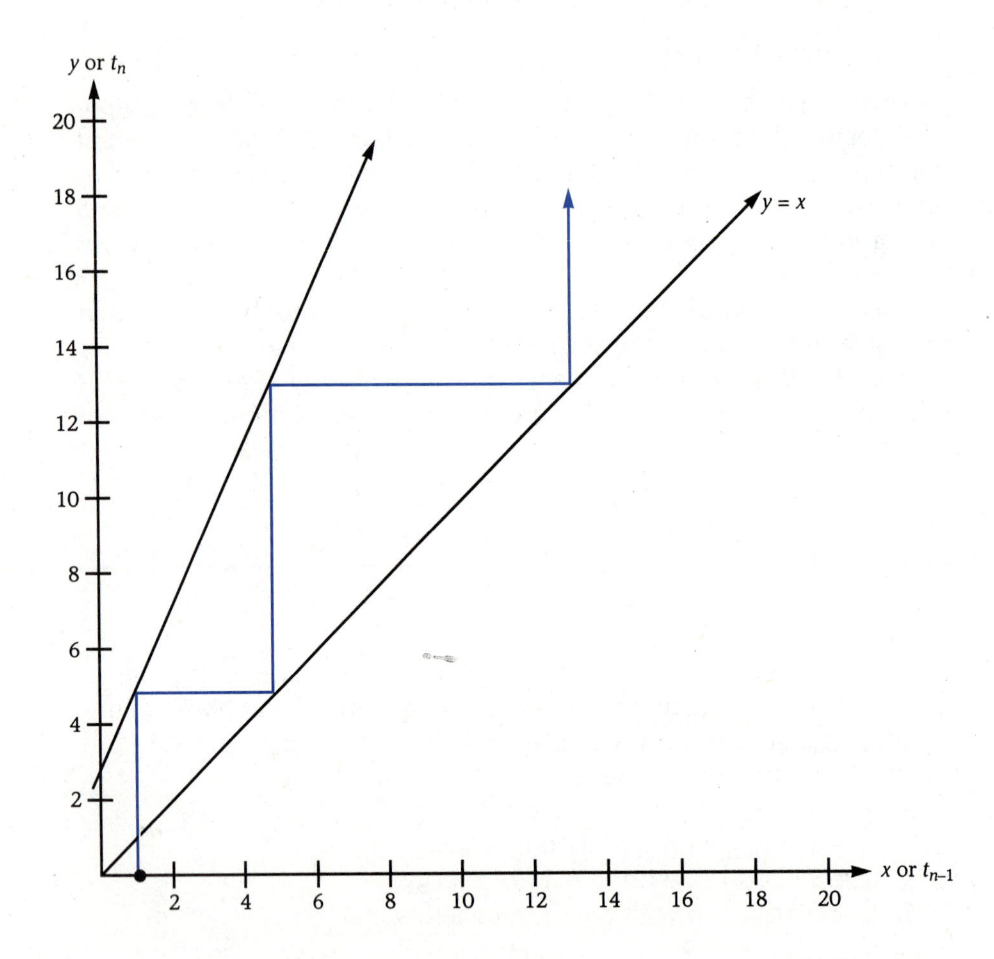

a. What is the first term of the recurrence relation?
b. How many terms can be determined from the diagram?
c. Make a table showing all terms that can be read from the diagram.
d. Write the recurrence relation.
e. Find the fixed point of the recurrence relation both algebraically and graphically.

5. Consider the second-degree recurrence relation $t_n = 4 - 0.5(t_{n-1})^2$ with $t_1 = 0$.
   a. Make a table showing the first 6 terms.
   b. Draw a cobweb diagram displaying the first 6 terms.
   c. Change the first term to 1. Use a spreadsheet or calculator to explore the behavior of the first 20 terms. Use a program to explore the related cobweb diagram.
   d. Change the first term to 2. What happens?
   e. Change the first term to 5. What happens?
   f. Use algebra to find the fixed points. Because this recurrence relation is second degree, it has 2.
   g. You have tried this recurrence relation with first terms of 0, 1, 2, and 5. In which cases were the terms attracted to a fixed point, and in which cases were they repelled? Were there any cases in which neither seemed to happen or in which it wasn't possible to tell?

6. A model that is sometimes used to represent the growth of a population in an environment that is capable of supporting only a limited number of the species says that if the uninhibited growth rate of the population is r (in decimal form) and the maximum number that the environment can support is m, then the recurrence relation that describes the total number of the species t_n in a given time period is $t_n = (1 + r(1 - t_{n-1}/m))t_{n-1}$. Suppose that a population of a particular species of animal has an annual growth rate of 10% and that the environment is capable of supporting 10 (in thousands) of the animals.
   a. Write the recurrence relation for the number of animals after n years.
   b. Suppose that the current animal population is 5,000. Use a spreadsheet or a cobweb diagram program to examine the growth of the population. Describe the results.
   c. Use your spreadsheet or cobweb diagram program to experiment with different initial populations. Be sure to include one that is over 10. Describe the results.
   d. What are the fixed points of the recurrence relation? Explain their significance in this situation.

Projects

7. The *order* of a recurrence relation is the difference between the highest and the lowest subscripts. In this chapter, you studied first-order recurrence relations. The recurrence relation $t_n = t_{n-1} + t_{n-2}$ is second order because the difference between n and $n - 2$ is 2. Investigate some recurrence relations of order higher than 1. Try to find some applications to include in your report.

8. Some recent mathematical topics related to recurrence relations include *chaos* and *fractals*. Investigate and report on either of these.

1. For each of the following, write a recurrence relation to describe the pattern, find a closed-form solution, and find the one-hundredth term.
   a. 2, 6, 10, 14, ...
   b. 3, 8, 23, 68, ...
   c. 3, 6, 12, 24, ...

2. Which of the sequences in the previous exercise are arithmetic? Which are geometric?

3. Find the fixed point for each of the following recurrence relations if one exists. Check your fixed point by using it and the recurrence relation to write the first four terms.
   a. $t_n = 5t_{n-1}$.
   b. $t_n = 5t_{n-1} + 3$.
   c. $t_n = t_{n-1} - 3$.

4. In the previous exercise, find the closed-form solution for each recurrence relation if the first term is 2. Use the closed form to find the one-hundredth term.

5. Which of the recurrence relations in Exercise 3 are arithmetic? Which are geometric?

6. Consider the sequence of squares shown on the next page.
   a. Let S_n represent the total number of squares of all sizes in the figure whose sides are n units long. For example, $S_2 = 5$ because there are four small squares and one large one in the second

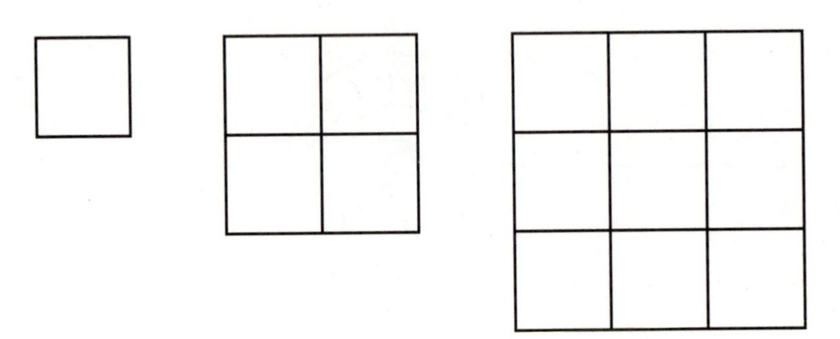

figure. Complete the following table by counting squares and making additional drawings.

n	S_n
1	1
2	5
3	
4	
5	

b. Add difference columns to your table.
c. Find a closed-form formula for S_n. Use your formula to find the total number of squares of all sizes on a checkerboard.

7. Use finite differences to determine the degree of the closed-form formula that was used to generate this sequence and find the closed form by solving an appropriate system.

$$-5, -2, 3, 10, 19, 30, 43, \ldots$$

8. The following table shows the number of gifts given on the nth days of Christmas and the total number of gifts given through the first n days as described in the song "The Twelve Days of Christmas."

Day	Gifts on That Day	Total Number of Gifts
1	1	1
2	1 + 2 = 3	1 + 3 = 4
3	1 + 2 + 3 = 6	4 + 6 = 10
4		
5		
6		

a. Complete the table through the sixth day.
b. Write a recurrence relation for the number of gifts given on the nth day, G_n, and a recurrence relation for the total number of gifts given through the nth day, T_n.
c. Find a closed-form formula for G_n and box T_n.

9. In 1994 the cost of a first-class letter was $0.29 for the first ounce and $0.23 for each additional ounce or fraction thereof.
a. Write a recurrence relation to describe the amount of postage, P_n, on a letter that weighs between $n - 1$ and n ounces.
b. Find a closed-form formula for P_n.

10. Roberto deposited $1,000 in a savings account paying 7.2% annual interest compounded monthly.
a. Complete the following table showing the balance in Roberto's account at the end of the first few months.

Month	Balance
0	$1,000
1	
2	
3	

b. Write a recurrence relation for the balance in Roberto's account at the end of the nth month.

c. Find the closed form for the balance at the end of the nth month.
d. Determine the number of years it will take for the amount in Roberto's account to double.

11. Joan has \$5,000 in an account to which she has decided to make \$100 monthly additions. The account pays 6.4% annual interest compounded monthly.
    a. Complete the following table showing the balance in Joan's account at the end of the nth month:

Month	Balance
0	\$5,000
1	
2	
3	

    b. Write a recurrence relation for the amount in Joan's account at the end of the nth month.
    c. Find the closed form for the amount in Joan's account at the end of the nth month.
    d. Find the balance in Joan's account at the end of 5 years.
    e. How long will it take Joan's account to grow to \$50,000?

12. Martha has borrowed \$11,000 to buy a car. Her loan carries an annual interest rate of 9.6%, compounded monthly, and her monthly payments are \$230.
    a. Write a recurrence relation for the unpaid loan balance at the end of the nth month.
    b. Write the closed form for the unpaid balance at the end of the nth month.
    c. How long will it take Martha to pay for her car?
    d. What is the total amount of interest that Martha will have paid?
    e. If Martha wants to pay off the loan in 3 years, what monthly payments should she make?

13. The following data indicate the change in value of a painting over a period of years.

Year	Value
1	$8,000
2	$16,000
3	$28,000
4	$46,000

a. Write a recurrence relation to describe the value of the painting after n years.
b. Find the closed form for the value of the painting after n years.

14. Create a cobweb diagram that displays the first four terms of $t_n = 0.5\, t_{n-1} + 12$, with $t_1 = 12$.

Bibliography

Arganbright, Deane. 1984. *Mathematical Applications of Electronic Spreadsheets.* New York: McGraw-Hill.

Cannon, Lawrence O., and Joe Elich. 1993. "Some Pleasures and Perils of Iteration." *Mathematics Teacher,* March, pp. 233–239.

Devaney, Robert L. 1990. *Chaos, Fractals, and Dynamics: Computer Experiments in Mathematics.* Menlo Park, CA: Addison-Wesley.

Gleick, James. 1987. *Chaos: Making a New Science.* New York: Viking Penguin.

Peitgen, Heinz-Otto, Hartmut Jurgens, and Dietmar Saupe. 1992. *Fractals for the Classroom, Part One: Introduction to Fractals and Chaos.* New York: Springer-Verlag.

Sandefur, James T. 1990. *Discrete Dynamical Systems: Theory and Applications.* Oxford: Oxford University Press.

Seymour, Dale, and Margaret Shedd. 1973. *Finite Differences: A Pattern-Discovery Approach to Problem Solving.* Palo Alto, CA: Dale Seymour.

Stewart, Ian. 1989. *Does God Play Dice? The Mathematics of Chaos.* Cambridge, MA: Basil Blackwell.

Selected Answers to the Exercises

CHAPTER 1

Lesson 1-1

10. 120 720
11. The number of schedules possible when there are n choices is n times the number of schedules possible when there are $n - 1$ choices.
12. 8 12 17

Lesson 1-2

5.
   a.

A	30.8%	69.2%
B	19.2%	0.0%
C	23.1%	0.0%
D	26.9%	30.8%

6. Plurality: B Borda: D Runoff: C Sequential runoff: C.
8. $C_n = C_{n-1} - 1$.

Lesson 1-3

3. A and B because A would defeat B, but A would lose to C.
4.
   a. A with 242 points. (B has 238, C has 120.)
   b. B with 238 points. (A has 204, C has 158.)
7.
   a. A defeats B by 68 to 32.
   b. C defeats A by 55 to 45.
8.
   a. A (280), B (260), C (190), D (170).
   b. With C removed: A, B, D.
      With a 3–2–1 Borda count: B (220), A (190), D (130).

10.

a. There are two new comparisons. A total of three comparisons must be made.

b. There are three new comparisons. A total of six comparisons must be made.

c.

1	0	0
2	1	1
3	2	3
4	3	6
5	4	10
6	5	15

Lesson 1-4

1. Nondictatorship.
2. If the method were repeated, the same ranking might not result. Therefore, condition 5 is violated. Nondictatorship is also violated.
4. Condition 4.
7. None.
11.

a. { } {A} {B} {C} {A, B} {A, C} {B, C} {A, B, C}.

b. { } {A} {B} {C} {D} {A, B} {A, C} {A, D} {B, C} {B, D} {C, D} {A, B, C} {A, B, D} {A, C, D} {B, C, D} {A, B, C, D}.

c. $V_n = 2\,V_{n-1}$.

13. 4 5
14. $V1_n = V1_{n-1} + 1$ or $V1_n = (n/(n-1))V1_{n-1}$.

Lesson 1-5

1.

a. The possible coalitions: { ; 0} {A; 3} {B; 2} {C; 1} {A, B; 5} {B, C; 3} {A, C; 4} {A, B, C; 6}.
The winning coalitions: {A, B; 5} {A, C; 4} {A, B, C; 6}.
The minimal winning coalitions: {A, B; 5} {A, C; 4}.

b. A:3 B:1 C:1.

c. A:2 B:2 C:0.

5. $C_n = 2C_{n-1}$.
6. { } {A} {B} {C} {D} {A, B} {A, C} {A, D} {B, C} {B, D} {C, D} {A, B, C} {A, B, D} {A, C, D} {B, C, D} {A, B, C, D}.
7.

a. The winning coalitions are {A, B; 51%} {A, C; 51%}

{A, B, C; 76%} {A, B, D; 75%} {A, C, D; 75%} {B, C, D; 74%} {A, B, C, D; 100%}.
Of these, A is essential to 5, B to 3, C to 3, and D to 1.

b. The winning coalitions are {A, B; 88%} {A, C; 54%} {A, D; 52%} {A, B, C; 95%} {A, B, D; 93%} {A, C, D; 59%} {B, C, D; 53%} A, B, C, D; 100%}.
Of these, A is essential to 6, B to 2, C to 2, and D to 2.

c. In part a, D has 24% of the stock and one-twelfth of the power. In part b, D has 5% of the stock, but two-twelfths of the power.

CHAPTER 2

Lesson 2-2

2.

a. $35,000. $30,000.
b. $35,000.
c. $35,000.
d. $30,000. Because there is $5,000 left in the estate, each receives $2,500. Garfield's share is $70,000 − $35,000 + $2,500 = 37,500. Marmaduke's is $30,000 + $2,500 = $32,500.
f. Marmaduke would receive $35,000 instead of $32,500. Garfield's fair share would be $35,000 instead of $37,500.

3.

	Amy	Bart	Carol
Fair share	3,833.33	4,033.33	3,766.67
Items received	Car, ticket	Painting	
Cash	−666.67	−966.67	3,766.67
	+955.56	+955.56	+955.56
	+288.89	− 11.11	
Final settlement	4,788.89	4,988.89	4,722.23

6.

a. $\begin{bmatrix} 8{,}000 & 58{,}000 & 0 \\ 6{,}000 & 64{,}000 & 0 \\ 7{,}000 & 59{,}000 & 0 \end{bmatrix}$

b. The value to Alan of the items that Betty receives.
c. The value to Betty of the items that Betty receives.

Lesson 2-3

1.

   b. 42.857.

   c. Sophomore quota: 10.83; junior quota: 5.6; senior quota: 4.57.

   d. Sophomore seats: 11; junior seats: 6; senior seats: 4.

   e. Although the size of the council increased and the class sizes didn't change, the seniors lost a seat.

2.

   a. Sophomore adjusted ratio: $464 \div 11 = 42.18$; junior adjusted ratio: $240 \div 6 = 40$; senior adjusted ratio: $196 \div 5 = 39.2$.

   b. Decrease the ratio until it drops below 40, but not below 39.2. Sophomore seats: 11; junior seats: 6; senior seats: 4.

5.

   a.

4.545	4.762
10.454	10.952

   When the ideal ratio is 22, the decimal part of the 100-member class is larger than the decimal part of the 230-member class. The situation is reversed when the ideal ratio drops to 21.

   b. For a small class. Decreases in a divisor result in larger changes in the decimal part of the quotient when dividing a small number than when dividing a large one.

Lesson 2-4

1.

   a.

10.83	10	10	11	11
5.6	5	5	6	6
4.57	4	4	5	5

   b. Sophomores: 11; juniors: 6; seniors: 4.

   c.

43.63	43.82
43.56	43.83

   d. 43.56 (Sr) 43.63 (Jr) 44.19 (So).
   Sophomore seats: 11; junior seats: 6; senior seats: 4.

   e. 43.82 (Jr) 43.83 (Sr) 44.24 (So).
   Sophomore seats: 11; junior seats: 5; senior seats: 5.

   f. The method favored by a given class can be seen in the following table of final apportionment results:

	Hamilton	Jefferson	Webster	Hill
Sophomore	11	11	11	11
Junior	6	6	6	5
Senior	4	4	4	5

3.
   a. 50
   b. 3.7 4
      2.6 3
      1.6 2
   c. 52.8571
      52
      53.3333
   d. Increase the ratio until it passes 51.3953 but stays below 52. Freshman: 21; sophomore: 4; junior: 3; senior: 2.
   e. The freshman quota is 22.1, so the number of seats should be either 22 or 23.

Lesson 2-5

1. Ann will feel she has exactly one-third. Bart and Carl each could feel he received more than one third.
3. One-sixth or 0.16.
4.
   a. One-sixth.
   b. Four-sixths or 0.67.
   c. One-third or 0.33.
5.
   a. $2 \times 3 = 6$.
   b. $k(k + 1)$ or $k^2 + k$.
6.
   a. Yes. No.
   b. Yes. Yes.
   c. Probably not. Yes.

Lesson 2-6

1.
   a. $k + 1$ $k - 1$.
   b. $k + 2$ k $2k + 1$ $2k - 1$.

2.
   a. New handshakes: 3. Total handshakes: 6.
   b. New handshakes: 4. Total handshakes: 10.

3.
   a. 7.
   b. k.
   c. $H_n = H_{n-1} + (n - 1)$.

4.
   a. 45.
   b. $k(k - 1)/2 \quad 2k(2k - 1)/2 \quad (k + 1)k/2$.
   c. $k(k - 1)/2$.
   d. $(k + 1)k/2$.
   e. k.
   f. $k(k - 1)/2 + k = (k^2 - k)/2 + 2k/2 = (k^2 + k)/2 = (k + 1)k/2$.

5.
   a. Example: With 10 people there are 45 potential conflicts, with 20 people there are 190. The number of potential disputes more than doubles.
   b. The demand for police and other legal services would probably grow more rapidly than the population because the number of potential problems outpaces the growth of the population.

6.
   a. $V_{k+1} = 2^{k+1}$.
   b. That the number of ways of voting doubles.
   c. $V_{k+1} = 2(2^k) = 2^1(2^k) = 2^{k+1}$.

CHAPTER 3

Lesson 3-1

3.

	Jackets	Shirts	Pants
Boutique 1	25	75	75
Boutique 2	30	50	50
Boutique 3	20	40	35

4. a. $A_{21} = \$1.09$, $A_{12} = \$10.86$, $A_{32} = \$3.89$.
   b. A_{21} represents the cost of drinks at Vin's.
   A_{12} represents the cost of pizza at Toni's.
   A_{32} represents the cost of salad at Toni's.
   c. S_3 represents the cost of pizza at Sal's.

8.

$$A + B = \begin{bmatrix} 10.10 + 1.15 & 10.86 + 1.10 & 10.65 + 1.25 \\ 3.69 + 0.00 & 3.89 + 0.45 & 3.85 + 0.50 \end{bmatrix}$$

$$C = \begin{array}{l} \\ \text{Pizza} \\ \text{Salad} \end{array} \begin{array}{c} \begin{array}{ccc} \text{Vin's} & \text{Toni's} & \text{Sal's} \end{array} \\ \begin{bmatrix} \$11.25 & \$11.96 & \$11.90 \\ \$3.69 & \$4.34 & \$4.35 \end{bmatrix} \end{array}$$

12.

	At bats	Runs	Hits	HRs	RBIs	Pct.	Avg.
C. Ripken	24	2	9	0	1	.002	.375
E. Martinez	16	2	5	0	3	−.003	.313
P. Molitor	21	6	8	2	5	.003	.381

14.

b. The greatest increase in times is the same for both boys and girls in the 14 to 17 age group. The smallest increase is for girls in the 10 to 11 age group.

15.

a. The ones along the diagonal represent the relationship of a variable with itself. (Answers may vary.)
b. No. If the matrix is symmetrical, the number of rows must equal the number of columns. In a matrix that is symmetrical, the entry C_{ij} will be the same as the entry C_{ji}.
c. Answers will vary. The values are symmetrical because they represent the relationship between variables. The relationship between ACT math and college GPA, for example, must be the same as the relationship between college GPA and ACT math. These values are in row 1, column 6, and in row 6, column 1.
d. ACT composite.
e. ACT math.

Lesson 3-2

1.

a. T represents the cost of four pizzas with additional toppings and four salads with a choice of two dressings from each of the three pizza houses.
b. $47.60.
c. T_{12} represents the cost of four pizzas at Toni's.
d. T_{21} represents the cost four salads at Vin's.

2.

a, b.

$$J = \begin{array}{l} \text{Pearl} \\ \text{Jade} \end{array} \overset{\begin{array}{cccc} \text{e} & \text{p} & \text{n} & \text{b} \end{array}}{\begin{bmatrix} 16 & 8 & 12 & 10 \\ 40 & 20 & 24 & 18 \end{bmatrix}}$$

c. 24.

d. J_{21} represents the number of jade earrings that Nancy expects to sell in June.

e. J_{12} represents the number of pearl pins that Nancy expects to sell in June.

5.

a. The dimensions of P times Q are 5 by 5.

$$\begin{array}{ccc} P & \times & Q \\ (5 \times 1) & & (1 \times 5) \end{array} \qquad (5 \times 5)$$

Same

Dimensions of the product.

b. $$\begin{bmatrix} 30 \\ 35 \\ 50 \\ 20 \\ 75 \end{bmatrix} [4 \;\; 5 \;\; 1 \;\; 3 \;\; 2] = \begin{bmatrix} 120 & 150 & 30 & 90 & 60 \\ 140 & 175 & 35 & 105 & 70 \\ 200 & 250 & 50 & 150 & 100 \\ 80 & 100 & 20 & 60 & 40 \\ 300 & 375 & 75 & 225 & 150 \end{bmatrix}.$$

7.

$$\begin{array}{l} \\ \text{Rate} \end{array} \overset{\begin{array}{ccc} \text{CD} & \text{CU} & \text{Bond} \end{array}}{[0.073 \;\; 0.065 \;\; 0.075]} \quad \begin{array}{l} \text{CD} \\ \text{CU} \\ \text{Bd} \end{array} \overset{\$}{\begin{bmatrix} 10{,}000 \\ 17{,}000 \\ 12{,}000 \end{bmatrix}} = \$2{,}735.$$

9.

a. The transpose of a row matrix is a column matrix, and the transpose of a column matrix is a row matrix.

b. $$M^T = \begin{array}{l} \text{e} \\ \text{p} \\ \text{n} \\ \text{b} \end{array} \overset{\begin{array}{cc} \text{p} & \text{i} \end{array}}{\begin{bmatrix} 8 & 20 \\ 4 & 10 \\ 6 & 12 \\ 5 & 9 \end{bmatrix}}$$

c. Answers will vary. A possible answer: It may be necessary to use the transpose of a matrix when performing matrix multiplication (see Exercise 10).

10.

a. $$\text{hours} \overset{\begin{array}{cccc} \text{e} & \text{p} & \text{n} & \text{b} \end{array}}{[2 \;\; 1 \;\; 2.5 \;\; 1.5]}$$

b. hours

	e	*p*	*n*	*b*
hours	[2	1	2.5	1.5]

	p	i
e	8	20
p	4	10
n	6	12
b	5	9

c.

	i	p
hours	[42.5	93.5]

d. It takes Nancy 42.5 hours to make the pearl jewelry and 92.5 hours to make the jade jewelry.

Lesson 3-3

1.

a.

	Mike	Liz	Kate
Mike	$261,000	0	$200
Liz	$235,000	0	$180
Kate	$255,000	0	$250

b. The entries in row 1, column 1, represent the value to Mike of the items that he, Liz, and Kate received. Row 2 represents the values to Liz, and row 3 the values to Kate.

2.

a. Matrix *Q*:

	Burger	Special	Potato	Fries	Shake
Emma	0	1	0	1	1
Ken	1	0	1	0	1

b. Matrix *C*:

	Cal.	Fat	Chol.
Burger	450	40	50
Special	570	48	90
Potato	500	45	25
Fries	300	30	0
Shakes	400	22	50

c. The dimensions of *Q* are 2 by 5, and those of *C* are 5 by 3.

d. The dimensions of the product will be 2 by 3.

$$\underset{(2 \times 5)}{Q} \times \underset{(5 \times 3)}{C} \qquad (2 \times 3)$$

Same

Dimensions of the product.

e. The dimensions of *C* can be described as Foods by Contents, and the dimensions of *Q* times *C* can be described as Persons by Contents.

Persons by Foods × Foods by Contents = Persons by Contents

Same

Dimensions of the product.

f.

	Cal.	Fat	Chol.
Emma	1,270	100	140
Ken	1,350	107	125

3.

a. The number of columns of A must equal the number of rows of B.
b. A and B both must be square matrices of the same dimensions.
c. Answers will vary. A possible answer follows:

$$A = \begin{bmatrix} 1 & 0 \\ 0 & 1 \end{bmatrix} \quad B = \begin{bmatrix} 1 & 2 \\ 3 & 4 \end{bmatrix} \quad AB = BA = \begin{bmatrix} 1 & 2 \\ 3 & 4 \end{bmatrix}$$

$$A = \begin{bmatrix} 1 & 2 \\ 0 & 4 \end{bmatrix} \quad B = \begin{bmatrix} 1 & 1 \\ 2 & 0 \end{bmatrix} \quad AB = \begin{bmatrix} 5 & 1 \\ 8 & 0 \end{bmatrix} \quad BA = \begin{bmatrix} 1 & 6 \\ 2 & 4 \end{bmatrix}$$

8. The diagram below shows the polygons plotted for parts a through h.

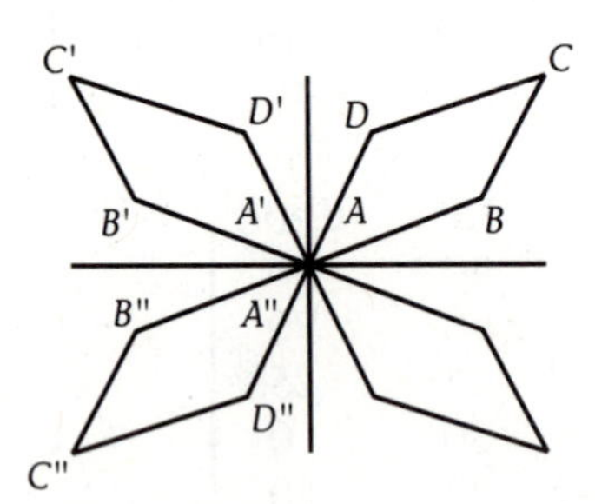

b.

$$T_1P = \begin{bmatrix} 0 & -6 & -8 & -2 \\ 0 & 2 & 6 & 4 \end{bmatrix}$$

c. Polygon $A'B'C'D'$ is the reflection of polygon $ABCD$ in the y-axis.

d.

$$T_2P = \begin{bmatrix} 0 & -6 & -8 & -2 \\ 0 & -2 & -6 & -4 \end{bmatrix}$$

e. Polygon $A''B''C''D''$ is the reflection of polygon $A'B'C'D'$ in the x-axis.

f.

$$T_1T_2 = \begin{bmatrix} -1 & 0 \\ 1 & -1 \end{bmatrix} = R$$

$$RP = \begin{bmatrix} 0 & -6 & -8 & -2 \\ 0 & -2 & -6 & -4 \end{bmatrix}$$

The effect of R on P is to rotate $ABCD$ 180 degrees about the origin.

g.

$$T_3 = \begin{bmatrix} -1 & 0 \\ 0 & 1 \end{bmatrix}$$

$$T_3P'' = \begin{bmatrix} 0 & 6 & 8 & 2 \\ 0 & -2 & -6 & -4 \end{bmatrix}$$

h.

$$T_4 = \begin{bmatrix} -1 & 0 \\ 0 & -1 \end{bmatrix} = T_2T_3$$

$$T_4P = \begin{bmatrix} 0 & 6 & 8 & 2 \\ 0 & -2 & -6 & -4 \end{bmatrix}$$

Lesson 3-4

1.
   a. 18.97.
   b. 9.96, 8.1, 7.29, 9.36, 2.4.
   c. 56 rats. 18.97, 9.96, 8.1, 7.29, 9.36, 2.4.
   d. 9 months: 18.32, 11, 0.38, 8.96, 7.29, 5.83, 5.62; total 57.
   12 months: 18.02, 10.99, 10.24, 8.07, 5.83, 3.5, total 57.
   e. Answers may vary. The population continues to grow, but the rate of growth seems to have slowed.
   f. The population growth may continue to slow or even become constant. Answers may vary.
2.
   a. 118.4.
   b. The product is the number of newborn deer after one cycle.
   c. 30, 24, 21.6, 21.6, 8.4.
   d.

i. [50 30 24 24 12 8] $\begin{bmatrix} 0.6 \\ 0 \\ 0 \\ 0 \\ 0 \\ 0 \end{bmatrix}$ ii. $\begin{bmatrix} 0 \\ 0.8 \\ 0 \\ 0 \\ 0 \\ 0 \end{bmatrix}$ iii. $\begin{bmatrix} 0 \\ 0 \\ 0.9 \\ 0 \\ 0 \\ 0 \end{bmatrix}$ iv. $\begin{bmatrix} 0 \\ 0 \\ 0 \\ 0.9 \\ 0 \\ 0 \end{bmatrix}$ v. $\begin{bmatrix} 0 \\ 0 \\ 0 \\ 0 \\ 0.7 \\ 0 \end{bmatrix}$

3.
   a. 0, 21, 0, 0, 0, 0; total 21.
   b. 11, 3, 4.5, 4.5, 4, 3; total 30.

Lesson 3-5

1.
   a. P_5 = [19.47 10.81 9.89 9.22 6.45 3.50].
   b. Total 59.35.
   c. P_7 = [20.47 12.11 10.51 8.76 7.12 4.43]; total 63.41.

2. To reach 250 females:

	Cycles	Population	Years
a.	61	253.2	15.25
b.	69	250.9	17.25
c.	76	252.9	19.00
d.	41	257.2	10.25
	42	257.2	10.5

3. Four cycles, 56.65.
   Five cycles, 59.35.
   Six cycles, 61.76.

4.
   a. 17.6%, 13.5%, 2.35%, −1.31%, 4.77%, 4.06%.
   b. The population appears to decline and then increase again.
   c. $P_{19} = 90.63$, $P_{20} = 93.38$, $P_{21} = 96.23$; Growth rates: 3.03%, 3.05%.

5. The long-term growth rate of the total population is 3.04% in each case. The initial population does not affect the long-term growth rate.

CHAPTER 4

Lesson 4-1

1.

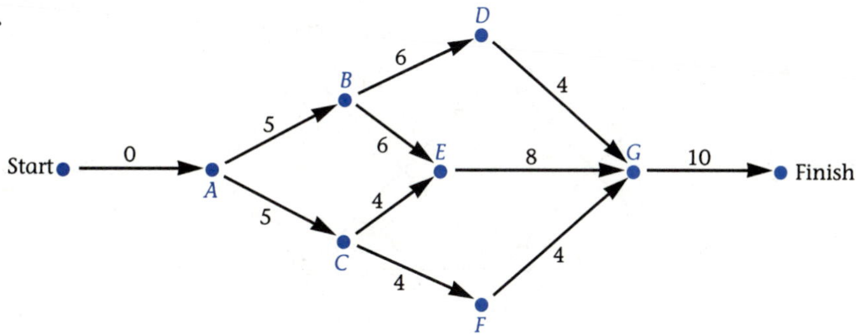

2.

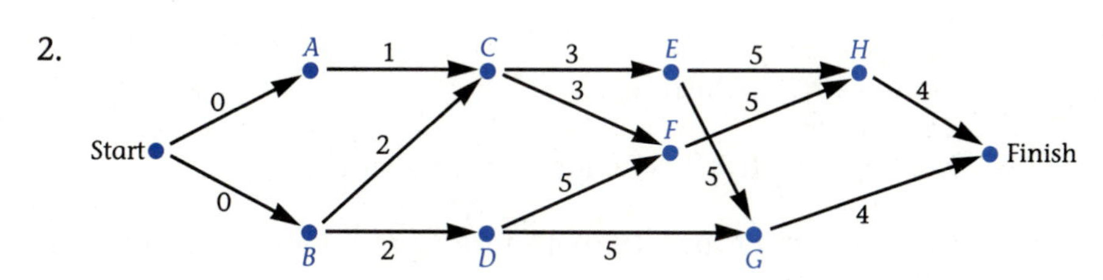

7. A: 2, none; B: 4, none; C: 3, A; D: 3, C and B; E: 2, B; F: 1, E; G: 4, D and F.

Lesson 4-2

1. EST for *C* through *G*: 7, 10, 11, 16, 23.
   Min. project time: 26.
   Critical path: Start—*ACEFG*—Finish.
3. EST for *D* through *I*: 6, 9, 8, 13, 11, 18.
   Min. project time: 26.
   Critical path: Start—*CFI*—Finish.

4. Min. project time: 20.
   Critical path: Start—*BCFG*—Finish.
5. a.

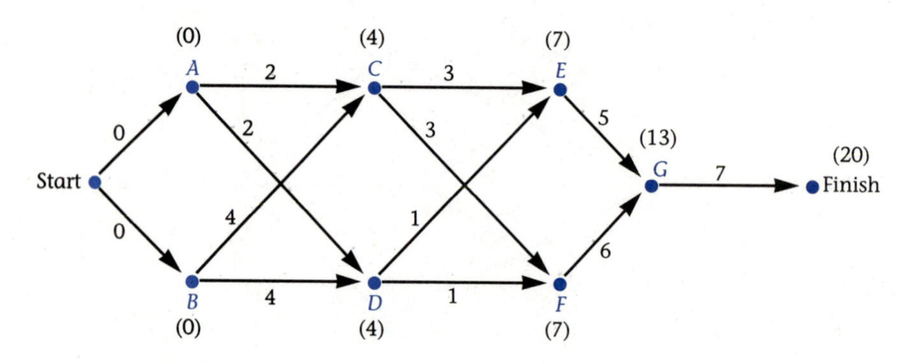

   b. 22.
   c. Start—*ADG*—Finish.
   d. 21, 20.
   e. No, below 8 days *A* is no longer on the critical path.
8.
   a. Day 16, day 17, day 18, both task *G* and the project will be delayed.
   b. Day 11.
   c. Day 5, day 6, day 5.

Lesson 4-3

1.

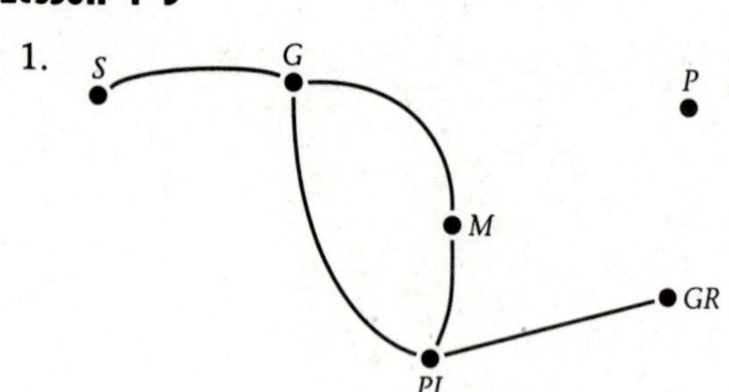

3.
   a. *A* and *F*, *B* and *F*, *B* and *C*, *A* and *C*, *A* and *D*, *A* and *E*, *B* and *E*, *B* and *D*, *F* and *C*, or *F* and *D*.
   b. *FEDC*.

c. No.
d. No.

5.

a.

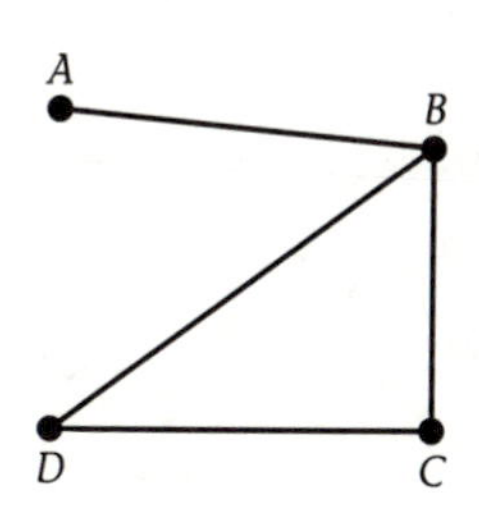

6.

a. $\begin{bmatrix} 0 & 1 & 1 & 0 \\ 1 & 0 & 1 & 1 \\ 1 & 1 & 0 & 1 \\ 0 & 1 & 1 & 0 \end{bmatrix}$

7. $\begin{bmatrix} 0 & 1 & 1 & 0 & 0 \\ 1 & 0 & 1 & 0 & 0 \\ 1 & 1 & 0 & 1 & 0 \\ 0 & 0 & 1 & 0 & 1 \\ 0 & 0 & 0 & 1 & 0 \end{bmatrix}$

a. All zeros.
b. Main diagonal.
c. The vertex is adjacent to itself; there are two edges between two vertices.

9. V:3, X:2, Y:2, Z:1.

10.

a. $\begin{bmatrix} 1 & 0 & 1 & 0 & 0 \\ 0 & 0 & 2 & 0 & 0 \\ 1 & 2 & 0 & 1 & 2 \\ 0 & 0 & 1 & 1 & 0 \\ 0 & 0 & 2 & 0 & 0 \end{bmatrix}$

b. A:3, B:2, C:6, D:3, E:2.

11. 4, 5, 6; 12, 20, 30;
$T_4 = T_3 + 6$ $\quad T_5 = T_4 + 8$ $\quad T_6 = T_5 + 10$ $\quad T_N = T_{N-1} + 2(N-1)$.

14. 4, 5, 6; 6, 10, 15;
$S_4 = S_3 + 3$ $\quad S_5 = S_4 + 4$ $\quad S_6 = S_5 + 5$ $\quad S_N = S_{N-1} + (N-1)$.

Lesson 4-4

1.
   a. Both.
2. New circuit: *S, b, e, a, g, f, S.*
   Final circuit: *S, e, f, a, b, c, S, b, e, a, g, f, S.*
3. *e, d, f, h, d, c, h, b, c, g, a, h, g, f, e.*
8.
   a.

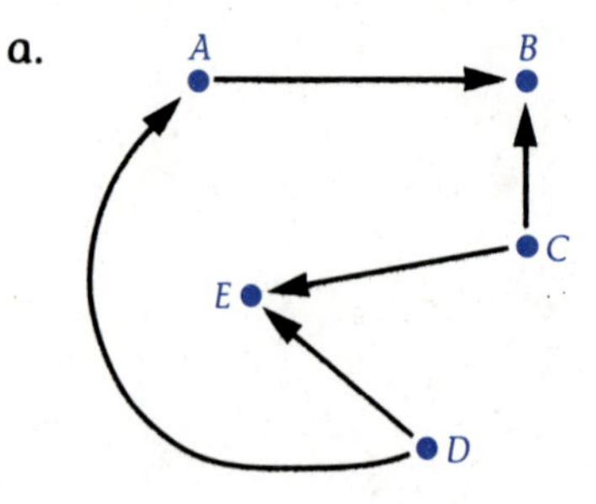

9. Yes. No. Yes.
11.
   a.

	a	*b*	*c*	*d*	*e*
a	0	0	1	0	1
b	1	0	0	1	0
c	0	1	0	0	0
d	0	0	1	0	0
e	0	1	0	1	0

Lesson 4-5

1.
   a. Yes.
   b, c. The theorem does not apply.
6.
   a. *BCAD.*
   b. *DBCA BCDA CDBA.*

8.

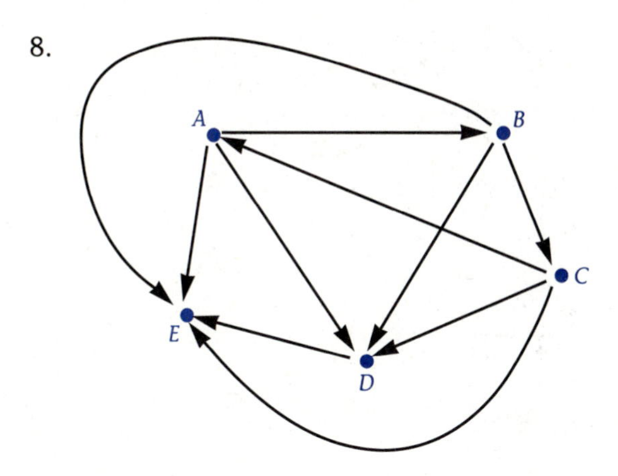

11. 6, 10, 15 $S_N = S_{N-1} + (N - 1)$.

14.

$$M = \begin{bmatrix} 0 & 0 & 1 & 1 & 1 \\ 1 & 0 & 1 & 1 & 0 \\ 0 & 0 & 0 & 1 & 1 \\ 0 & 0 & 0 & 0 & 1 \\ 0 & 1 & 0 & 0 & 0 \end{bmatrix}.$$

$$M^2 = \begin{bmatrix} 0 & 1 & 0 & 1 & 2 \\ 0 & 0 & 1 & 2 & 3 \\ 0 & 1 & 0 & 0 & 1 \\ 0 & 1 & 0 & 0 & 0 \\ 1 & 0 & 1 & 1 & 0 \end{bmatrix}.$$

The winner would be B.

$$M^3 = \begin{bmatrix} 1 & 2 & 1 & 1 & 1 \\ 0 & 2 & 0 & 1 & 3 \\ 1 & 1 & 1 & 1 & 0 \\ 1 & 0 & 1 & 1 & 0 \\ 0 & 0 & 1 & 2 & 3 \end{bmatrix}.$$

Lesson 4-6

1. 4, 3, 2.
3. List the vertices from the ones with the greatest degree to the ones with the least.

4. Those not adjacent to it or adjacent to another vertex with that color.
5. The vertex with the next highest degree of those not already colored. Any vertex not adjacent to that one or to one receiving the second color.
6. When all of the vertices are colored.
9. 2, 3, 4, N.
12. 4.
15.

 a.

CHAPTER 5

Lesson 5-1

1. Planar.

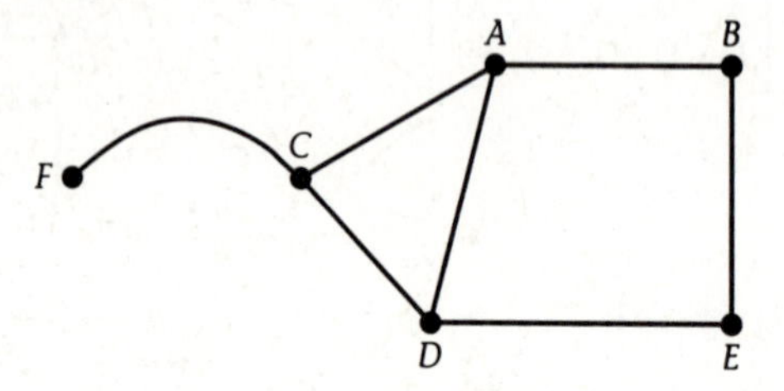

4. Hint: Move A and B around.

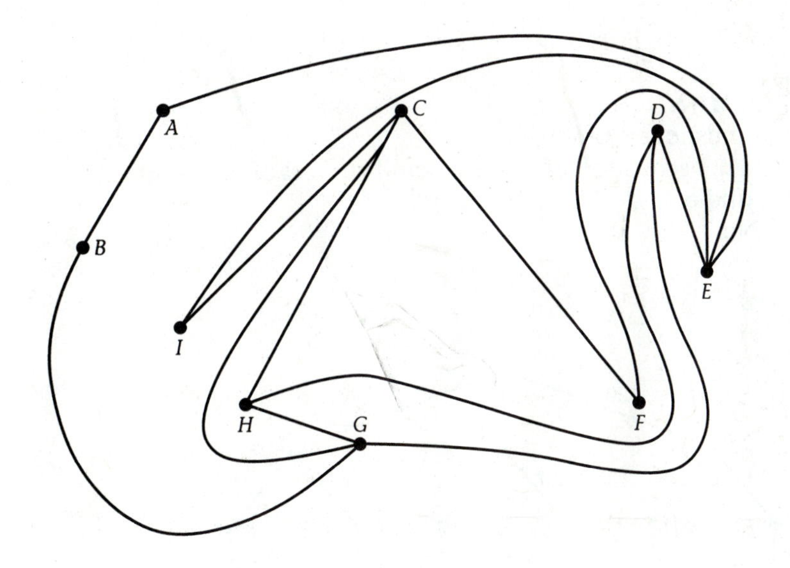

7.

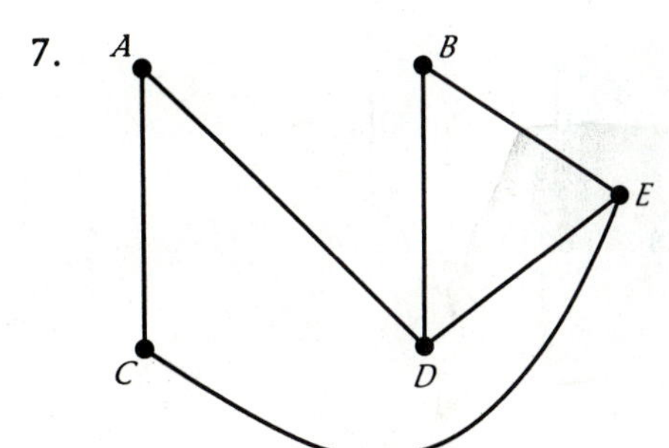

10.

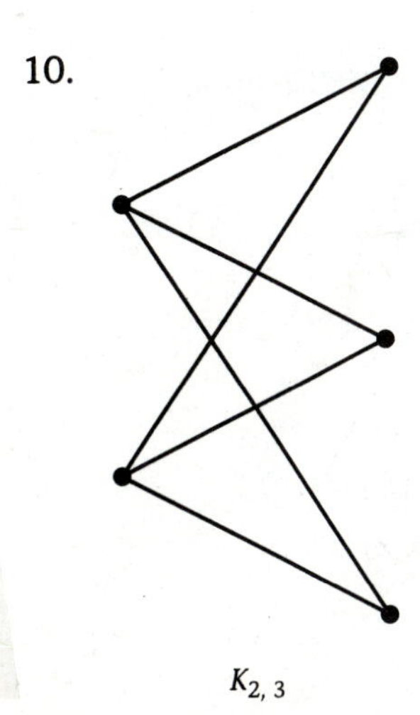

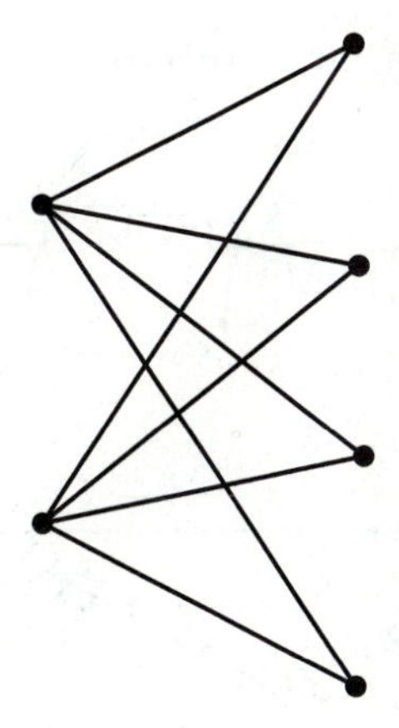

$K_{2,3}$ $K_{2,4}$

11. a. $\{A, B, C, D, E, F\}$
    $\{G\}$
14. 6, 12, $M \cdot N$.
16. 30 handshakes, bipartite.
19. a and d, because all of the edges and vertices of the original graph are in a and d.
20. No. No.

Lesson 5-2

1. a and b.

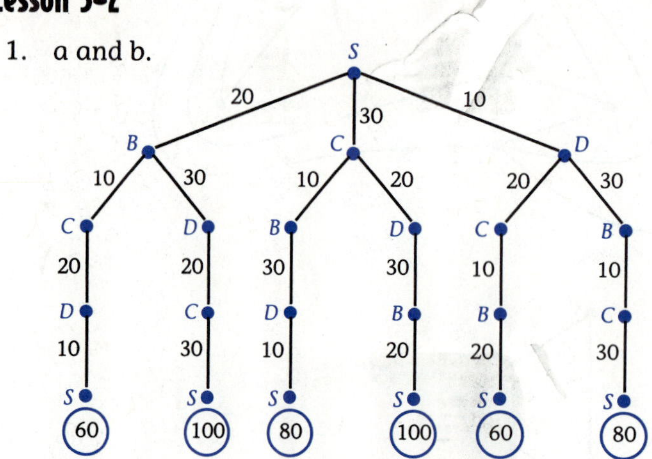

5. 0.36 seconds, 24 hours.

Lesson 5-3

1. 6, 12, 11, C, BC; the shortest path from A to E is AHGE.

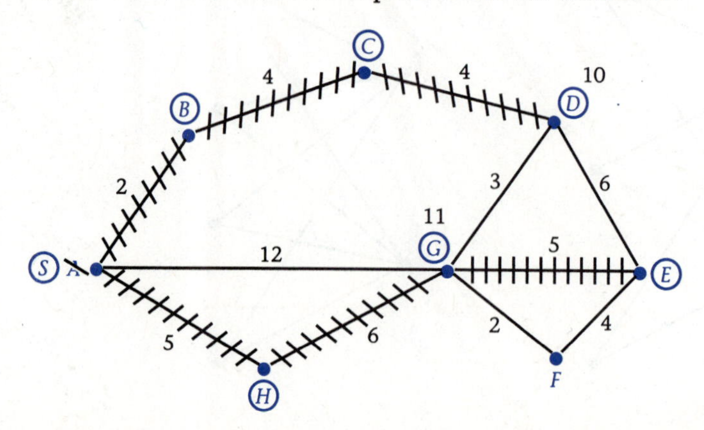

2. *ABECDF* (11).
5. a. Albany, *CEH*, Ladue.
   b. Albany, *BD*, Fenton, *GK*, Ladue. The algorithm was used from Albany to Fenton, and then the algorithm was begun again with Fenton as Start.
8.

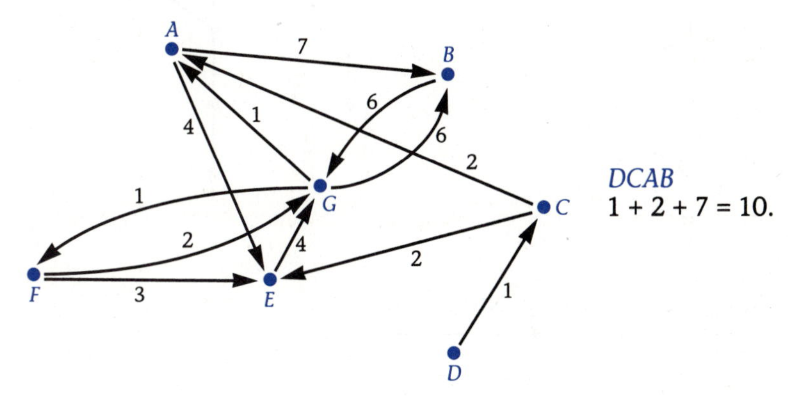

Lesson 5-4

1. *BCEFB, CDEC, BCDEFB, BCFB, CEFC, CDEFC.*
3. For five vertices there are three trees:

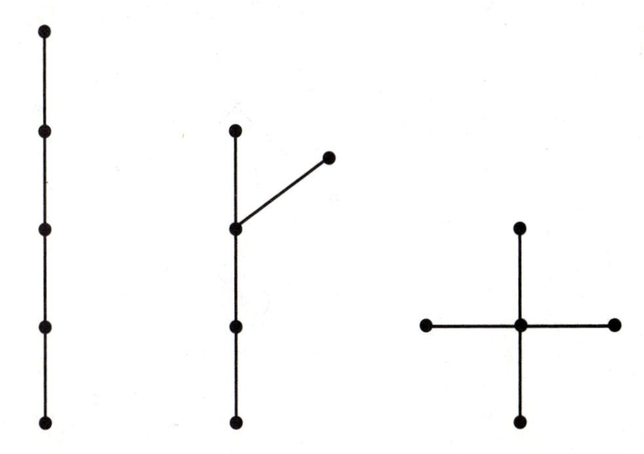

For six vertices, there are six different trees possible:

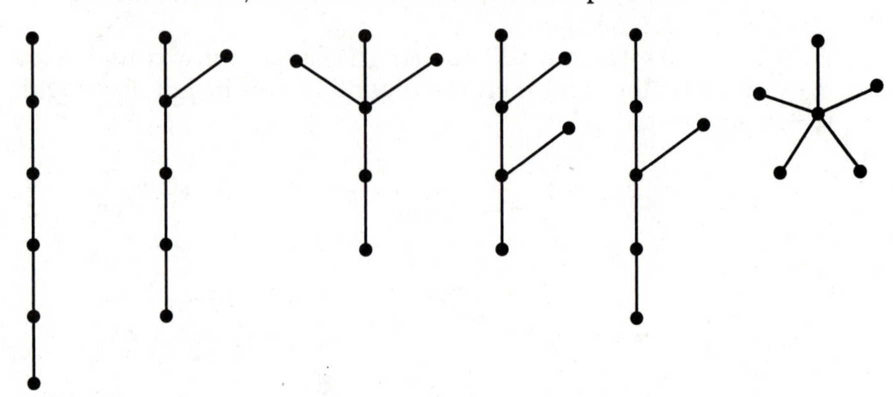

4.

Number of vertices	Number of edges
1	0
2	1
3	2
4	3
N	$N - 1$

18 edges, 16 vertices.

7. 3 4 $S_3 = S_2 + 2$
   4 6 $S_4 = S_3 + 2$
   5 8 $S_5 = S_4 + 2$
   6 10 $S_6 = S_5 + 2$
   $S_N = S_{N-1} + 2$

9.

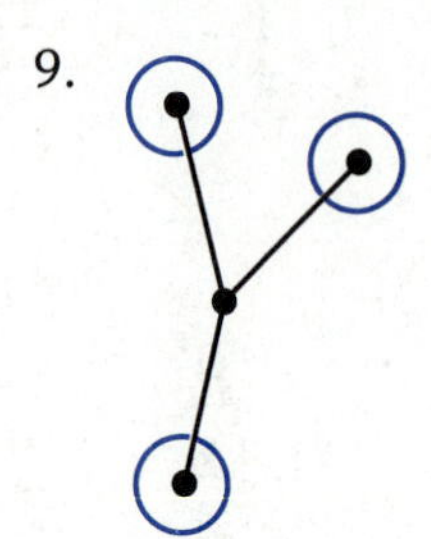

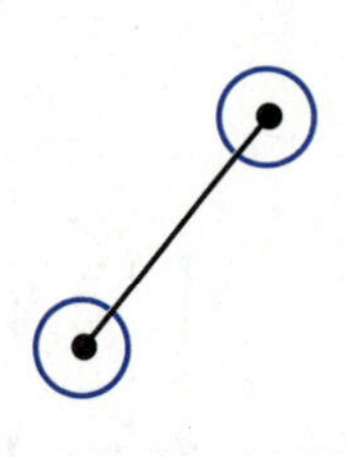

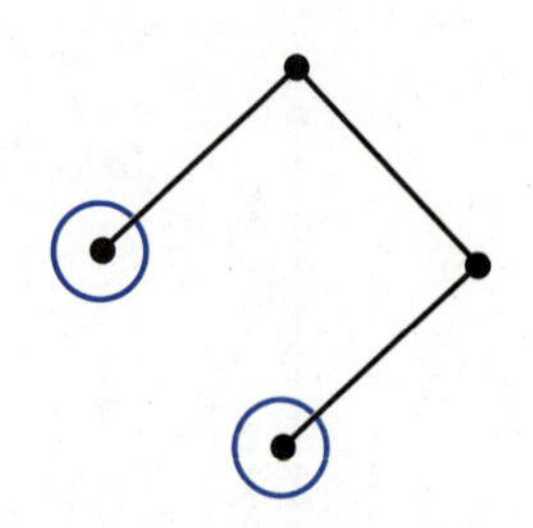

11. Rhombus.

Lesson 5-5

1.

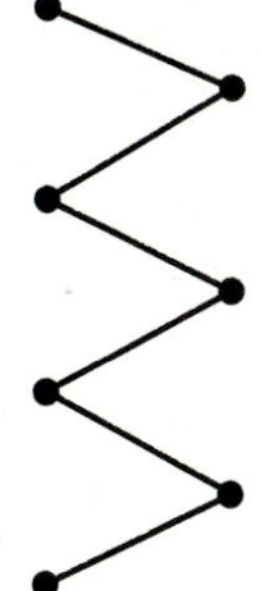

4.

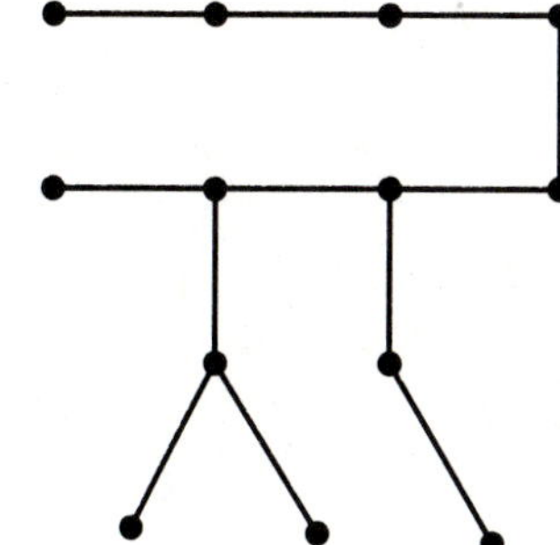

6.

a. Yes. b. *E* and *J*. c. *DE* and *DJ*. d. *F*, *K*, *I*.

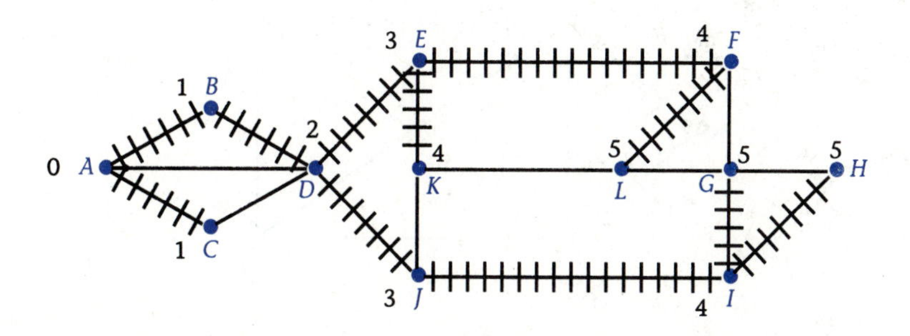

7. This is one of many possibilities:

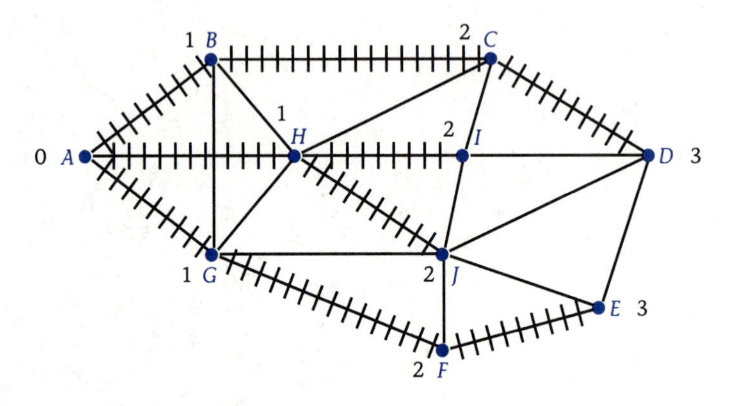

9. The minimum weight is 10.
10. The minimum weight is 18.
12. $2,100.
16.
   a. Weight 28.

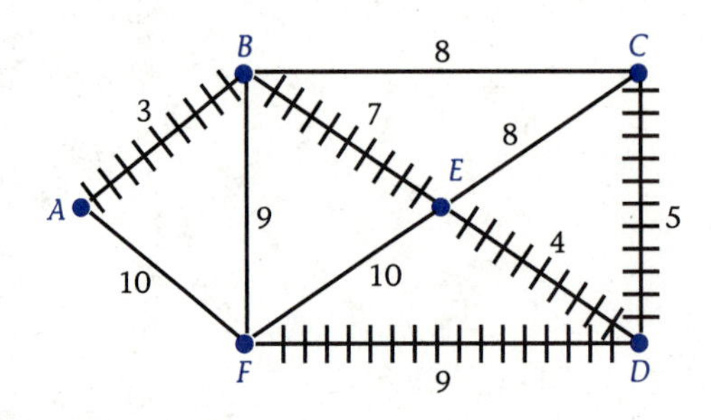

   b.

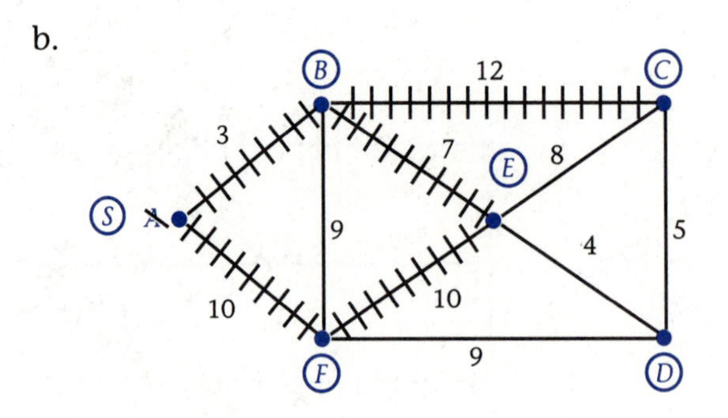

A to *F* – 10
B – 3
C – 15
E – 10
D – 14

   c. No.

Lesson 5-6

1.

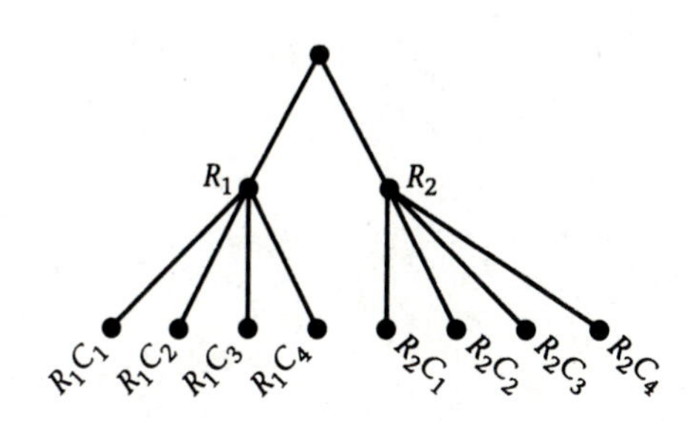

3. Binary tree. a. *V* is level 2. b. *C* is the parent. c. *G* and *H* are children.
6. Binary tree. a. *V* is level 3. b. *E* is the parent. c. No children.
7.

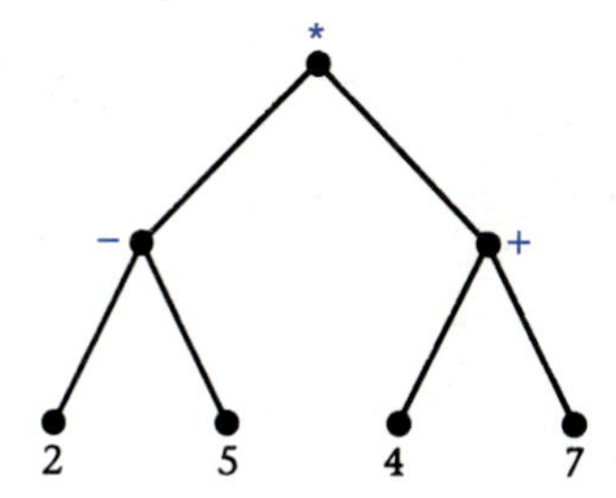

10.

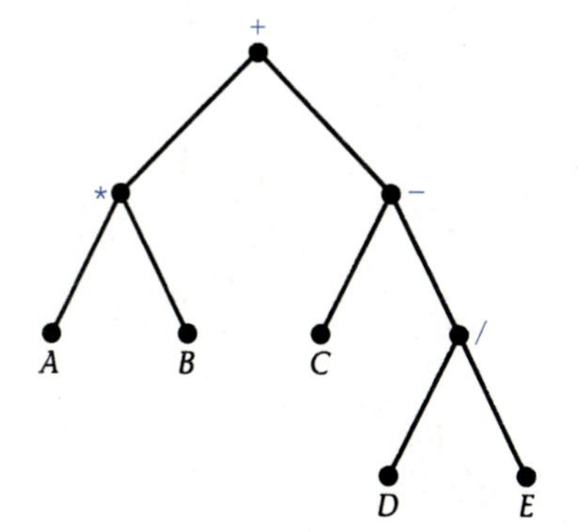

11. 3 2 * 8 2 3 * − +.
12. 5 1 * 13 + 2 9 3/* +.
14. a. 33. b. 7. c. 9. d. 14.
15. a. 2 3 6 * + 4 1 + −
    b. 5 3 − 2 * 7 6 2/− +.
18. *ABDEGHCFI*.
20. 17.
21. a. 8. b. 9.

CHAPTER 6

Lesson 6-1

1. li, lo, ln, ls, il, io, in, is, ol, oi, on, os, nl, ni, no, ns, sl, si, so, sn.
2. 5, 4, $5 \times 4 = 20$, $6 \times 5 = 30$.
3. (1,2) (1,3) (1,4) (1,5) (1,6) (1,7) (1,8) (1,9) (2,3) (2,4) (2,5) (2,6) (2,7) (2,8) (2,9) (3,4) (3,5) (3,6) (3,7) (3,8) (3,9) (4,5) (4,6) (4,7) (4,8) (4,9) (5,6) (5,7) (5,8) (5,9) (6,7) (6,8) (6,9) (7,8) (7,9) (8,9).
4. 9, 8, $9 \times \frac{8}{2} = 36$.
9. $9 \times 8 = 72$. Modify by dividing by 2.
13.

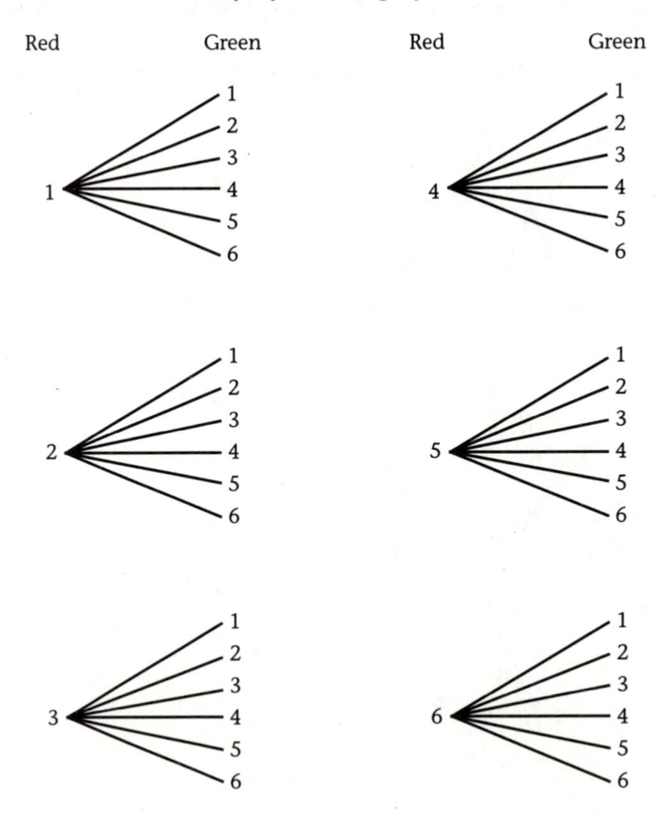

15. Win \$1: 10 ways; win \$2: 1 way; lose \$1: 25 ways; you'd lose money.

Lesson 6-2

1. 10!/6! 5040.
3.
   a. 15.
   b. A particular front sprocket and a particular rear sprocket.
5.
   a. $23^3 10^3 = 12{,}167{,}000$.
   b. $1/23^3 = 1/12617$.
7.
   a. $6 \times 3 \times 2 \times 2 \times 1 \times 1 = 72$.
   b. $\frac{1}{2}$.
   c. 0.
   d. 1.
10. $52 \times 52 = 2{,}704$; $52 \times 51 = 2{,}652$.
13.
   a. 30! or about 2.6525×10^{32}.
   b. About 1.0827×10^{28}; the number of seating arrangements is about 24,500 times as large.
14.
   a. $3 \times 2 = 6$.
   b. A road from Claremont to Upland and a road from Upland to Pasadena.
   c. $3 \times 3 = 9$.
18.
   a. $2^{20} = 1{,}048{,}576$.
   b. About 10 years; about 262 feet.
19. $10^5 = 100{,}000$ manufacturers; $10^5 = 100{,}000$ products.

Lesson 6-3

3.
   a. 672.
   b. 504.
   c. 2,380. They are the same.
4.
   a. $C(10,4) = 210$.
   b. $C(10,6) = 210$.
   c. $2^{10} = 1{,}024$.

5.
   a. C(52,2) = 1,326.
   b. C(26,2) = 325.
   c. 325/1,326, or about 0.245.

8.
   a. C(5,2) = 10.
   b. C(5,3) = 10.

9.
   a. C(44,6) = 7,059,052.
   b. About 245 weeks, or a little less than 5 years.
   c. About 353 feet.
   d. C(49,6) = 13,983,816.
   e. 80,000/13,983,816, or about .00572.
   f. The probability of winning in Virginia is nearly twice as high.

10.
   a. C(6,5) × C(38,1) = 228.
   b. C(6,4) × C(38,2) = 10,545.
   c. C(6,3) × C(38,3) = 168,720.

14. 255, which can be found by evaluating C(8,1) + C(8,2) + · · · + C(8,8) or by evaluating $2^8 - 1$.

19.
   a. 7.
   b. C(7,2) = 21.
   c. 7 + C(7,2) = 28.
   d. $\frac{7}{28}$ or $\frac{1}{4}$.
   e. 13 + C(13,2) = 91.

Lesson 6-4

1.
   a. 520/1,000 = .52.
   b. 196/360, or about .544.
   c. No, but they are fairly close.
   d. 360/1,000 = .36.
   e. 196/1,000 = .196. The product is .1872.
   f. .544 × .36 = 196. They are the same.
   g. No.

2.
   a. 684/1,000 = .684.
   b. .36 + .52 = .88, which is larger than .684.
   c. No.

3.

a.

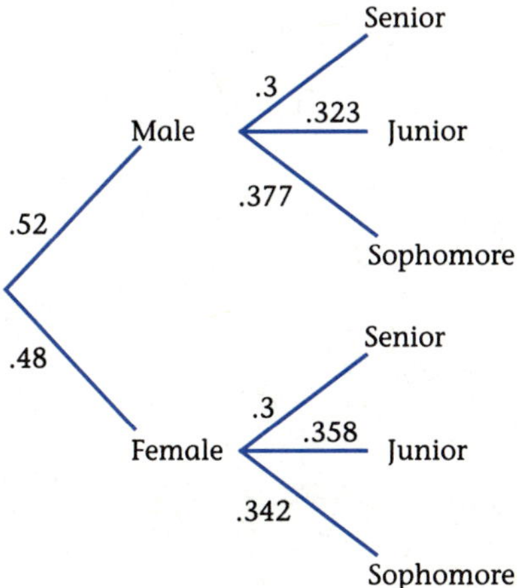

p(Male and Senior) = .52 × .3 = .156

p(Male and Junior) = .52 × .323 = .168

p(Male and Sophomore) = .52 × .377 = .196

p(Female and Senior) = .48 × .3 = .14

p(Female and Junior) = .48 × .358 = .172

p(Female and Sophomore) = .48 × .342 = .164

b. 1.

7.

a. .9 × .6 = .54.

b. That the outcomes of the two games are independent. Probably not—although answers will vary.

8.

a.

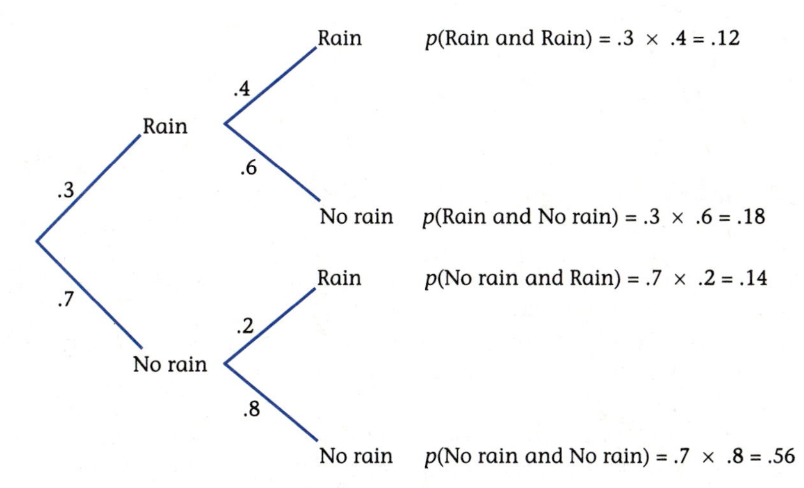

b. .12.

c. $.18 + .14 = .32$.
d. .56.
e. No, because the probability of rain tomorrow depends on today's weather.

11. $\frac{1}{6} \times \frac{1}{6} \times \frac{1}{6} = \frac{1}{216}$.

14.
a. $.977^6$, or about .870.
b. $\frac{1}{6}$.
c. The first is .870, and the second is $1 - \frac{1}{6} = \frac{5}{6}$, or about .83.
d. $.023 \times .023 = .000529$.
e. $1 - .000529 = .999471$, $.999471^6 = .99683$.

17.
a. $\frac{1}{8} \times \frac{1}{10} = \frac{1}{80}$.
b. That the two events are independent.
c. The probability of selecting a man with red hair is .1; the probability of selecting a man who owns a blue car is .125; and the probability of selecting a man who has red hair and owns a blue car is about .0124. The product of the first two probabilities is .0125, so the events are quite close to being independent.

18.
a. $6 \times 6 = 36$.
b. $\frac{1}{36}$.
c. $\frac{1}{36} \times \frac{1}{36} = \frac{1}{1296}$.
d. $\frac{35}{36}$.
e. $\frac{35}{36} \times \frac{35}{36} = \frac{1225}{1296}$.
f. $(\frac{35}{36})^{21}$, or about .5534.

Lesson 6-5

1.
a. The probability of each outcome is $\frac{1}{2}$, and successive applications are independent.
b. The probability of choosing the third answer is $\frac{1}{2}$, and the probability of choosing each of the others is $\frac{1}{4}$; successive applications are independent.
c. Each possibility probably has a $\frac{1}{4}$ chance of occurring, but successive applications may not be independent if, for example, one finger is damaged.
d. Each possibility has the same chance of occurring, and successive applications are independent.

2.

a. $C(5,3) = 10$.
b. $C(5,3)(.5)^3(.5)^2 = .3125$.
c. .03125, .15625, .3125, .3125, .15625, .03125.
d. .00243, .02835, .1323, .3087, .36015, .16807.

3.

a. .2051.
b. .1172.
c. .0439.
d. .0098.
e. .00098.
f. .3770.

5.

a. $C(6,3)(.4)^3(.6)^3 = .2765$.
b.

Number of women	0	1	2	3	4	5	6
Probability	.0466	.1866	.3110	.2765	.1382	.0369	.0041

c. The probability of fewer than two women is .2332, which probably isn't unlikely enough for an accusation of discrimination.

8.

a. 1/7,059,052; 7,059,051/7,059,052.
b. $2.82, but this assumes that the jackpot is not shared with another party.
c.

Amount won	27,000,000	−5,000,000
Probability	5,000,000/7,059,052	2,059,052/7,059,052

The expectation here is $19,124,380, but again this assumes that the jackpot is not shared.

10.

a.

Amount won	−1	1	20
Probability	$\frac{21}{36}$	$\frac{14}{36}$	$\frac{1}{36}$

b. $0.36.
c. No, the council would lose about 36 cents per play. One way of correcting this would be to give no prize for matching a single number. In fact, the jackpot could then be increased but should be kept under $35.

13.

a. $\frac{1}{4}$
b.

Amount won	−$0.50	$1.00
Probability	$\frac{3}{4}$	$\frac{1}{4}$

c. −$0.50.
d. Yes, about 50 cents per play.

CHAPTER 7

Lesson 7-1

1. 25; 475; 45; 455; 2,000; 1,900; 5,000; 4,750; 2,500; 125; 7,500; 375; $0.05P$; $P - 0.05P$.
2.
   a. 2%.
   b.
   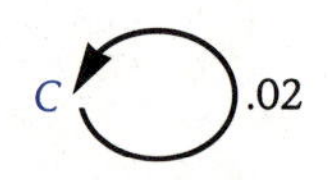

   c. $P - 0.02P = D$.
   d. $P = 20{,}408$.
3.
   a.
   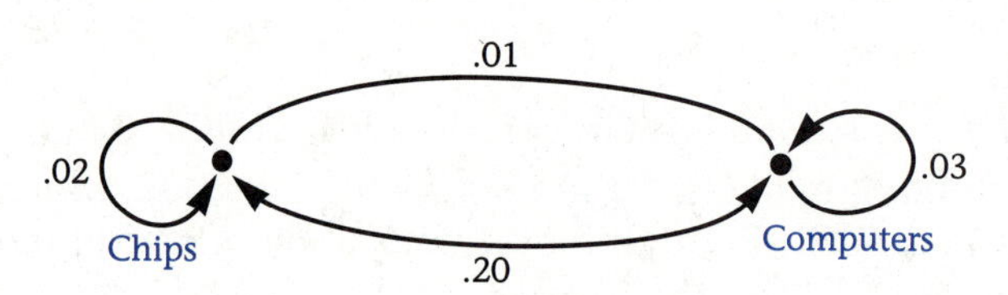

   b.

	Chips	Computers
Chips	0.02	0.20
Computers	0.01	.03

   c. $20 chips, $10 computers.
   d. $150 computers, $1,000 chips.
   e. $20,408.
   f. $51,546.
4.
   a.
   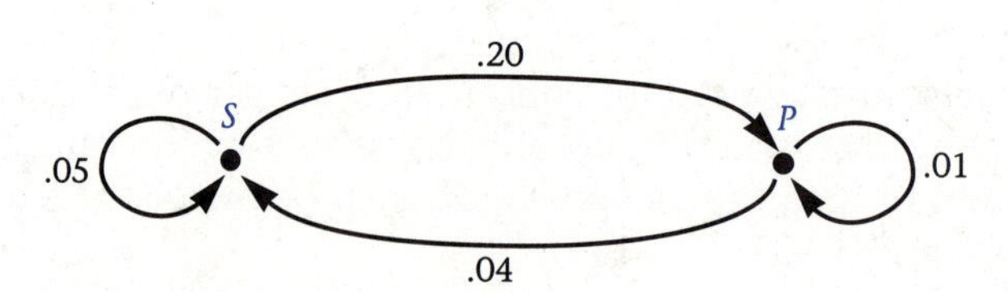

   b. 5 cents, 20 cents, 1 cent, 4 cents.
   c. $1 million, $0.8 million.
   d. $0.4 million, $0.8 million.
   e. $11 million, $38.8 million.

Lesson 7-2

1.

a. $T = \begin{matrix} \text{chips} \\ \text{computers} \end{matrix} \overset{\begin{matrix} \text{Chips} & \text{Computers} \end{matrix}}{\begin{bmatrix} 0.02 & 0.20 \\ 0.01 & 0.03 \end{bmatrix}}.$

b. $P = \begin{matrix} \text{chips} \\ \text{computers} \end{matrix} \overset{\text{Dollars}}{\begin{bmatrix} 40{,}000 \\ 50{,}000 \end{bmatrix}}.$

c. $TP = \begin{matrix} \text{chips} \\ \text{computers} \end{matrix} \overset{\text{Dollars}}{\begin{bmatrix} 10{,}800 \\ 1{,}900 \end{bmatrix}}.$

d. $D = \begin{matrix} \text{chips} \\ \text{computers} \end{matrix} \overset{\text{Dollars}}{\begin{bmatrix} 29{,}200 \\ 48{,}100 \end{bmatrix}}.$

e. $P = \begin{matrix} \text{chips} \\ \text{computers} \end{matrix} \overset{\text{Dollars}}{\begin{bmatrix} 51{,}444 \\ 52{,}077 \end{bmatrix}}.$

4.

a.

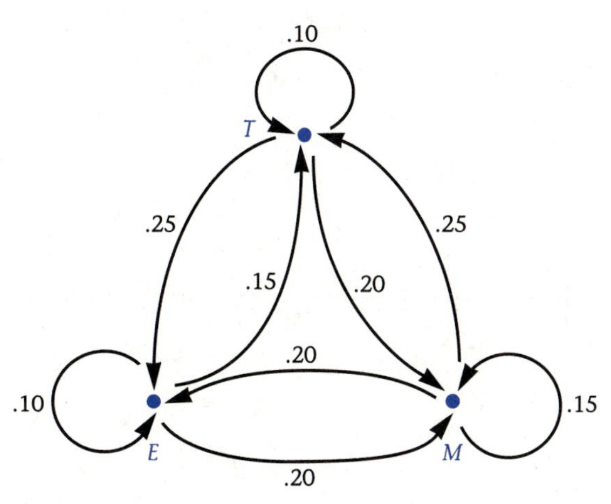

b. $T = \begin{matrix} \text{T} \\ \text{E} \\ \text{M} \end{matrix} \begin{bmatrix} 0.10 & 0.25 & 0.20 \\ 0.15 & 0.10 & 0.20 \\ 0.25 & 0.20 & 0.15 \end{bmatrix}$

c. $P = \begin{matrix} \text{T} \\ \text{E} \\ \text{M} \end{matrix} \begin{bmatrix} 150 \\ 200 \\ 160 \end{bmatrix}$

d. $TP = \begin{matrix} T \\ E \\ M \end{matrix} \begin{bmatrix} 97.00 \\ 74.50 \\ 101.50 \end{bmatrix}$

e. $D = \begin{matrix} T \\ E \\ M \end{matrix} \begin{bmatrix} 53.00 \\ 125.5 \\ 58.50 \end{bmatrix}$

f. $T = \begin{matrix} T \\ E \\ M \end{matrix} \begin{bmatrix} 218.6 \\ 195.2 \\ 239.6 \end{bmatrix}$

5.

a.

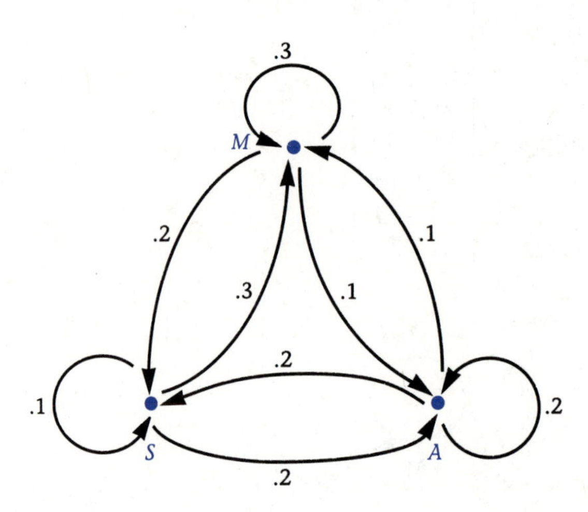

b. Most dependent on itself and least dependent on agriculture.

c. $8 million, $8 million.

d. $\begin{matrix} S \\ M \\ A \end{matrix} \begin{bmatrix} 12.50 \\ 13.00 \\ 9.50 \end{bmatrix}$

e. $\begin{matrix} S \\ M \\ A \end{matrix} \begin{bmatrix} 11.03 \\ 11.61 \\ 9.21 \end{bmatrix}$

7.

a. Yes. Answers will vary.

b. $P = \begin{bmatrix} 184 \\ 124 \end{bmatrix}.$

c. Three times the production for part a: $\begin{bmatrix} 276 \\ 186 \end{bmatrix}$.

Lesson 7-3

1.

a. $D_0 = [.75 \quad .25]$ $D_1 = [.625 \quad .375]$
$D_2 = [.5875 \quad .4125]$
$D_3 = [.57625 \quad .42375]$ $D_4 = [.572875 \quad .427125]$.

b. $D_{10} = [.571430 \quad .428570]$
$D_{15} = [.571429 \quad .428571]$.

d. No.

e. $T_{15} = \begin{bmatrix} 0.571429 & 0.428571 \\ 0.571429 & 0.428571 \end{bmatrix}$.

2.

a. $[.571429 \quad .428571]$.

b. They are the same.

3.

a. $D_0 = [1 \quad 0]$ $D_1 = [.7 \quad .3]$ $D_2 = [.61 \quad .39]$
$D_3 = [.583 \quad .417]$ $D_4 = [.5749 \quad .4251]$.

b. $D_{10} = [.571431 \quad .428569]$
$D_{15} = [.571429 \quad .428571]$.
After several weeks. 57% of the students will be eating in the cafeteria.

4.

a. No. Row 2 adds to more than 1.

b. No. Not square.

c. No. $1.2 > 1$ or $-.4$ is not a probability.

d. Not square or $.3 + .3 + .3$ is not equal to 1.

e. Yes.

f. Yes.

5.

a. .42.

b. .08.

c. $[.6 \quad .4]$.

d. $\begin{bmatrix} .7 & .3 \\ .2 & .8 \end{bmatrix}$.

e. $D_7 = [.40 \quad .60]$.

10.

a. $D_0 = [1 \quad 0 \quad 0]$.
b. $D_1 = [.8 \quad .2 \quad 0]$ $D_2 = [.66 \quad .28 \quad .06]$
$D_3 = [.556 \quad .300 \quad .144]$
$D_4 = [.4748 \quad .2912 \quad .2340]$.
After one month: 47% in min., 29% in max., 23% dead.
c. He or she will be dead.

Lesson 7-4

1.

Best strategies	Player 1	Player 2	Strictly determined	Saddle point
a.	row 1	column 2	yes	8
b.	row 1	column 1	yes	0
c.	row 2	column 1	no	
d.	row 1	column 2	no	
e.	row 3	column 3	yes	4
f.	row 1	column 1	yes	0

2.

a. Best strategies: Row 2, column 2; saddle point is 3.
b. Best strategies stay the same. Adds 4 to the saddle point.
c. Best strategies stay the same. Doubles the saddle point.
d. A hypothesis: When a constant is added to each value in the payoff matrix of a strictly determined game, the best strategies stay the same, and the saddle point is increased by the constant. If each value is multiplied by a constant, the best strategies stay the same, and the saddle point is also multiplied by the constant.

4.

a. Every other row dominates row B. Eliminate row B. Column E dominates column G. Eliminate column G. Row C dominates rows A and D. Eliminate rows A and D. Column F dominates column E. Eliminate column E. Best strategies: Row C and column F.

7. $\begin{array}{c} \\ 1 \\ 2 \\ 3 \end{array} \begin{array}{c} \begin{array}{ccc} 1 & 2 & 3 \end{array} \\ \begin{bmatrix} 10 & -10 & -10 \\ -20 & 20 & -20 \\ -30 & -30 & 30 \end{bmatrix} \end{array}$ No saddle point.

Lesson 7-5

1.
   a. $P(\text{HH}) = 21/80$.
   b. $P(\text{TT}) = 15/80$.
   c. $P(\text{HT or TH}) = 44/80$.
   d. Probability distribution:

Amount won	3	1	−2
Probability	21/80	44/80	15/80

   e. Expectation: $3(21/80) + 1(15/80) + (-2)(44/80) = -10/80$.
   f. He expects to lose 1/8 cents each time he plays.
2.
   a. Sol's best strategy is to play heads 40% of the time and tails 60% of the time.
   b. Tina's best strategy is to play heads 30% of the time and tails 70% of the time.
   c. Probability distribution:

Amount won	4	−2	−3	1
Probability	.12	.28	.18	.42

   d. Expectation: −0.2 cents per play, or −2 cents in 10 plays.
   e. Lose 20 cents in 100 plays.
5. Payoff matrix:

$$\begin{array}{cc} & \text{Player 2} \\ & \begin{array}{cc} 1 & 2 \end{array} \\ \text{Player 1} \begin{array}{c} 1 \\ 2 \end{array} & \begin{bmatrix} 2 & -3 \\ -3 & 4 \end{bmatrix}. \end{array}$$

   Best strategies: Both row and column players play 1 finger seven-twelfths of the time and 2 fingers five-twelfths of the time. The row player can expect to lose 1 cent in 12 plays. Not a fair game.
7. Row 2 dominates row 1. Eliminate row 1 (making phone calls). Column 2 dominates column 1. Eliminate column 1 (making phone calls). Group against should send out mailings three-fourths of the time and go door to door one-fourth of the time. Group in favor should send out mailings one-half of the time and go door to door one-half of the time. Opposing group gets only 250 signatures (not enough to get the issue on the ballot).

CHAPTER 8

Lesson 8-1

1.

b. 1, 4, 9.

c.

Number of couples	Number of handshakes	Recurrence relation
1	1	—
2	4	$H_2 = H_1 + 3$
3	9	$H_3 = H_2 + 5$
4	16	$H_4 = H_3 + 7$
5	25	$H_5 = H_4 + 9$

2.

a.

Couples	Handshakes
1	0
2	2
3	6
4	12

b. $2n - 2$.

c. $H_n = H_{n-1} + 2n - 2$.

3.

a.

i. $H_n = H_{n-1} + 3$.
ii. $H_n = (H_{n-1}) \cdot 2$.
iii. $H_n = H_{n-1} + n$.
iv. $H_n = (H_{n-1}) \cdot n$.

b. 16, 64, 36, 4320.

4.

a. 4.
b. −1.
c. 0.
d. None.

6.

a.

Term number	Number of handshakes	Differences	Differences
1	0	—	—
2	1	1	—
3	3	2	1
4	6	3	1
5	10	4	1
6	15	5	1
7	21	6	1
8	28	7	1

b. Second degree. 2.

9.

a.

Term number	Number of bees
0	5,000
1	5,600
2	6,272
3	7,024.64
4	7,867.60

b. $B_n = 1.12B_{n-1}$.

c. After 27 years, in 2014.

Lesson 8-2

2.

a. $H_n = 2n^2 - 5n$.

b. $H_n = 0.29 + 0.23(n - 1)$.

c. $H_n = n^2 - 5n + 4$.

d. $H_n = 3^{n-1}$.

3.

a. $T_n = T_{n-1} + n - 1$.

b. 0.

c. $T_n = n(n - 1)/2$.

4. a. 1, −2, −11, −38, −119, −362.
   b. 2.5.
6.
   a.

Row number	Number of seats	Total seats
1	24	24
2	26	50
3	28	78
4	30	108
5	32	140
6	34	174

   b. $S_n = S_{n-1} + 2$.
   c. $S_n = 24 + 2n - 2$.
   d. 37.
   f. $T_n = T_{n-1} + 24 + 2n - 2$.
   g. $T_n = n^2 + 23n$.
7.
   a.

Year	Value
1980	38,000
1981	41,040
1982	44,323.20
1983	47,869.06
1984	51,698.58
1985	55,834.47
1986	60,301.22
1987	65,125.32
1988	70,335.35
1989	75,962.18
1990	82,039.15
1991	88,602.28
1992	95,690.46
1993	103,345.70

   b. $V_n = (V_{n-1}) \cdot 1.08$.

Lesson 8-3

1.
  b.
    i. $H_n = H_{n-1} + 3.$
    ii. $H_n = (H_{n-1})/2.$
    iii. $H_n = (H_{n-1}) \cdot 1.2.$
    iv. $H_n = H_{n-1} + H_{n-2}.$
    v. $H_n = (H_{n-1}) \cdot 0.1.$
    vi. $H_n = H_{n-1} + n - 1.$
  c.
    i. $H_n = 2 + 3(n - 1).$
    ii. $H_n = 64(0.5^{n-1}).$
    iii. $H_n = 10(1.2^{n-1}).$
    v. $H_n = 0.3(0.1^{n-1}).$

5.
  a. $M_n = (M_{n-1}) \cdot 1.084.$
  b. $M_n = 5{,}000(1.084^n).$
  c. $6,368.80.
  d. 0.7%.
  e. $M_n = (M_{n-1}) \cdot 1.007.$
  f. $M_n = 5{,}000(1.007^n).$
  g. $6,427.34.
  h. Yearly = $6,368.80.
     Monthly = $6,427.34.

6. 7.2% monthly: $5,372.12, $5,771.94, $6,201.51; 8% yearly: $5,400, $5,832, $6,298.56.

7.
  a. 74, 585.
  b. 63.25, 817.5

10.
  a. $38,265.77.

11.
  a. Approximately 6.5%.

15.
  a. $88,048.25.
  b. $15,710.53.

17.
  a. 1.14471.
  b. 2,318.
  c. After 33 hours.

Lesson 8-4

1.
   a. \$93,070.22.
   b. \$48,000.
   c. \$45,070.22.
   d. \$202,107.52, \$93,070.22.
   e. \$118,789.44, \$93,070.22.
   f. \$139,605.33, \$93,070.22.

4.
   a.

T (in months)	T_n
0	12,000
1	11,802
2	11,602.22
3	11,400.64

   b. $A_n = (A_{n-1}) \cdot 1.009 - 306$.
   c. It takes 49 months.
   d. 34,000; 34,000; 34,000; the fixed point.

6.
   a. $M_n = (M_{n-1}) \cdot 1.2 - 1{,}500$.
   b. \$3,180, \$2,316.
   c. 20%.
   d. \$7,500. The interest would equal the payments, and the loan amount would never decrease.

7. \$10,206.75.

Lesson 8-5

1.
   a. -3; $T_n = 4(2^{n-1}) - 3$; 2.5353×10^{30}.
   b. 3.5; $T_n = 1.5(3^{n-1}) + 3.5$; 2.5769×10^{47}.
   c. $\frac{5}{3}$; $T_n = (\frac{1}{3})(4^{n-1}) + \frac{5}{3}$; 1.3391×10^{59}.

2.

a. $B_n = (B_{n-1})(1 + 0.08/12) + 150.$

b.

Month	Balance
0	150
1	301
2	453.01
3	606.03
4	760.07

c. −22,500.

d. $22{,}801(1 + 0.08/12)^{n-1} - 22{,}500$, or $22{,}650(1 + 0.08/12)^{n} - 22{,}500$.

e. $225,194.28.

f. 472 months.

g. About $333.

3.

a. $B_n = (B_{n-1}) \cdot 1.01 - 220.$

b.

N	B_n
0	10,000
1	9,880
2	9,758.80
3	9,636.39
4	9,512.75

c. 22,000.

d. $B_n = -12{,}120(1.01^{n-1}) + 22{,}000$ or $B_n = -12{,}000(1.01^{n}) + 22{,}000$.

e. $6,763.18.

f. 61 months.

g. $13,420, $3,420.

h. $332.14.

Lesson 8-6

1.

a.

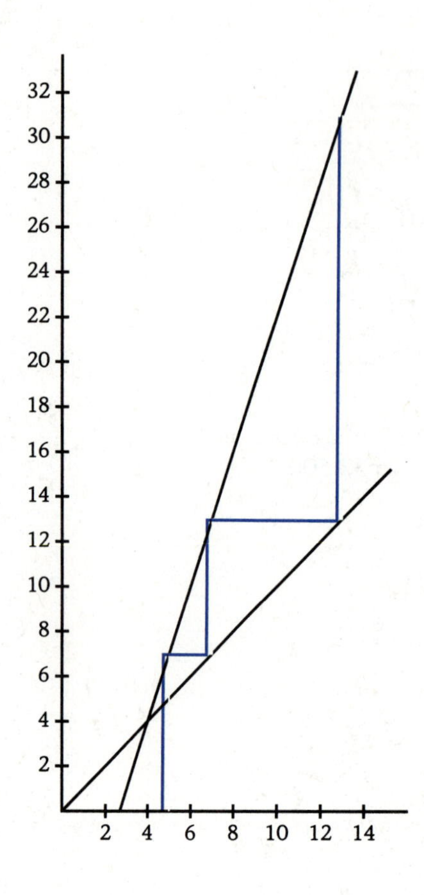

b.

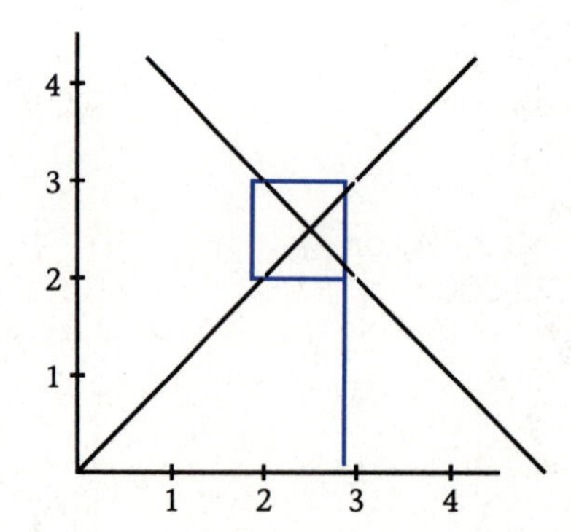

c.

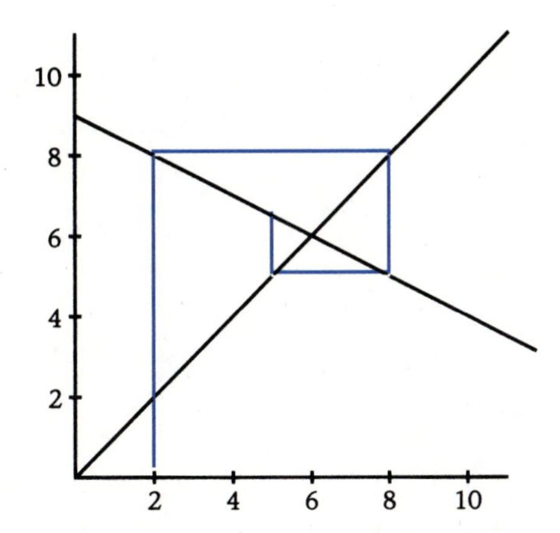

d.

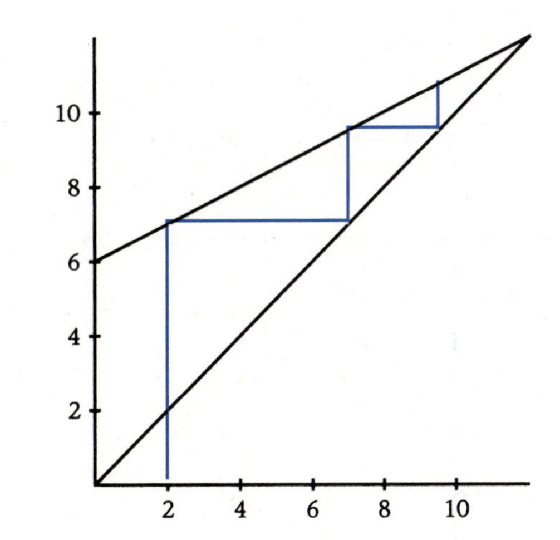

4.

a. 1.

b. 3.

c.

n	t_n
1	1
2	5
3	13

d. $t_n = 2t_{n-1} + 3$.

e. -3.

5.

a.

n	t_n
1	0
2	4
3	−4
4	−4
5	−4
6	−4

b.

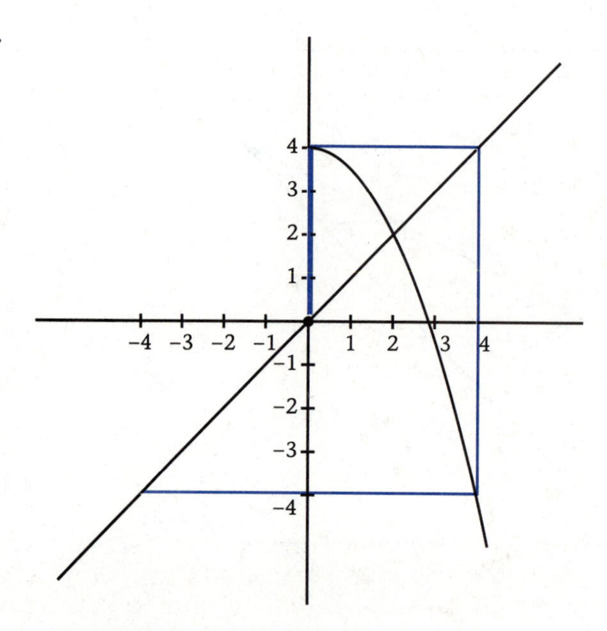

c. The behavior is unpredictable.
d. 2 is a fixed point.
e. The terms diverge.
f. 2, −4.
g. Attracted when $t_1 = 0$ or 2; repelled when $t_1 = 5$; unpredictable when $t_1 = 1$.

Answers to the Chapter Review Exercises

CHAPTER 1

1.
   a. D.
   b. B.
   c. A.
   d. E.
   e. C.
   f. C.
2. 19, 42, 89.
3.
   a. It can occur in either the run-off or sequential run-off method.
4.
   a. Wilson, no.
   b. They rated him last.
   c. The voters in the last group could have switched to Roosevelt, their second choice, and thereby have prevented Wilson from winning.
   d. Answers can vary, but Roosevelt will probably win by most other methods.
5. Conditions 2, 3, and 5.
6. Arrow proved that no group ranking method that ranks three or more choices will always adhere to his five fairness conditions.
7.
   a. {A, B, C, D; 12}, {A, B, C; 10}, {B, C, D; 8}, {A, C, D; 9}, {A, B, D; 9}, {A, B; 7}, {A, C; 7}.
   b. A: 5, B: 3, C: 3, D: 1.
   c. No, A's power is disproportionately high, while D's is low.
   d. All voters now have equal power.

CHAPTER 2

1. Answers are rounded to the nearest dollar.

	Joan	Henry	Sam
Fair share	$8,880	$8,600	$8,433
Items received	Lot	Computer Stereo	Boat
Cash	$880	$6,000	$1,733
Final settlement	$9,342	$9,062	$8,895

2. Answers are rounded to the nearest dollar.

	Anne	Beth	Jay
Fair share	$1,800	$1,567	$1,617
Items received	Car, computer		Stereo
Cash	−$2,600	$1,567	$ 417
Final settlement	$2,005	$1,772	$1,822

3.
   a. 10.
   b. 64.7, 24.7, 10.6.
   c. 65, 25, 10.
   d. 64, 24, 10.
   e. 9.9538, 9.8800, 9.6364.
   f. 65, 25, 10.
   g. 65, 25, 11.
   h. 10.0310, 10.0816, 10.952.
   i. 64, 25, 11.
   j. 65, 25, 11
   k. 10.0313, 10.0837, 10.1067.
   l. 64, 25, 11.
   m. 64, 25, 11.
   n. State A gained population and State C lost, but A lost a seat to C.
4. Balinski and Young proved that any apportionment method will sometimes produce one of three undesirable results: violation of quota, the loss of a seat when the size of the legislative body is increased even if population doesn't decrease, and the loss of a seat by one state whose population has increased to another whose population has decreased.
5. Arnold and Betty.
6. Have each of the original four divide his or her piece of cake into five pieces that he or she considers equal. Have the new person select a piece from each of the others.
7. $(k^2 - k)/2$ or $k(k - 1)/2$.

CHAPTER 3

1. No. Answers will vary.
2.
   a. $L = \begin{bmatrix} 35 & 6 & 6 & 12 \end{bmatrix}$ (columns M, C, S, D).
   b. L_2 = number of bags of chips ordered.
      L_4 = number of six-packs of drinks.
   c. $216.60.
3.
   a.

	Lodging	Food	Rec.
Crystal	13.00	20.00	5.00
Springs	12.50	19.50	7.50
Bear	20.00	18.00	0.00
Beaver	40.00	0.00	0.00

   b. $C_{22} = 19.50$.
      $C_{43} = 0$.
   c. C_{13} = cost for recreation at Crystal Lodge.
      C_{31} = cost for lodging at Bear Lodge.
4.
   a.

	System	Cart.	Case
Z-Mart	39.50	24.50	8.50
Base	49.90	29.95	12.50

   b.

	System	Cart.	Case
Z-Mart	35.55	22.05	7.65
Base	39.92	23.96	10.00

   c.

	System	Cart.	Case
Z-Mart	3.95	2.45	0.85
Base	9.98	5.99	2.50

   d.

	System	Cart.	Case
Z-Mart	142.20	88.20	30.60
Base	159.68	98.84	40.00

5.
   a.

	Plate	Large	Small
No.	[5	3	7].

   b.

	Ebony	Walnut	Rose	Maple
Plate	100	800	600	400
Large	200	1200	1000	800
Small	50	500	450	400

   c.

Ebony	Walnut	Rose	Maple
[1450	11,000	9,150	7,200].

d.
$$\text{No.}\begin{matrix} \text{Plate} & \text{Large} & \text{Small} \\ [\,5 & 3 & 7\,] \end{matrix} \begin{matrix} \\ \text{Plate} \\ \text{Large} \\ \text{Small} \end{matrix} \overset{\text{Weeks}}{\begin{bmatrix} 3 \\ 4 \\ 2 \end{bmatrix}} = \text{No.}\ \overset{\text{Weeks}}{[15 + 12 + 14]} = [41].$$

6.
$$\begin{matrix} \text{Tennis} & \text{Golf} & \text{Soccer} \\ [\$50{,}000 & \$100{,}000 & \$75{,}000] \end{matrix} \begin{matrix} \text{Tennis} \\ \text{Golf} \\ \text{Soccer} \end{matrix} \overset{\text{Return}}{\begin{bmatrix} 0.082 \\ 0.065 \\ 0.075 \end{bmatrix}} = \overset{\text{Return}}{[\$16{,}225]}.$$

7.

c.
$$\begin{matrix} & \text{Jazz} & \text{Symp.} & \text{Orch.} \\ \$[& 300.00 & 335.00 & 373.50]. \end{matrix}$$

8.
$$A^T = \begin{bmatrix} 4 & 5 \\ 2 & 1 \\ 6 & 3 \end{bmatrix}.$$

9.

a. 3×2.
b. 4×3.
c. Not possible.
d. 4×3

10.

a. $M^2 = \begin{bmatrix} 2 & 2 \\ 2 & 2 \end{bmatrix}$ $M^3 = \begin{bmatrix} 4 & 4 \\ 4 & 4 \end{bmatrix}$ $M^4 = \begin{bmatrix} 8 & 8 \\ 8 & 8 \end{bmatrix}$.

b. $M^5 = \begin{bmatrix} 16 & 16 \\ 16 & 16 \end{bmatrix}$.

c. $M^n = \begin{bmatrix} 2^{n-1} & 2^{n-1} \\ 2^{n-1} & 2^{n-1} \end{bmatrix}$.

d. $M^2 = \begin{bmatrix} 1 & 0 \\ 8 & 9 \end{bmatrix}$ $M^3 = \begin{bmatrix} 1 & 0 \\ 26 & 27 \end{bmatrix}$ $M^4 = \begin{bmatrix} 1 & 0 \\ 80 & 81 \end{bmatrix}$

$M^5 = \begin{bmatrix} 1 & 0 \\ 242 & 243 \end{bmatrix}$ $M^n = \begin{bmatrix} 1 & 0 \\ 3^n - 1 & 3^n \end{bmatrix}$.

11. Identity, a square matrix with ones along the diagonal and zeros elsewhere.

12.

a. Yes. $AB = BA = I$.
b. Yes. $AB = BA = I$.
c. No. Not square matrices.

13. a.

$$\begin{matrix} \\ \text{A} \\ \text{B} \\ \text{C} \end{matrix} \begin{matrix} \text{Male} & \text{Female} \\ \end{matrix}\begin{bmatrix} 189 & 196 \\ 175 & 180 \\ 251 & 254 \end{bmatrix}$$

b.
$$\begin{matrix} \text{A} \\ \text{B} \\ \text{C} \end{matrix} \begin{bmatrix} 385 \\ 355 \\ 505 \end{bmatrix}$$

14.

a. 24 months.

b.

$$L = \begin{bmatrix} 0.0 & 0.6 & 0.0 & 0.0 & 0.0 & 0.0 \\ 0.5 & 0.0 & 0.8 & 0.0 & 0.0 & 0.0 \\ 1.1 & 0.0 & 0.0 & 0.9 & 0.0 & 0.0 \\ 0.9 & 0.0 & 0.0 & 0.0 & 0.8 & 0.0 \\ 0.4 & 0.0 & 0.0 & 0.0 & 0.0 & 0.6 \\ 0.0 & 0.0 & 0.0 & 0.0 & 0.0 & 0.0 \end{bmatrix}.$$

c. 10%. d. 28 months.

CHAPTER 4

1.

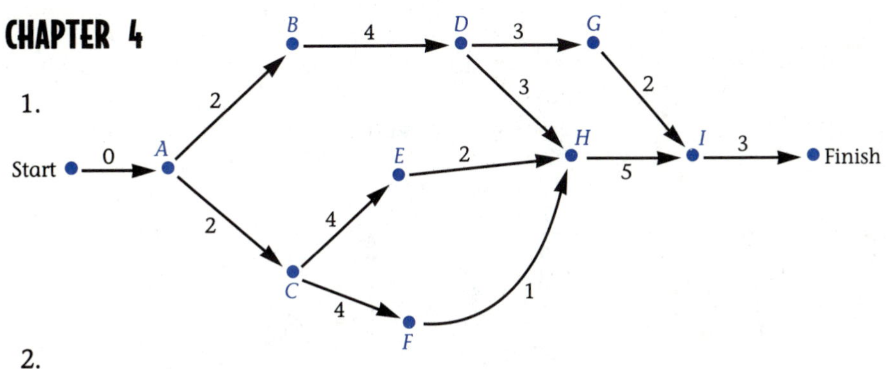

2.

Task	Time	Prerequisites
Start	0	—
A	2	None
B	3	None
C	4	A
D	4	A, B
E	2	B
F	3	C
G	5	D, E
H	7	F, G
Finish		

3. A, (0); B, (4); C, (4); D, (7); E, (6); F, (11); G, (10); H, (15); I, (16); J, (18). Minimum project time: 23.

4. A, (0); B, (2); C, (2); D, (6); E, (6); F, (6); G, (9); H, (9); I, (14). Critical path: Start–$ABDHI$–Finish. Minimum project time: 17.

5.

a. Yes, a path exists from each vertex to every other vertex.
b. No, not every pair of vertices are adjacent.
c. A, D, or C.
d. $BCDE$ or $BCAE$.
e. Deg(C) = 4.
f.

$$\begin{array}{c} \\ A \\ B \\ C \\ D \\ E \end{array} \begin{array}{c} \begin{array}{ccccc} A & B & C & D & E \end{array} \\ \begin{bmatrix} 0 & 0 & 1 & 0 & 1 \\ 0 & 0 & 1 & 0 & 0 \\ 1 & 1 & 0 & 1 & 1 \\ 0 & 0 & 1 & 0 & 1 \\ 1 & 0 & 1 & 1 & 0 \end{bmatrix} \end{array}.$$

6.

a. Euler path. Two vertices have odd degrees, and the remaining vertices have even degrees.
b. Euler circuit. All vertices have even degrees.

7.

a.

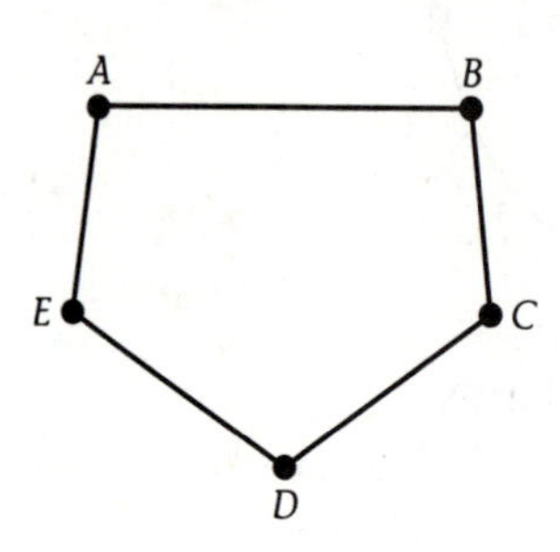

b.

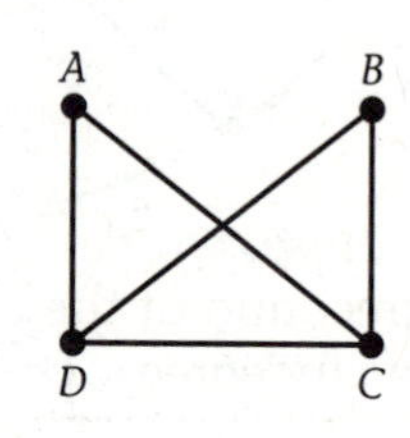

8.

a. No.
b. Yes.
c.

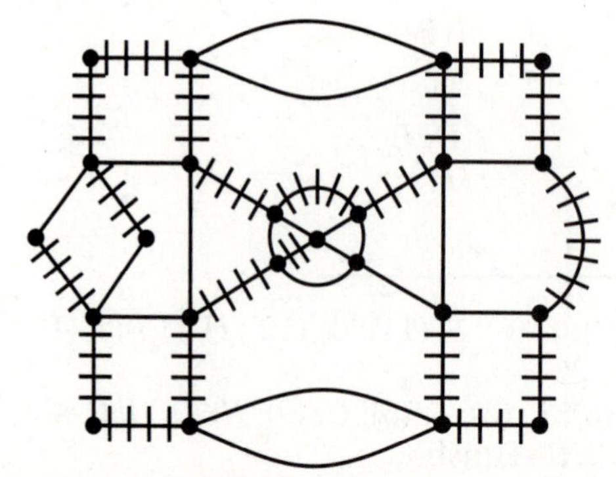

9.
   a. Yes.
   b. No, the graph has exactly two odd vertices. You must begin at one of the vertices with an odd degree and end at the other.

10.
   a.

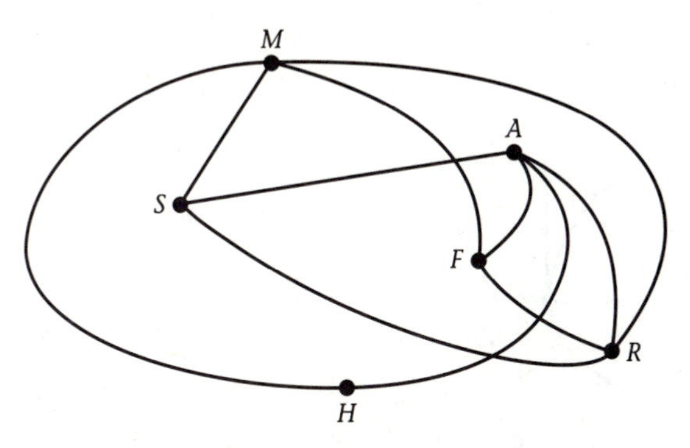

   b. Three time slots. One possible schedule:
   Time 1—Math and Art   Time 2—Reading and History
   Time 3—Science and French.

11.
   a. No, the outdegrees are not equal to the indegrees at each vertex.
   b. Yes, the outdegree equals the indegree at all vertices but two. At one of those two vertices, the indegree is one greater than the outdegree, and at the other vertex, the outdegree is one greater than the indegree.
   Answers will vary, but the paths must begin at B and end at D.

12.
   a, b.

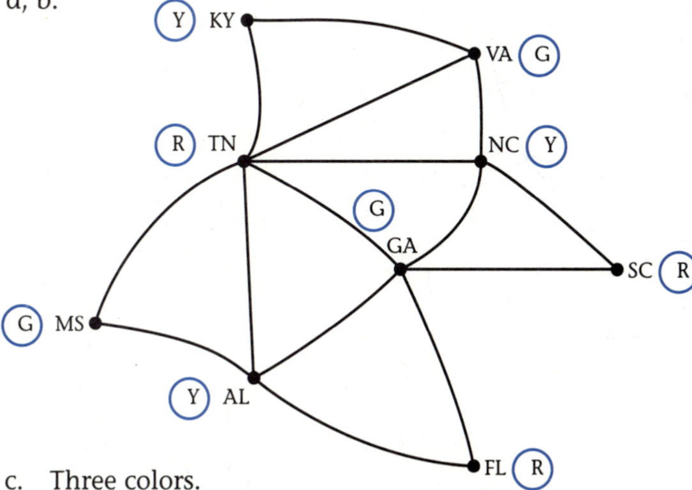

   c. Three colors.

CHAPTER 5

1.

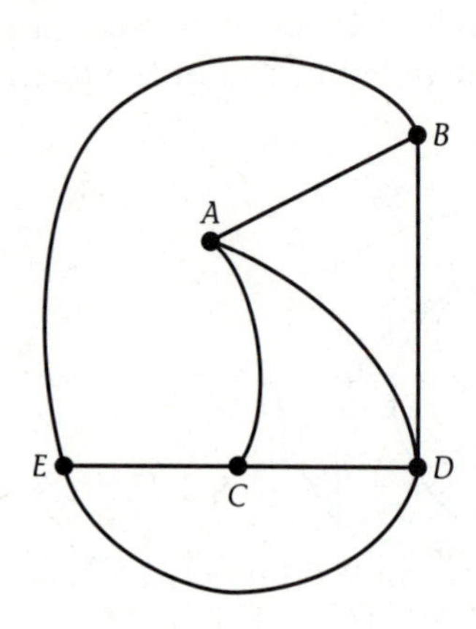

2. 2. 2. 2. 2.
3.
   a. Yes, the vertices of the graph can be divided into two sets so that each edge of the graph has one endpoint in each set.
   b. Yes, all possible edges from one set of vertices to the other are drawn.
   c.

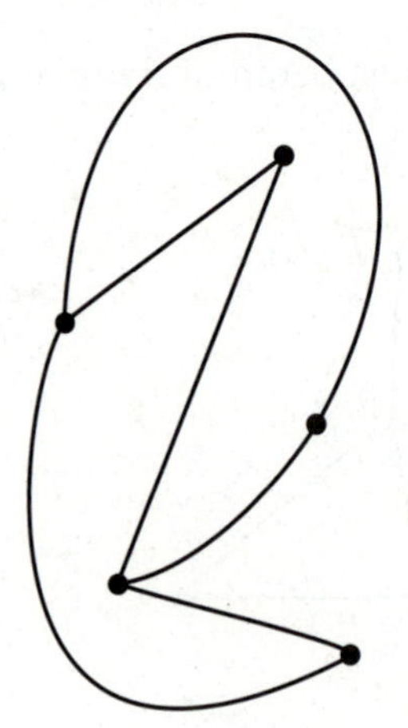

   d. 2.

4.
   a. O–SCM–O.
   b. 314 ft.
   c. Hamiltonian circuit.

5.

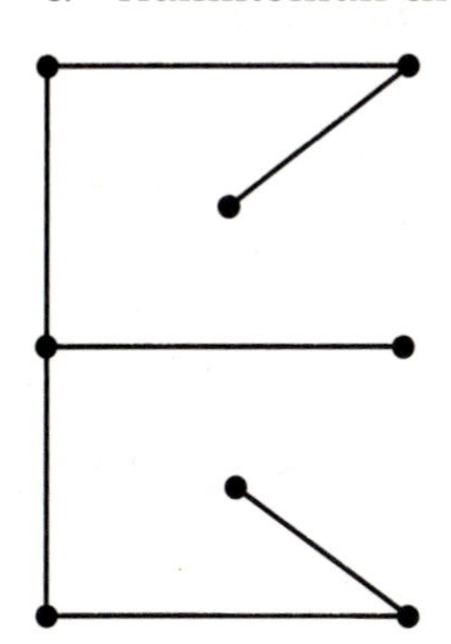

6.
   a. Home-T-P-G-H-BB.
   b. 14 miles.
   c.

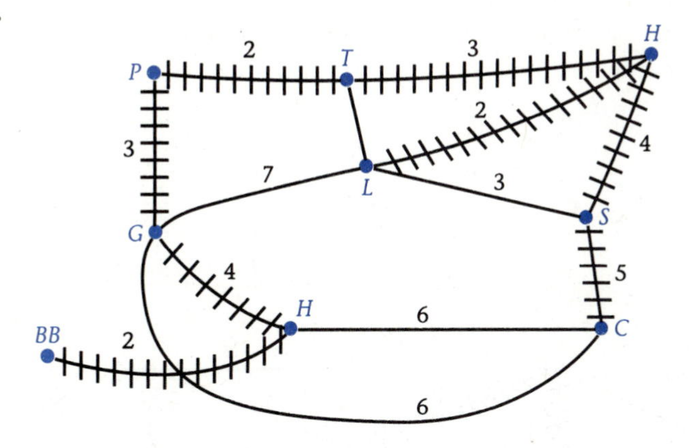

7. The total weight of a minimum spanning tree for the graph is 23 miles.

8.
   a. Yes.
   b. Yes.
   c. No.

9.

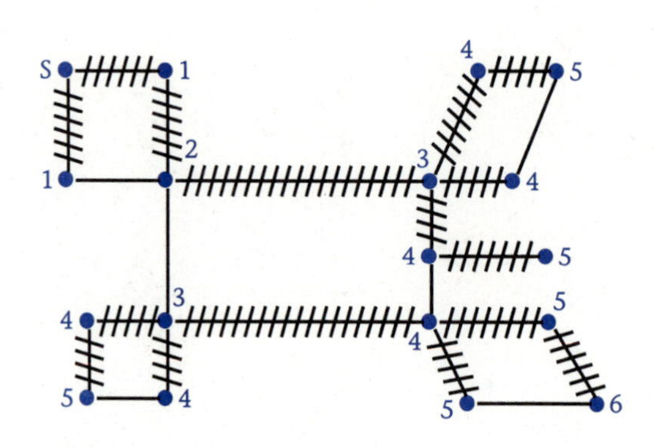

10. One possible solution.

11.

a.

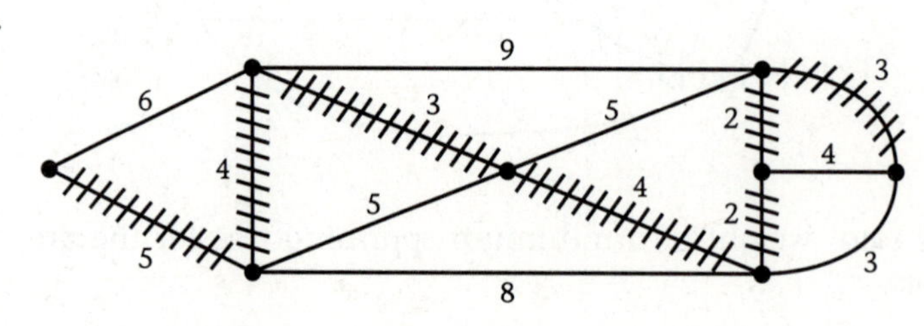

b. Total weight = 23.

12.

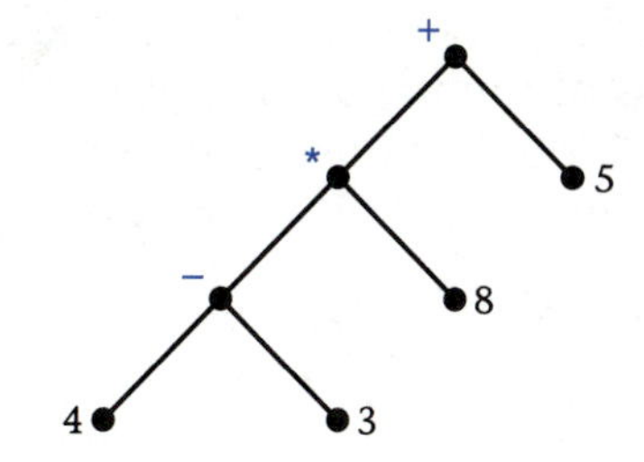

13. 18.

CHAPTER 6

1.
   a. 7,900/46,900, or about .168.
   b. 2,300/13,700, or about .168.
   c. 13,700/46,900, or about .292.
   d. 2,300/46,900, or about .049.
   e. 19,300/46,900, or about .412.
   f. Yes.
   g. No.
2. $C(5,1) = 5$, $C(5,2) = 10$, $C(5,3) = 10$, $C(5,4) = 5$, $C(5,5) = 1$.
3.
   a. $(1/2)^4 = 1/16$.
   b. $(1/2)^4 = 1/16$.
   c. $1/16 + 1/16 = 1/8$.
4.
   a. $P(6,6) = 6! = 720$.
   b. $2 \times 1 \times 4 \times 3 \times 2 \times 1 = 48$.
   c. $48/720 = 1/15$.
   d. The math books can be in positions 1 and 2, positions 2 and 3, positions 3 and 4, positions 4 and 5, or positions 5 and 6. $5 \times 48 = 240$.
   e. $240/720 = 1/3$.
5.
   a.

Amount won	\$2	\$1	−\$1
Probability	1/4	1/4	1/2

   b. $2(1/4) + 1(1/4) - 1(1/2) = 1/4$.
   c. Win about \$25.
6. $3!5! = 720$, about 28.

7. There are $C(39,5) = 575,757$ different winning tickets possible in the first and $C(36,6) = 1,947,792$ ways of winning in the second, so the probability of winning in the first is between three and four times as great as in the second. About five in the first and one or two in the second.
8.
   a. $13/52 = 1/4$.
   b. $12/51 = 4/17$.
   c. $1/4 \times 4/17 = 1/17$.
   d.

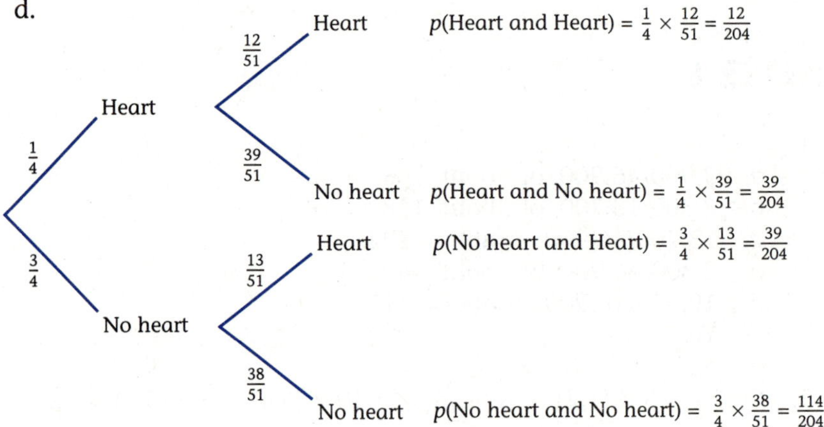

   e. No.
9.
   a. $C(11,3) = 165$.
   b. $C(5,1) \times C(6,2) = 75$.
   c. $75/165$.
   d. $C(5,1) \times C(6,2) + C(5,2) \times C(6,1) = 135$.
   e. $135/165$.
10.
   a.

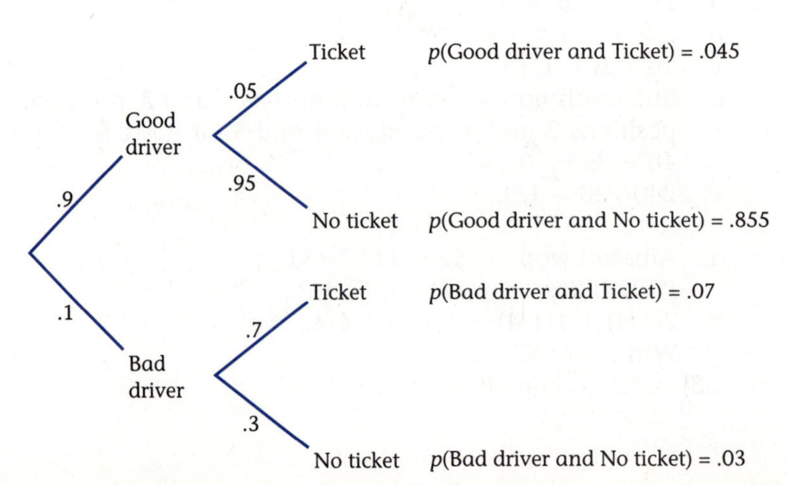

b. $.045 \times 50{,}000 + .07 \times 50{,}000 = 5{,}750$.
c. 3,500.
d. 3,500/5,750.

11.
a. $(1/10)^3 = 1/1{,}000$.
b. $(9/10)^3 = 729/1{,}000$.
c. $9/10 \times 8/10 \times 7/10 = 504/1{,}000$.

12.
a. $10 \times 10 = 100$.
b. $P(10,2) = 90$, $C(10,2) = 45$.

13.
a. $C(5,2)(1/6)^2(5/6)^4$, or about .134.
b. $C(5,0)(1/6)^0(5/6)^6 + C(5,1)(1/6)^1(5/6)^5 + C(5,2)(1/6)^2(5/6)^4$, or about .804.

14. $C(4,1) + C(4,2) + C(4,3) + C(4,4) = 15$.

CHAPTER 7

1.
a.

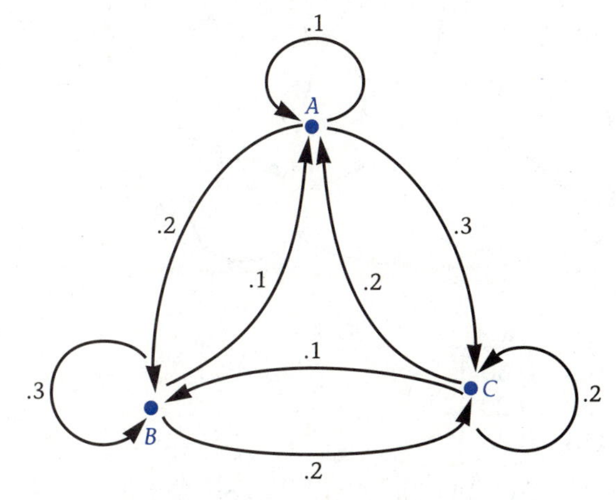

b. $TP = \begin{matrix} A \\ B \\ C \end{matrix} \begin{bmatrix} 7.7 \\ 7.4 \\ 5.8 \end{bmatrix} \quad D = \begin{matrix} A \\ B \\ C \end{matrix} \begin{bmatrix} 0.3 \\ 4.6 \\ 9.2 \end{bmatrix}.$

c. $P = \begin{matrix} A \\ B \\ C \end{matrix} \begin{bmatrix} 18.6 \\ 20.4 \\ 22.2 \end{bmatrix}.$

2. Dan should bluff one-eleventh of the time.
3.
   a. Probability of another quiz on Friday is 29%.
   b. Start class with a quiz one-third of the time.
   c. Review: 40% of the time.
4.
   a.
$$\begin{array}{c} \\ O \\ I \\ B \end{array} \begin{array}{c} \begin{array}{ccc} O & I & B \end{array} \\ \begin{bmatrix} 0.20 & 0.25 & 0.55 \\ 0.45 & 0.35 & 0.20 \\ 0.20 & 0.25 & 0.55 \end{bmatrix} \end{array}.$$
   b. 28.5%.
   c. O: 27%, I: 28%, B: 45%.
   d. Answers will vary. Example: The clerk will know how many sandwiches to order.
5.
   a.

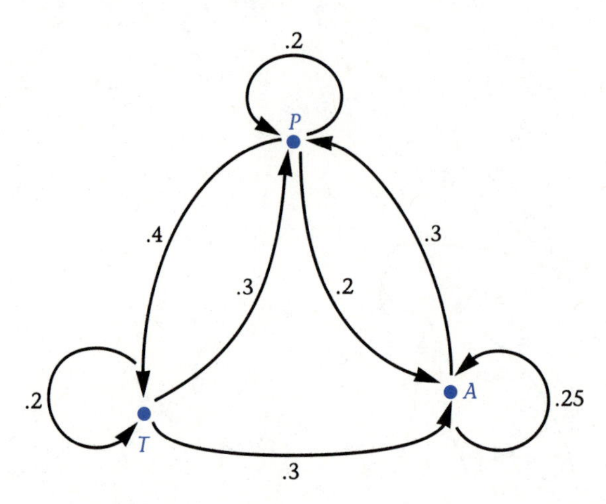

   b.
$$\begin{array}{c} \\ T \\ P \\ A \end{array} \begin{array}{c} \begin{array}{ccc} T & P & A \end{array} \\ \begin{bmatrix} 0.2 & 0.3 & 0.30 \\ 0.4 & 0.2 & 0.20 \\ 0.0 & 0.3 & 0.25 \end{bmatrix} \end{array}.$$
   c. Transportation is most dependent on petroleum. Transportation is least dependent on agriculture.
   d. \$1.08 million from petroleum. \$1.35 million from agriculture.
   e. $TP = \begin{array}{c} T \\ P \\ A \end{array} \begin{bmatrix} 16.00 \\ 16.00 \\ 11.25 \end{bmatrix} \quad D = \begin{array}{c} T \\ P \\ A \end{array} \begin{bmatrix} 4.00 \\ 9.00 \\ 3.75 \end{bmatrix}.$

f. $P = \begin{matrix} \text{T} \\ \text{P} \\ \text{A} \end{matrix} \begin{bmatrix} 16.4 \\ 17.5 \\ 11.0 \end{bmatrix}$ (in millions of dollars).

6. The best strategy for both companies is to focus on school district A.
7. The Democrats' best strategy is to go with strategy A one-fourth of the time and strategy B three-fourths of the time. The Republicans' best strategy is to go with strategy C one-half of the time and strategy D one-half of the time. Expectation for the Democrats: 45% of undecided voters joining them.

8.

a.

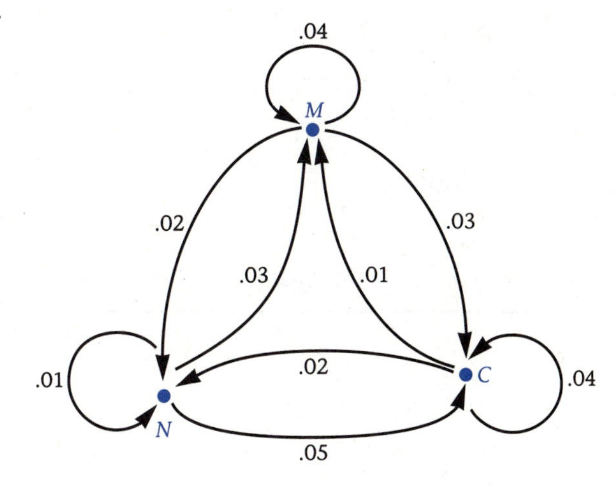

b. $P = \begin{bmatrix} 54.14 \\ 34.11 \\ 42.94 \end{bmatrix}.$

c. $\begin{bmatrix} 4.14 \\ 4.11 \\ 2.94 \end{bmatrix}.$

d. Internal consumption change: $\begin{bmatrix} 1.01 \\ 1.06 \\ 0.80 \end{bmatrix}.$ Production change: $\begin{bmatrix} 11.01 \\ 9.06 \\ 12.80 \end{bmatrix}.$

9. Store A: Lower prices. Store B: No change.

CHAPTER 8

1.
   a. $H_n = H_{n-1} + 4$; $H_n = 2 + 4(n - 1)$; 398.
   b. $H_n = 3H_{n-1} - 1$; $H_n = \frac{5}{2}\, 3^{n-1} + \frac{1}{2}$; 4.29×10^{47}.
   c. $H_n = (H_{n-1}) \cdot 2$; $H_n = 3(2^{n-1})$; $1.9014759 \cdot 10^{30}$.
2.
   a. Arithmetic. b. Neither. c. Geometric.
3.
   a. 0; 0, 0, 0, 0.
   b. −0.75; −0.75, −0.75, −0.75, −0.75.
   c. No fixed point.
4.
   a. $H_n = 2(5^{n-1})$, $3.1554436 \cdot 10^{69}$.
   b. $H_n = 2.75(5^{n-1}) - 0.75$, $4.338735 \cdot 10^{69}$.
   c. $H_n = 2 + (-3)(n - 1)$, −295.
5.
   a. Geometric. b. Neither.
   c. Arithmetic.
6.
   a.

N	S_n	Differences	Differences	Differences
1	1	—	—	—
2	5	4	—	—
3	14	9	5	—
4	30	16	7	2
5	55	25	9	2

   c. $n^3/3 + n^2/2 + n/6$, 204.
7. Second degree, $H_n = n^2 - 6$.
8.
   a.

Day	Gifts that day	Total gifts
1	1	1
2	3	4
3	6	10
4	10	20
5	15	35
6	21	56

b. $G_n = G_{n-1} + n.$
$T_n = T_{n-1} + n^2/2 + n/2.$
c. $G_n = n^2/2 + n/2.$
$T_n = n^3/6 + n^2/2 + n/3.$

9.
a. $P_n = P_{n-1} + 0.23.$
b. $P_n = 0.29 + 0.23(n - 1).$

10.
a.

Month	Balance
0	$1,000
1	$1,006
2	$1,012.04
3	$1,018.11

b. $B_n = 1.006(B_{n-1}).$
c. $B_n = 1{,}000(1.006)^n$, or $1{,}006(1.006)^{n-1}$.
d. 9 years, 8 months.

11.
a.

Month	Balance
0	$5,000
1	$5,126.67
2	$5,254.01
3	$5,382.03

b. $B_n = (B_{n-1})(1 + 0.064/12) + 100.$
c. $B_n = 23{,}750(1 + 0.064/12)^n - 18{,}750.$
d. $13,928.99.
e. 200 months.

12.
a. $B_n = 1.008(B_{n-1}) - 230.$
b. $B_n = -17{,}750(1.008)^n + 28{,}750.$
c. 61 months.
d. $3,030.
e. $352.88.

13.
a. $V_n = 1.5V_{n-1} + 4{,}000.$
b. $V_n = 16{,}000(1.5)^{n-1} - 8{,}000.$

SOURCES OF ILLUSTRATIONS

Facing page 1
Miriam White © 1987

page 10
Bettmann

page 19
Bettmann

page 39
UN Photo 182120/M. Tzovaras Doc. 10150

page 44
Various Cakes, 1981. Wayne Thiebaud, private collection, courtesy: Alan Stone Gallery, New York, NY

page 69
UPI/Bettmann Newsphotos

page 96
UPI/Bettmann Newsphotos

page 102
Archive Photos/White

page 118
©Eve Arnold-Magnum

page 125
Annabelle Breakey

page 141
Chip Clark/Smithsonian Institution

page 150
From *Cartography and Geographic Information Systems: A Career Guide* by the American Cartographic Association

page 159
Archive Photos

page 160
San Francisco Chronicle. Photographer: Gary Fong

page 173
Bettmann

page 183
Bettmann

page 206
Robert Patterson and Donna Cox, NCSA

page 221
K. Wiles Stabile/Newsday

page 239
Michael Macor, *The Oakland Tribune*

page 254
Courtesy of Hewlitt-Packard Company

page 264
New York State Lottery/George Kanatous

page 277
Bob Thomason/Tony Stone Images

page 288
© 1960 Billboard Publications. Used with permission from BILLBOARD.

page 315
Chip Henderson/Tony Stone Images

page 330
Bettmann

Page 333
Bettmann

page 352
NOVOSTI/SOVFOTO

page 388
© The Stock Market/Tom Sanders

page 417
Wilson North © 1983 International Stock Photo

page 419
Bernice Abbott

INDEX